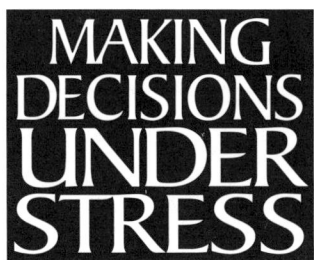

MAKING
DECISIONS
UNDER
STRESS

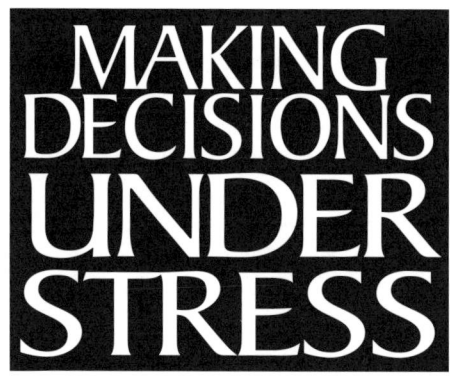

MAKING DECISIONS UNDER STRESS

IMPLICATIONS FOR INDIVIDUAL AND TEAM TRAINING

EDITED BY

JANIS A. CANNON-BOWERS
EDUARDO SALAS

AMERICAN PSYCHOLOGICAL ASSOCIATION
WASHINGTON, DC

Published by
American Psychological Association
750 First Street, NE
Washington, DC 20002

Copies may be ordered from
APA Order Department
P.O. Box 92984
Washington, DC 20090-2984

Typeset in Century by EPS Group Inc., Easton, MD

Printer: Edwards Brothers, Ann Arbor, MI
Cover designer: Minker Design, Bethesda, MD
Technical/production editor: Tanya Y. Alexander

Library of Congress Cataloging-in-Publication Data
Making decisions under stress : implications for individual and team training /
    edited by Janis A. Cannon-Bowers, Eduardo Salas.
        p.   cm.
    Includes bibliographical references and index.
    ISBN 1-55798-525-1 (hardcover : alk. paper)
    1. Leadership. 2. Command of troops.   3. Group problem solving—
United States. 4. Naval tactics—Decision making.   5. United States.
Navy—Management. I. Cannon-Bowers, Janis A.   II. Salas, Eduardo.
VB203.D43   1998
359.3'3041—dc21   98-25041
                                                                          CIP

Printed in the United States of America
First edition

To Marty Tolcott, who guided us masterfully with wisdom, patience, and diplomacy. His warmth, good nature, and quiet persistence were an insipiration to us all.

Contents

Contributors

Bruce Acton, Afloat Training Group Pacific, San Diego, CA

Scott J. Behson, State University of New York Albany, Albany, NY

Elizabeth Blickensderfer, Naval Air Warfare Center Training Systems Division, Orlando, FL

John J. Burns, Sonalysts, Inc., Orlando, FL

Janis A. Cannon-Bowers, Naval Air Warfare Center Training Systems Division, Orlando, FL

Marvin S. Cohen, Cognitive Technologies, Inc., Arlington, VA

Stanley C. Collyer, Office of Naval Research, Arlington, VA

James E. Driskell, Florida Maxima, Orlando, FL

Elliot E. Entin, Alphatech, Burlington, MA

Arthur D. Fisk, Georgia Institute of Technology, Atlanta, GA

Jared T. Freeman, Cognitive Technologies, Inc., Arlington, VA

James H. Hicinbothom, CHI Systems, Inc., Lower Gwynedd, PA

William C. Howell, American Psychological Association, Washington, DC

Susan G. Hutchins, Naval Post Graduate School, Monterey, CA

Joan H. Johnston, Naval Air Warfare Center Training Systems Division, Orlando, FL

Richard T. Kelly, Pacific Science & Engineering Group, Inc., San Diego, CA

Alex Kirlik, Georgia Institute of Technology, Atlanta, GA

Steve W. J. Kozlowski, Michigan State University, East Lansing

Gerald S. Malecki, Office of Naval Research, Arlington, VA

James A. McPherson, Summit Technologies, Orlando, FL

Ronald A. Moore, Pacific Science & Engineering Group, Inc., San Diego, CA

Jeffrey G. Morrison, Naval Command, Control and Ocean Surveillance Center, RDT&E Division (NRaD), San Diego, CA

Stephanie C. Payne, Naval Air Warfare Center Training Systems Division, Orlando, FL

John Poirier, Sonalysts, Inc., Virginia Beach, VA

John S. Pruitt, Naval Air Warfare Center Training Systems Division, Orlando, FL

J. Thomas Roth, Micro Analysis and Design, Inc., Butler, PA

Ling Rothrock, Georgia Institute of Technology, Atlanta, GA

Joan M. Ryder, CHI Systems, Inc., Lower Gwynedd, PA

Eduardo Salas, Naval Air Warfare Center Training Systems Division, Orlando, FL

Daniel Serfaty, Aptima, Inc., Burlington, MA

Kimberly A. Smith-Jentsch, Naval Air Warfare Center Training Systems Division, Orlando, FL

Scott I. Tannenbaum, State University of New York at Albany

Bryan Thompson, Cognitive Technologies, Inc., Arlington, VA
Neff Walker, Georgia Institute of Technology, Atlanta, GA
Wayne W. Zachary, CHI Systems, Inc., Lower Gwynedd, PA
Rhonda L. Zeisig, Naval Air Warfare Center Training Systems Division, Orlando, FL

Foreword

Shortly after I reported to the AEGIS Training Center in Dahlgren, Virginia, as commanding officer in 1991, scientists from the Naval Air Warfare Center Training Systems Division (NAWCTSD) called me to ask if they could get an hour on my calendar to brief me on the Tactical Decision Making Under Stress, or TADMUS, program.

I had just finished a 3-year tour of duty as captain of the AEGIS cruiser USS *Princeton* CG 59. Early in 1988, I had assembled a new crew in the shipyard in Pascagoula, Mississippi, where the ship was built, and in February 1989, when the ship was commissioned, we sailed for our new home port in Long Beach, California. In the ensuing 2½ years that I was aboard, the crew was constantly on the go, continually training, maintaining the complex equipment, and conducting operations, often simultaneously. We had the newest and most capable ship in the Navy. We had a bright enthusiastic crew with a wardroom of 28 officers, and all were excited at the prospect of bringing a new ship to life.

This was the most complex warship in the world and the first that used digital technology to integrate the multitude of sensors, weapons, and control systems. Our first major event was combat systems testing. To check out the final construction of this integrated AEGIS Combat System, each new crew is required to conduct a lengthy Combat System Ship Qualification Trial (CSSQT). During this fast-paced 6-week process, the crew carried out intensive maintenance and training and test-fired every weapon on the ship with the exception of Tomahawk under realistic operational conditions. From CSSQT it was back to the shipyard for completion of warranty work and other repairs during the period following sea trials.

Because the *Princeton* was the first ship to have the AN/SPY-1B radar, which was a major performance upgrade from the original AN/SPY-1A multifunction phased array radar, the crew of the *Princeton* was required to proof and test the radar, first during developmental tests under the direction of the AEGIS program manager PMS-400, and then during operational tests conducted by the commander of the Operational Test and Evaluation Force. During these testing periods we fired 25 of our standard surface-to-air missiles against a variety of targets in very stressful scenarios. On one occasion we were required to detect and engage six simultaneous targets: two supersonic and four subsonic sea-skimming drones, the most stressing targets for both operators and combat systems. By the spring of 1990 our testing was complete and we joined the Pacific Fleet for operations. We participated in a multinational exercise in the Hawaiian operational areas (OPAREAS) and then prepared for a unique assignment. The *Princeton* was selected as the flagship for the first U.S. Navy visit to the Soviet Union's Pacific port of Vladivostok since before World War II.

Before that port visit was completed, the crew received word that the remainder of our 4-month cruise around the rim of the Western Pacific was canceled. Saddam Hussein had invaded Kuwait, and we were directed to return to Long Beach and join Rear Admiral "Zap" Zlatoper's USS *Ranger* CV 61 Battle Group, which was in final preparations prior to sailing for the Persian Gulf.

The ship arrived in the Persian Gulf on January 13, 1991, just days before the beginning of Operation Desert Storm. For a month the USS *Princeton* was consumed with combat operations in the Persian Gulf: conducting air surveillance, controlling aircraft, providing air defense for the aircraft carriers and amphibious and mine-sweeping ships, attacking the Iraqi Navy, shooting Tomahawk missiles at targets within Iraq, searching for floating mines, and maintaining communications and data links with other ships and commanders.

On February 17, 1991, the *Princeton's* participation in the Persian Gulf War came to an abrupt end when an Iraqi influence mine exploded directly under the ship's stern. Although major damage was done to the ship and there were numerous injuries, miraculously there were no deaths.

After 6 weeks of repairs in Dubai, United Arab Emirates, the ship returned to the United States. Shortly thereafter, I was relieved of command of the *Princeton* and proceeded to the Naval Surface Warfare Center in Dahlgren, the site of the AEGIS Training Center. I present this travelogue to provide a portion of the context that those designing our complex combat systems must consider. With the exception of the mine strike, this nonstop pace of training and operations is typical for naval combatant ships. Even without the mine strike, the entire schedule for the ship and crew from precommissioning through combat operations was one of continual stress. Twelve- to 18-hour workdays, 6–7 days a week for extended periods, were the norm. In many respects the pace was grueling.

The key to this green crew's ability to execute the schedule successfully was the massive amount of training that was provided throughout the process. Formal team training both ashore and afloat, hundreds of individual schools, damage-control training, maintenance training, operator training, tactical training, CSSQT training, predeployment training, refresher training—the list goes on and on. The ship had a crew of 380, many of whom spent upward of 2 years in the training pipeline prior to reporting to the ship. After the crew was assembled, the training continued on a daily, nonstop basis and in fact drove the pace of the schedule more than did scheduled operations. My crew and I knew we were operating in a stressful environment, starting long before we got into combat.

Therefore, when TADMUS personnel indicated to me they would like to brief me on a program aimed at the understanding of stress, human factors, and training, and their potential implications for combat systems design and training technologies and methods, I was an enthusiastic listener. I had already come to the conclusion that the process I had just gone through with the *Princeton* was unnecessarily burdensome and difficult, although I had not yet figured out the reason why.

That original briefing for me by TADMUS personnel constituted the proverbial lightbulb being turned on in my head. Of course, there were better ways to conduct team training. Of course, cognitive modeling and shared mental model theory made sense. Of course, improved combat system displays based on human engineering research could improve tactical decision making. Of course, psychological research on team performance had produced data that should influence both training and ship system design.

Although the design concept for the integrated AEGIS Combat System was brilliant and revolutionary when developed in the 1970s, it did not incorporate the results of human factors engineering or psychological research, much of which was done after the AEGIS system was designed.

It became obvious to me that the fast paced and often grueling schedule we had on the *Princeton* was not the primary cause of the difficulties we had experienced; it was only a symptom. The root problem that we faced was one of engineering design. The human had not been integrated into the design. The computer program architecture was complex and difficult to understand, and the displays were not optimized for ease of use or ease of understanding. To the great credit of the AEGIS program manager, the means to overcome these problems was provided: massive and comprehensive training. As is often the case, engineering design problems required training solutions, but as good as this training was, the TADMUS program showed that the Navy was extremely inefficient in conducting training.

All is not lost, however, and this story is proceeding toward a happy ending. The Navy's AEGIS program manager has embraced the TADMUS program and has been a strong supporter of additional research and development leading to advanced training concepts and better human-engineered combat system designs. AEGIS ships will be the mainstay of the surface combatant force for the next 30 years, until they are replaced by the SC-21 family of ships, which are currently on the drawing board. The TADMUS team from NAWCTSD has been brought in as advisors on the next generation of ships at the beginning of the design process. One of the hallmarks of these ships will be reduced manning, efficient training, and advanced human factors engineering. The TADMUS program has provided a wake-up call to today's ship design community. The beneficiaries will be the captains and crews that man ships and take them to sea in the next century, confident in their ability to operate, maintain, and take their ship into harm's way.

Captain Edward B. Hontz, U.S. Navy (retired)

Preface

It is probably the case that in late 1989, when we were first confronted with the prospect of working on the Tactical Decision Making Under Stress (TADMUS) project, we did not realize that we were being presented with the opportunity of a lifetime (at least from the perspective of applied psychological researchers). Of course we realized that it was an uncommonly large research and development (R&D) investment and that it was likely to be a highly visible program owing to its origins and magnitude. Little did we know, however, that we would have the chance to meet and work with extremely talented and inspired colleagues, to learn as much as we have about the complexities and intricacies of Navy tactics, and to grow from the experience of leading a large-scale R&D effort. This is not to say that we lacked our share of problems: too many people who had to be pleased, squabbles among constituents and within the research team, people who doubted our competence or the program's intentions, as well as the typical complications encountered whenever complex applied research is attempted.

Despite the moments when it all seemed too hard or just plain impossible, we look back now and feel terribly rewarded for having endured. We clearly have not solved the problem of training for decision making under stress, but we have derived principles that can have great benefit to those who must operate in these most difficult and challenging environments. Perhaps more than any other accomplishment, we are most satisfied to report that we altered the way many in the surface Navy think about decision-making training. Now, everywhere we turn, we see and hear people confronting human-performance issues directly and ascribing to them the same degree of importance as hardware and software problems. If we had even the slightest hand in making this happen, it was all worth it.

There were many people along the way who contributed to our success; in fact, some of these people are in one way or another responsible for the program's accomplishments. Any challenge of the magnitude we confronted in TADMUS requires a team effort, and we were fortunate enough to be part of a great one. It would be impossible to recognize by name all of the people who contributed to the success of TADMUS; however, we feel compelled to highlight a few of our teammates.

The first category of people who deserve recognition are the decision makers at the Office of Naval Research, most notably, Phil Selwyn, who believed in us enough to fund the effort. Others who deserve our sincerest gratitude are Bill Vaughn, Steve Zornetzer, and of course our program manager, Jerry Malecki, who kept things moving forward despite seemingly insurmountable odds at times. Next, we recognize others in the Navy civilian workforce: Jeff Grossman and his team in San Diego; early believers Reuben Pitts and John Burrow at Dahlgren; our boss Bill Rizzo for letting us follow our hearts even when it made him a little uncomfortable; Mike Sovereign and Gary Porter at Navy Post Graduate School; P. A. Federico at the Navy Personnel Research and Development Center (NPRDC); and of course our own talented team at the Naval Air Warfare Center Training Systems Division, led most recently by our competent and loyal colleague, Joan Johnston. Others in our organization—management, support personnel, and fellow scientists—also contributed to our success.

When we reflect on the program, it is clear that our TADMUS Technical Advisory Board (TAB), led initially by Marty Tolcott and then by Bill Howell, also contributed significantly to our success. The list of individuals composing the TAB can be found in Appendix A; we are indebted to this most impressive group of individuals. In fact, it was Marty's wisdom, leadership, and patience early on that helped to ensure continuation of the program. We all suffered a great loss in his passing.

The next category of people who deserve our thanks is the group of highly motivated, dedicated, and often brilliant contractors who conducted most of the R&D. Our gratitude to these individuals cannot be overstated. In this group, we would like to recognize expressly Reid Benton at Lockheed Martin because he had enough vision and confidence not to be threatened by us; instead, he helped to open several doors.

Next is a large group: all of the active-duty and retired Navy personnel who in one way or another contributed to the research. We could not have done the job without the aid of these bright, dedicated, and hard-working people. We specifically call out Captain Gary Zwirschitz and his crew on the USS *Harry E. Yarnell* for patiently teaching the senior editor nearly everything she knows about Navy combat information centers. We also recognize those at the Surface Warfare Officer's School—Rear Admiral Dick West, Captain Mark Edwards, Captain Joe Volpe, Captain Mick McDonough, Captain Paul Surfass, Lieutenant Commander Ray Valez, Lieutenant Commander Steven Evans, Lieutenant Commander Deidre McLay, Lieutenant Mike McCarthy, and Dave Monroe—who were all instrumental in helping us establish credibility and collect research data over the years.

Others who deserve recognition for their help and support include Rear Admiral George Huchting, Rear Admiral Phil Balisle, Captain Al Koster, Captain Gary Storm, Captain Tom Bush, Captain Shelly Margolis, Captain Doug Armstrong, Commander Bruce Acton, Lieutenant Commander Don McConkey, Lieutenant Commander Don Sine, Lieutenant Tim Gilbride, and Petty Officer Mike Hazen. Two early TAB members—Captain Marty Chamberlain and Commander Bill Boulay—played an important role in keeping our detractors from destroying us in the beginning: They were early believers. Finally, in addition to their active support, Commander Lee Pitman, Captain El Halley, and Captain Ted Hontz are the kind of people who make others better simply by knowing and associating with them. We are fortunate to have met and worked with these individuals and honored to count them among our colleagues and friends.

We must also thank Bill Howell—another person who supported us even when the going got tough—and the American Psychological Association for agreeing to publish this book. We are grateful to all of the chapter authors, who made it easy and pleasurable to edit this volume. In addition, Sandra Richards and Danielle Merket deserve recognition for keeping us organized and on track. Of course, our spouses, Clint Bowers and Vickie Salas, helped us to stay sane through it all.

Above all, we are indebted to Stan Collyer for the role of his vision in making TADMUS a reality. He provided more than vision, however; he conceived of TADMUS and demonstrated unswerving conviction to the program's success. We hope that he realizes how that vision and conviction have so positively touched the lives of so many people. We also hope that when he looks back he can enjoy the same sense of accomplishment we do in achieving something that only a few people have the privilege to experience, that is, the successful application of psychological principles in solving real problems. He gave us all the opportunity to make this volume a reality, and in doing so he changed our lives permanently for the better.

Introduction

This volume represents the culmination of a 7-year research project called Tactical Decision Making Under Stress (TADMUS). The program, which was sponsored by the Office of Naval Research, was conceived in response to an incident involving the USS *Vincennes*, a warship that mistakenly shot down an Iranian airbus in July 1989 (see chapter 1, this volume, for more details). Briefly, the goal of the program was to develop training, simulation, decision support, and display principles that would help to mitigate the impact of stress on decision making. With the exception of chapter 16, this volume focuses on research directed at training and simulation solutions to the problem.

Our purposes in documenting the results of this important program are twofold. First, TADMUS represented a true *program* of research, driven by a common operational problem, based on a common theoretical framework, and employing a common method and task. As such, the significance of the work can be best understood as a series of related studies, each of which builds on and augments findings from prior experiments. This, we believe, is a viable approach to research and development (R&D) directed at a problem as difficult, complex, and multifaceted as decision making under stress.

The second reason that we decided to organize this volume was that many of the results documented herein would not have been accepted for publication in typical peer-reviewed psychological journals. This is because many of the experiments were field studies, using complex, realistic tasks and experienced Navy participants. As is typical in such investigations, the requirements of laboratory research (e.g., control groups, large sample sizes, random assignment) were not always met. However, attempts to use sound principles of quasi-experimentation were made, and the rigors of more controlled experiments were compromised only when absolutely necessary. Hence, we believe that the results of these quasiexperiments are valuable and should be disseminated throughout the psychological community.

This volume is organized into several sections. The first section describes the overall background, research approach, and paradigm employed in TADMUS. To begin with, the chapter by Collyer and Malecki describes the history, background, and origins of the TADMUS program. Next, the chapter by Cannon-Bowers and Salas lays out the theoretical underpinnings of TADMUS research; it is followed by a description of the experimental methods, paradigm, and task used in most of the TADMUS experiments in the chapter by Johnston, Poirier, and Smith-Jentsch. Rounding out this section is a chapter by Smith-Jentsch, Johnston, and Payne that describes the performance-measurement strategies used in the experiments.

The second section focuses on research in the training of decision-making skills at the individual level. This body of work concerns the provision of training that will prepare individuals to operate in complex, team environments, including investigations of individual level training interventions. The section begins with a chapter by Kirlik, Fisk, Walker, and Rothrock, who studied the impact of part-task training and feedback on decision-making performance. Next, Kozlowski documents work addressing the acquisition of expertise required to perform successfully in a decision-making team; following that, Cohen, Freeman, and Thompson describe a study showing the benefits of metacognitive training. Finally, the chapter by Driskell and Johnston details work regarding the positive impact of stress-exposure training on decision-making performance.

The third section of the volume describes efforts to investigate team training interventions. These chapters are all concerned with the formulation of principles concerning teams and team decision making and their application to the development of team-level training interventions. To begin with, Serfaty, Entin, and Johnston document the results of team coordination and adaptation training on team decision-making performance. Following that, Tannenbaum, Smith-Jentsch, and Behson describe successful attempts to improve decision making through team leader training. Next, Smith-Jentsch, Zeisig, Acton, and McPherson detail a team training intervention called *team dimensional training*, which guides performance measurement and feedback in decision-making teams. Finally, Blickensderfer and colleagues summarize research into cross training as a means to enhance team decision-making performance.

The fourth section of the volume documents TADMUS follow on and related applied research. Zachary, Ryder, and Hicinbothom begin this section by describing an approach to cognitive modeling initiated under TADMUS that has value as a means to capture decision-making performance in an executable format. The chapter by Roth focuses on the problem of training instructors, particularly in the operational environment, to impart sound decision-making skills in decision-making teams. Next, in the short chapter by Cannon-Bowers, Burns, Salas, and Pruitt, a brief description is offered of more advanced R&D being pursued as a means to develop further and apply the concepts and principles derived under TADMUS. Finally, Morrison and colleagues document a related effort conducted under TADMUS directed at developing an operational decision support system to improve tactical decision-making performance.

The final section in the book is best considered reflections on the program: what was learned from the experience of conducting a large applied R&D effort. First, Salas, Cannon-Bowers, and Johnston summarize the lessons learned as a result of conducting and managing the TADMUS program. Finally, Howell provides commentary about the program: what he believes were the successes of TADMUS as well as areas in need of improvement if such a program is attempted again in the future.

Throughout the volume we pressed authors not only to focus on the research implications of their efforts, but also to generate a series of lessons learned that could guide those interested in applying results of the research in operational environments. Whereas TADMUS focused on a military decision-making environment, the findings could be applicable across a variety of task environments that pose similar demands on human operators. We hope these lessons learned will help the interested reader to make use of the knowledge base we have amassed over the years. In addition, we are confident that many viable research questions and propositions can be generated to guide future research in this crucial area.

Part I

TADMUS History, Background and Paradigm

Tactical Decison Making Under Stress: History and Overview

Stanley C. Collyer and Gerald S. Malecki

> Since it appears that combat induced stress on personnel may have played a significant role in this incident, it is recommended the Chief of Naval Operations direct further study be undertaken into the stress factors impacting on personnel in modern warships with highly sophisticated command, control, communications and intelligence systems, such as AEGIS.—From the formal investigation into the circumstances surrounding the downing of Iran Air Flight 655 on 3 July 1988 (Fogarty, 1988)

Background

On July 3, 1988, the USS *Vincennes* mistakenly shot down Iran Air Flight 655 over the Persian Gulf, killing 290 people. Among the many consequences of this widely publicized catastrophe was the initiation of a research and development program called Tactical Decision Making Under Stress (TADMUS). In this introductory chapter we discuss the history of TADMUS, including some of its scientific antecedents; describe some features of the program that we believe have helped to make it successful; and briefly discuss some of the more significant accomplishments thus far.

The launching of the TADMUS program required a convergence of many factors, some operational in nature, some scientific and technical, and some bureaucratic. Along with the *Vincennes* disaster were other Navy incidents in the same time period that caused many to question whether improvements in training and decision support technology could have reduced their likelihood of occurrence. Meanwhile, recent scientific advances in decision theory and team training had matured to the point that they appeared ready for testing in an applied setting.

Operational Context

The *Vincennes*, a guided missile cruiser (CG-49) possessing the Navy's most sophisticated battle-management system (the AEGIS system), was designed to protect aircraft carrier battle groups against air attack in open ocean scenarios against an enemy force of hundreds of airplanes and missiles. On July 3, 1988 it was returning to the Persian Gulf after an escort mission through the Strait of Hormuz.

A great deal has been written about the *Vincennes* tragedy. The official report (Fogarty, 1988) and the endorsement of the report by then chief of naval operations Admiral Crowe (Crowe, 1988) provide good starting points for those wishing to understand the facts. Since the appearance of those reports, which were necessarily prepared quickly to satisfy the great demand for information and explanations, many articles have been published, both in technical journals and in the popular press, providing discussions of the incident from various perspectives and sometimes, new information. Some of the articles are quite controversial; suffice it to say that there will probably always be differing opinions as to all the reasons for this accident and the relative importance of those reasons. The following brief summary presents many of the key facts that contributed to the decision of the commanding officer to fire:

- The Iranian airbus had taken off from an airport (Bandar Abbas International) that served both military and civilian aircraft. Although it was a regularly scheduled flight, it departed approximately 27 minutes late.
- Intelligence reports had recently warned naval forces operating in the Persian Gulf that the Iranians might be planning an incident to embarrass the United States on or near Independence Day.
- Approximately 40 minutes before the shootdown, the helicopter operating from the *Vincennes* had reported being fired on by Iranian gunboats. As a result, the *Vincennes* and another U.S. warship, the frigate USS *Elmer Montgomery* (FF-1082), soon thereafter engaged in a gun battle against small armed boats manned by members of the Iranian Revolutionary Guard Corps. This resulted in high noise levels and some high-speed maneuvering, which undoubtedly produced greater than normal levels of stress in the combat information center (CIC) of the *Vincennes* (Fogarty, 1988). The gun battle also provided the context in which an approaching aircraft could be viewed as another component of a coordinated attack.
- A member of the *Vincennes* crew incorrectly identified the unknown aircraft as a probable F-14, apparently because of an error in operating the system that interrogates aircraft transponders for identification friend or foe (IFF) information.
- F-14s were known to be operating from Bandar Abbas.
- An Iranian patrol aircraft (P-3) was flying in the vicinity and was well positioned to be providing targeting information to any aircraft preparing to attack the *Vincennes*; furthermore, it had just turned toward the *Vincennes*.
- Although the unknown aircraft was within the 20-mile-wide commercial air corridor, it was not in the center of that corridor; commercial aircraft normally took pains to fly exactly on the centerline.
- The contact was heading directly toward the *Vincennes*.

- Although the aircraft was continuously climbing until it was shot down, two console operators in the CIC reported that it was descending. This report was passed along to the commanding officer without being independently confirmed by the officer responsible for coordinating anti-air warfare.
- The aircraft did not emit any electronic radiations that would have definitively identified it as a commercial airliner.
- Repeated radio warnings from the *Vincennes* to the unidentified aircraft were not answered, and no course changes were made.
- There was very limited time to make a decision; the elapsed time from takeoff to shootdown was only 7 minutes.

The Fogarty report concluded that the commanding officer had acted properly, given all the information (correct and incorrect) that he had received. The report said that "stress, task fixation, and unconscious distortion of data may have played a major role in this incident," presumably owing to "scenario fulfillment," or "expectancy bias" (given a belief that an attack was underway, participants may have misinterpreted facts and selectively attended to facts that supported this hypothesis).

The results thus were generally attributed to "human error," although many believe this explanation is simplistic. Klein (1989), for example, downplayed the role of bias, saying it was more likely that inadequate displays led to the mistaken belief that the aircraft was descending. Furthermore, although it may be true that the AEGIS system worked as designed (no system failures were found to have occurred), it clearly did not work perfectly in this scenario. It is important to note that it was not designed for low-intensity conflict scenarios, in which there is often a high degree of ambiguity concerning, for example, the identity and intentions of unidentified radar contacts.

Among the many consequences of the *Vincennes* incident, the following are particularly relevant here. Rear Admiral Fogarty, the chief investigating officer, recommended research be conducted on stress and CIC team training. Admiral Crowe, in his endorsement of the Fogarty report, recommended further human factors research in decision support systems and information display. The chairman of the House Armed Services Committee, Representative Les Aspin, sent a letter to Senator John Tower, President Bush's nominee for secretary of defense, encouraging further research on behavioral decision making. In addition, the House Armed Services Committee held hearings on the *Vincennes* incident in October 1988. Several psychologists were invited to provide testimony; among their recommendations, not surprisingly, was the advice to increase funding for behavioral research.

In addition to the *Vincennes* tragedy and its aftermath, there were other incidents that contributed to a favorable climate for supporting an applied research program to develop improved training and decision support technology in the Navy. Three months before the *Vincennes* incident, a Navy frigate, the USS *Samuel B. Roberts* (FFG-58), struck an Iranian mine in the gulf and was nearly sunk. One year before that, the USS *Stark*

(FFG-31) was struck by two Exocet missiles mistakenly fired by an Iraqi pilot, killing 37 sailors. In both incidents, crew training and shipboard displays were cited as areas in which improvements could be made.

Scientific Context

Overview of theories and models of decision making. Tolcott (1992) observed that the two primary components of military decision making are situation assessment (what is happening) and action selection (what to do about it). Subsumed under these components are other tasks, such as generating hypotheses to account for information being received and generating and evaluating alternate actions. To support military decision makers in complex, demanding environments, scientists have investigated the underlying cognitive processes in problem solving, judgment, and decision making, with the goal of exploiting strengths and compensating for weaknesses of individuals. The successes of decision support techniques and systems have been directly related to their relevance to the operational environment and their compatibility with attributes of human cognitive processes. The following sections will present an overview of the progression of selected decision theories and models relative to their use for military decision support.

Rational models. During the period approximately between 1955 and 1975, decision research was dominated by models of rational choice and behavior. The most frequently used models were subjective expected utility theory, multiattribute utility theory, and Bayesian inference models. These theories and models, which are still in use, involve decomposing the decision problem into its elements in a form that makes explicit the choices, the uncertainties that would affect the outcomes, and the outcomes themselves. These models have been successfully used in military organizations for evaluating alternative system designs, allocating financial resources, evaluating the merits of new military tactics, and analyzing intelligence information. There are at least two major limitations of the models of rational choice. The first is the relatively long time required to structure or model the decision problem and obtain judgments needed to derive a solution. Although the models can be used effectively in long-range planning problems, there are not well suited generally for high-tempo tactical decisions. One exception is the Bayesian inference model, which has been used successfully for certain target-classification tasks. Even that model has limitations in this context, however, particularly when the operational situation changes (e.g., a sensor becomes less reliable or the enemy introduces new tactics). The second shortcoming is that the explicit judgments required by the models are unnatural and difficult for many people to make. Expert decision makers tend to make these judgments implicitly without assigning numerical values and on a more intuitive basis than the rational models require.

Descriptive models. During the period approximately from 1965 to 1985, decision researchers produced compelling findings indicating that humans do not normally make decisions as described by rational models. In this work, researchers took the rational models as a starting point, assumed that these models furnished the correct answers, and regarded deviations from the prescribed procedures as heuristics and deviations from the correct responses as biases. Simon (1957) conducted pioneering research in this domain, and his findings indicated that human decision making consistently deviated from the processes and outcomes described by the models of rational choice. He introduced the concept of *bounded rationality* and demonstrated that in many situations people do not attempt to evaluate all available response choices to maximize their subjective expected utility, but rather consider only as many alternatives as needed to discover one that satisfies them. He referred to the process as *satisficing*.

These findings motivated performance of new studies that focused on comparisons of human judgments with those generated by formal models and produced a large body of experimental findings that identified a wide range of typical human decision biases. Examples are (a) biases in probability estimation, such as the availability bias and the confirmation bias; (b) biases in utility judgments, such as intransitivity of judgments and framing biases; and (c) inadequate generation of alternatives and hypotheses. Much of that literature is described in a book by Kahneman, Slovic, and Tversky (1982). The models that are based on these approaches are referred to as *normative descriptive models*.

This body of knowledge provided a robust basis for the development of interactive and adaptive decision support tools to reduce biases. Examples of these support tools are prescriptive computer-based aids, outcome calculators, expert systems, and graphic displays. These categories of decision support systems have been effectively used for mission planning, estimating enemy courses of action, and route planning, and as tactical decision aids that perform complex and burdensome calculations, such as calculating the best sonobuoy pattern for localizing a submarine target. A library of these decision support tools was developed to operate in conjunction with the Joint Operational Tactical System (JOTS 1990), which has since been renamed the Joint Maritime Command Information System (JMCIS). These support systems have a major limitation in that they are best suited for planning, replanning, and training and are not optimally suited for high-tempo and rapidly changing tactical operations. Nonetheless, they represent an important asset in the repertoire of tactical decision support tools.

Decision models in natural settings. During the 1980s decision researchers moved from the laboratory to the natural workplace of the decision makers of interest (Klein, Calderwood, & Clinton-Cirocco, 1986). College students were replaced by trained professional experts as study participants. An important characteristic of this new line of research is that the focus and scope of the choice and judgment tasks tend to encompass the total decision task. Investigators more frequently present the task as realistically as possible instead of breaking it down into its constituent parts. In addition, researchers recently have been making concerted efforts to understand the underlying cognitive processes through the use of cognitive task analysis, analysis of verbal protocols, interview data and recorded communications—techniques that were pioneered during the eras of rational models and descriptive models. This period of research has produced new and refined decision theories, such as recognition primed decision making (Zsambok & Klein, 1997), metacognitive and critical-thinking models (Cohen, Freeman, & Wolf, 1996; Cohen, Freeman, Wolf, & Militello, 1995), and schema-based models of problem solving (Marshall, 1995). These models are making significant contributions to the development of innovative decision support concepts and training strategies (Smith & Marshall, 1997).

Scientific endorsements. As mentioned earlier, the state of science and technology provided a credible foundation for undertaking the TADMUS research program. This state of affairs was articulated at least two highly regarded forums, the Naval Studies Board of the National Academy of Sciences and the American Psychological Association's Defense Research Roundtable. During 1988 and 1991, the Naval Studies Board (NSB) conducted two in-depth scientific reviews of the research programs in the Cognitive and Neural Sciences Division of the Office of Naval Research (ONR). The task was to identify promising research opportunities for consideration by ONR in planning the future directions of its programs and maintaining the strong science base needed to support the U.S. fleet in the 21st century. The report of the first NSB review included strong recommendations for accelerated and increased support in the areas of problem solving and decision making (Naval Studies Board, 1988). A second, similar review was conducted 3 years later; in that report high priority was assigned to research on the cognitive processes in learning and to the development of intelligent computer-based instruction and training (Naval Studies Board, 1991).

In addition, the American Psychological Association assembled leading scientists in 1988 in a workshop to assess the state of research on human cognition and decision making, to evaluate its relevance to the design of military decision support systems, and to recommend a research agenda to stimulate and contribute to the effectiveness of decision-aiding technologies. The report of the workshop (Tolcott & Holt, 1987) included the following assessments and recommendations:

- Research on human judgment and decision making has produced rich findings about cognitive abilities, but designers of decision-aiding systems often fail to take advantage of this knowledge.
- Decision-aiding research and development has been dominated by the push of popular technologies rather than by systematic study of user needs, resulting in widespread lack of confidence and frequent rejection by potential users.

- Improved methods are needed for testing and validating decision-aid effectiveness, and indeed for measuring the quality of decision performance.
- New research thrusts are needed in the areas of decision making in high-workload environments, option generation in unstructured situations, pattern recognition, the effective training of pattern recognition skills, and avoiding restriction of decision support systems applications to narrowly focused problems without requiring impossibly large knowledge bases.

It is not coincidental that the research agenda and plan for the decision support portion of the TADMUS program incorporated many of the tenets and recommendations of these two respected bodies, with a special emphasis on information processing and tactical decision making by shipboard command teams in air defense operations under conditions of short decision times, high operational workload, and ambiguous and incomplete information.

Training research. The preceding discussion focused on some of the scientific underpinnings for the decision research agenda of TADMUS. A similar treatment of the training research agenda is unnecessary because it is covered in depth in subsequent chapters. In short, the TADMUS training research program was based on the following three broad categories of interventions:

- *Increasing overall skill levels.* The hypothesis was that individuals and teams that are highly competent in executing a task will be more resilient to the negative effects of stress on performance.
- *Exposing trainees to stress during training.* This approach is based on the hypothesis that stress-exposure training can inoculate trainees against the impact of stress in actual task-performance situations.
- *Targeting skills that are particularly vulnerable.* This approach was based on data collected during initial baseline studies in which critical skills that are particularly vulnerable to stress effects were identified.

The specific strategies and interventions are described by Cannon-Bowers, Salas, and Grossman (1992). Briefly, the principal innovative strategies investigated were shared mental model training, teamwork behavior training, coordination and adaptability training, guided practice and feedback training, instructor and team-leader training, and interventions to train novice decision makers to employ strategies used by expert decision makers in skills such as pattern recognition and schema-based decision making.

The TADMUS Program

Objectives and Tasks

The TADMUS research program was begun in 1990. It is an interdisciplinary program involving the development of both training and human factors technologies. The objective of the program is to enhance the quality of tactical decision making in high-stress operational environments by applying recent developments in decision theory, simulation and training technology, and information display concepts. The collective body of research in these domains had recently incorporated important scientific innovations and was ready for harvesting and further refinement tailored to important operational military needs. These innovative theories and models of decision performance provided a technically credible basis for undertaking the program.

Cannon-Bowers et al. (1992) described the objectives and approach of the TADMUS project and highlighted the new and emerging technologies that were being used as a basis for the research. They stated succinctly:

> The TADMUS program was designed to: define the decision problems facing Navy tactical teams and develop measures of tactical decision performance; collect empirical data to document the impact of stress on decision making; and develop principles for decision support, information display, training system design and simulation that will mitigate these stress effects. Several emerging areas of research are being exploited to accomplish these objectives, including: recognition primed decision theory, shared mental models, human performance modeling and team training. (p. 49)

The program originally was scheduled to end in 1996. Its duration was later extended to 1999, and a major new technical task was added. The extension came about primarily because of the strong operational support and the high quality of the technical products. Although work is continuing on the refinement of the original five tasks, the principal focus is presently on Task 6. The six tasks may be summarized as follows:

1. *Definition and measurement of critical decision tasks.* To evaluate properly any new technology being developed, it is important to provide the context of a realistic operational scenario. This task involves defining the operational tasks, setting up laboratories in which to study those tasks, developing a strong performance measurement capability, and developing knowledge of the decision-making processes for that operational environment.
2. *Examination of stress effects on decision making.* The objective of this task is to understand how combat-related stress affects tactical decision making. Prior to the evaluation of the technology under development in this program, a baseline was needed that described how performance degrades under various stressors.

Work in this task involved selecting a number of stressors for investigation, determining which stressors should be used as approximations to actual combat stress, and determining how to quantify their effects.

3. *Development of decision support principles and an experimental prototype.* An experimental decision support system was produced that incorporates sufficient flexibility to permit extensive exploration of alternative concepts and architectures. The design of the prototype was guided by cognitive models of decision making with a focus on the repertoire of decision tasks related to situation assessment and management of tactical responses. The principal attributes of these models are schema-based feature matching and explanation-based reasoning. The prototype is being evaluated in simulated tactical environments using experienced shipboard command teams as study participants. Future tests will be conducted during shipboard sea trials and exercises.

4. *Development of training and simulation principles.* The focus of this task is on developing and demonstrating a variety of individual and team training strategies and techniques to minimize the adverse effects of stress. Products of this task are strategies, principles, and prototype training systems for automaticity training and pattern recognition instruction, training to foster shared mental models, training to improve the team's ability to adapt and configurally restructure during periods of high stress, team leader skills training, stress-exposure training, and cross training.

5. *Development of display principles.* Work was undertaken to examine man–machine interface concepts that maximize the effectiveness of tactical decision aids under stressful conditions. Products of this task include display principles for target–contact deconfliction, threat assessment, hypothesis and option generation, resolution of conflicting or ambiguous information, response management, and adherence to rules of engagement.

6. *Integration of training and decision support strategies.* The goal of this task (begun in 1997) is to develop principles and guidelines for embedding training strategies into the decision support system and then to produce and evaluate experimentally the prototype under both laboratory and shipboard conditions. Research is currently underway to combine the training and decision support strategies. Data will be collected to determine how and to what extent performance enhancements found separately will scale-up when the strategies are combined as well as how and to what extent the laboratory findings will transfer to shipboard operations.

Program Features

TADMUS is an applied research program in Department of Defense parlance (this is also known as Budget Category 6.2, formerly called exploratory development). This category of funding supports the transition of basic research results into defense-related applications. It thus serves as a bridge between pure science and highly specific system development activities. Although it is funded modestly by the standards of many applied research programs in the Navy, the program has had a substantial budget compared to other training or human factors efforts funded by the Navy. (Its total budget, from 1990 through 1997, was approximately $18 million, about half of which went to decision support research and half to training research.)

The program was initially managed by the Office of Naval Technology (ONT). Although it was funded in part from already programmed training and human factors funds, those resources were substantially augmented. A reorganization in 1995 resulted in the merging of ONT into the Office of Naval Research, which has had management responsibility for TADMUS since that time.

From a management standpoint, the TADMUS program has several features that we believe have contributed significantly to its success. Foremost among these is the unusually strong connection to the operational Navy, which was present from the beginning of the program. This was accomplished initially by the formation of a technical advisory board (TAB) composed of both "blue suiters" and scientists. This board, consisting of about a dozen members, has met semiannually since program inception to review progress and make recommendations.

The Navy members of the TAB were senior officers (mostly commanders and captains) with extensive operational experience; in some cases they were former commanding officers of AEGIS ships. These officers provided a crucial "reality check," particularly in the early days of the program, when many scientifically interesting but operationally impractical (or unsalable) ideas were under consideration. The officers also provided important entries into various fleet organizations, facilitating data collection and marshaling support for the program.

The TAB also contained respected scientists and research and development (R&D) managers from all three services, as well as one member from the Canadian defense research establishment. These members offered technical advice and provided important linkages to basic research and related applied programs conducted elsewhere.

The TAB meetings were not always serene. Many controversial issues were raised, and the arguments sometimes became heated. Underlying everything, however, was the mutual respect that the operational and technical communities had for each other, along with the realization that this program was an opportunity to accomplish something of potentially great value. Most important, the chairman of the TAB, Dr. Martin A. Tolcott, provided the glue that held the group together. His composed and courteous demeanor, combined with a remarkable incisiveness and a wealth of knowledge, was usually sufficient to broker a compromise acceptable to all. Marty chaired the TAB from 1990 until his death in 1996. Since then, the TAB has been ably chaired by Dr. William C. Howell.

Another unusual feature of TADMUS was the joint participation of representatives of two Navy laboratories. The Naval Air Warfare Command's Training Systems Division (NAWCTSD) in Orlando was responsible for the training and simulation portion of the program, and the Naval Command, Control and Ocean Surveillance Center's Research, Development, Test & Evaluation Division (NRaD) in San Diego conducted the decision support and human systems interface R&D. Some of the functions (e.g., development of testbed capabilities) were performed jointly. Although it is not unusual for multiple laboratories to be involved in a major program, the customary arrangement is for one to assume the lead and the others to be subordinate. In this case neither was secondary to the other, a situation that can be a recipe for conflict and other unproductive behaviors. Although a certain amount of competitiveness manifested itself from time to time, this was generally kept within reasonable bounds.

Products and Transitions

The TADMUS program has produced abundant technical and operationally relevant accomplishments during the past 7 years. Perhaps the most significant has been the rapport and vital working relationships that developed between the research teams at NAWCTSD and NRaD and the operational and training commands of both the Atlantic and Pacific Fleets, in particular AEGIS cruisers and destroyers, the AEGIS training center, and its afloat and shore-based training groups. Those relationships have provided mechanisms for gaining insights and understanding of the most critical operational issues and problems as well as providing support for the transition of research products to other programs. The most noteworthy transitions to date are (a) an advanced embedded training development demonstration project and (b) an advanced development project on integrated decision support for command and control enhancement.

Publications are the products easiest to tabulate and catalogue. A conservative tally of all publications emanating from the TADMUS program yields a count of more than 250 publications in the form of journal articles, technical reports, book chapters, and symposium proceedings.

A compelling body of data based on laboratory experiments with experienced naval command teams convincingly has demonstrated that significant improvements, often at levels of 40%–50%, in both the quality and timeliness of tactical decision making occurred when the command teams used either the innovative training or decision support interventions. Performance improvements have been observed consistently in the quality and efficiency of communications, anticipating team members' information needs and other aspects of crew coordination, planning response actions to threats, maintaining an awareness of tactical situations, reducing occasions of blue-on-blue (engaging friendly forces) and blue-on-white (engaging neutral forces) incidents, and adhering to rules of engagement. Many of these results are discussed in detail in the following chapters.

Conclusion

The TADMUS program has been successful for many reasons, including the national attention focused on some high-profile naval incidents, the high quality of the researchers involved in the program, and the availability of adequate funding. Of paramount importance is the fact that the products of TADMUS are due to an unusual degree of cooperation between scientists and the operational navy. This synergy has produced some remarkably positive results for both communities.

References

Cannon-Bowers, J. A., Salas, E., & Grossman J. D. (1992). Improving tactical decision making under stress: Research directions and applied implications. In S. C. Collyer (Ed.), *The 27th International Applied Military Psychology Symposium: A focus on decision making research* (ONR Europe Report 92-4-W, pp. 49–71). London: Office of Naval Research.

Cohen, M. S., Freeman, J. T., & Wolf, S. P. (1996). Meta-recognition in time stressed decision making: Recognizing, critiquing, and correcting. *Human Factors, 38*, 206–219.

Cohen, M. S., Freeman, J. T., Wolf, S. P., & Militello, L. (1995). *Training metacognitive skills in naval combat decision making* (Cognitive Technologies Technical Report). Arlington, VA: Cognitive Technologies.

Crowe, W. J., Jr. (1988). *Second endorsement on Rear Admiral Fogarty's letter of 28 July 1988*. Unclassified Letter Ser. CM-1485-88 of 18 August 1988, to Secretary of Defense.

Fogarty, W. M. (1988). *Formal investigation into the circumstances surrounding the downing of a commercial airliner by the U.S.S. Vincennes (CG 49) on 3 July 1988*. Unclassified Letter Ser. 1320 of 28 July 1988, to Commander in Chief, U.S. Central Command.

House Armed Services Committee. (1989). *Iran Air Flight 655 compensation*. Hearings before the Defense Policy Panel, held August 3 and 4, September 9, and October 6, 1988 (HASC No. 100–119). Washington, DC: U.S. Government Printing Office.

Kahneman, D., Slovic, P., & Tversky, A. (Eds.). (1982). *Judgment under uncertainty: Heuristics and biases*. Cambridge, England: Cambridge University Press.

Klein, G. A. (1989). Do decision biases explain too much? *Human Factors Society Bulletin, 32*, 1–3.

Klein, G. A., Calderwood, R., & Clinton-Cirocco, A. (1986). Rapid decision making on the fire ground. In *Proceedings of the 30th Annual Human Factors Society meeting* (Vol. 1, pp. 576–580). Dayton, OH: Human Factors Society.

Marshall, S. P. (1995). *Schemas in problem solving*. Cambridge, England: Cambridge University Press.

Naval Studies Board. (1988). *Research opportunities in the behavioral sciences* (Naval Studies Board Report). Washington, DC: National Academy of Sciences.

Naval Studies Board. (1991). *ONR research opportunities in the behavioral sciences* (Naval Studies Board Report). Washington, DC: National Academy of Sciences.

Simon, H. A. (1957). *Models of man: Social and rational. Mathematical essays on rational human behavior in a social setting*. New York: Wiley.

Smith, D. E., & Marshall, S. P. (1997). Applying hybrid models of cognition in decision aids. In C. E. Zsambok & G. Klein (Eds.), *Naturalistic decision making* (pp. 331–341). Mahwah, NJ: Erlbaum.

Tolcott, M. A. (1992). Understanding and aiding military decisions. In S. C. Collyer (Ed.), *The 27th International Applied Military Psychology Symposium: A focus on decision making research* (ONR Europe Report 92-4-W, pp. 33–48). London: Office of Naval Research.

Tolcott, M. A., & Holt, V. E. (1987). *Impact and potential of decision research on decision aiding* (APA technical report). Washington, DC: American Psychological Association.

Zsambok, C. E., & Klein G. (Eds.). (1997). *Naturalistic decision making*. Mahwah, NJ: Erlbaum.

2

Individual and Team Decision Making Under Stress: Theoretical Underpinnings

Janis A. Cannon-Bowers and Eduardo Salas

As previously discussed, the Tactical Decision Making Under Stress (TADMUS) program was designed to (a) define the decision problem facing Navy tactical teams, (b) develop measures of tactical decision making (TDM) performance, (c) collect empirical data to document the impact of stress on TDM, and (d) develop and test principles for decision support, information display, training system design, and simulation that would mitigate these stress effects. Given the complexity of the environment (see Collyer & Malecki, chap. 1, this volume), the project team was presented with a formidable challenge: how to weed through the universe of things that we *could* do to improve performance and select only those interventions that would yield the most significant gains. We were motivated both practically and scientifically to answer this question. On the practical side, it was clear that a program afforded the resources and visibility of TADMUS needed to provide some demonstrable results fairly quickly. More important, we were committed from the outset to develop and test interventions that would actually be implemented in operational settings —we did not have the time or money to suffer many failures. From the scientific side, we were determined to make best use of what we considered to be a tremendous opportunity—the mandate and resources to study a difficult, real-world problem in depth. This was an opportunity that does not present itself very often.

Given this dilemma, we fell back to an old adage that traces back to Kurt Lewin: "There's nothing more practical than a good theory" (Marrow, 1969). Our reasoning was that if we adopted a few strong theoretical positions and generated promising hypotheses from these, our chances of success would be greatly enhanced. Therefore, several emerging areas of theory and research were exploited as a means to drive empirical work. The purpose of this chapter is to review briefly these theoretical perspectives. To do this, we first provide a description of the boundaries of the problem we were dealing with and offer definitions of key concepts. Next, we discuss the way we approached the problem of selecting theories that we believed could drive the specification of training interventions. Following this, we describe the major theoretical perspectives we adopted for the program. Finally, we summarize the major hypotheses associated with these theories and describe briefly the training interventions that were developed from them.

Boundaries of the Problem

To select theoretical perspectives appropriate for the problem with which we were dealing, it was necessary to establish the boundaries of the problem and delineate important definitions. In the following sections, we provide a description of the decision environment and task, the nature of stressors characteristic of the environment, and major definitions.

Defining the Decision Environment

Demands on the human decision maker in military tactical environments are becoming more complicated with advances in technology and changes in the world order. Modern combat scenarios are often characterized by rapidly evolving and changing conditions, severe time compression, and high degrees of ambiguity and uncertainty. In addition, such situations often present the decision maker with an overwhelming amount of data and require the coordinated performance of a team of operators who must gather, process, integrate, communicate, and act on these data in support of a tactical decision. A variety of other stressors (both physical and psychological) also exist in the operational setting, not the least of which is the catastrophic costs of making an error, which mitigate against effective individual and group performance. Coupled with the fact that the modern military scenario is likely to be complex and multinational, these factors have provided unprecedented demands on the human decision maker.

An example may help to illustrate this contention (see Johnston, Poirer, & Smith-Jentsch, chap. 3, this volume, for details). One of the tasks facing a team of operators in a Navy combat information center (CIC) is to defend the ship against hostile aircraft. This task is accomplished by a hierarchically structured team of operators/decision makers, with final decision authority retained by a single individual. Team members perform a variety of functions in support of the final decision: They operate sensor consoles to detect aircraft, integrate information collected regarding the aircraft's intent, make decisions about how and when to seek additional information, and make decisions about how and when to transmit pertinent situation assessment information. Once information is passed to a final decision maker, it must be considered against the situational constraints (e.g., rules of engagement) and potential consequences before a decision can be reached or action taken.

To function in such a situation, team members must understand the system at several levels. First, they must understand the dynamics and control of the equipment (both hardware and software) with which they are interacting to extract information. Second, they must understand the demands of the task and how to accomplish them (e.g., the significance of information, types of information required, strategies to combine information, necessary procedures, etc.). They must also understand how various facets of the environment affect the task and task demands (e.g., when workload increases as a function of air traffic in the area, or when radar reception is affected by weather conditions). Third, they must understand sound decision-making process so that they optimize the use of available information and avoid errors. Fourth, they must understand their role in the task, that is, what their particular contribution is, how they must interact with other team members, who requires particular classes of information, and so forth.

Finally, to perform optimally, tactical team members must be familiar with the knowledge, skills, attitudes, preferences, and other task-relevant attributes of their teammates. This is because expectations for the behavior of their teammates will vary as a function of the individuals who comprise the team. When working with a particularly competent teammate, for example, a team member may alter his or her behavior so that it is consistent with how he or she thinks that teammate will perform.

Defining Stressors in the Environment

There are many definitions of stress in the literature (e.g., Hogan & Hogan, 1982; Ivancevich & Matteson, 1980; Janis & Mann, 1977). For our purposes, we adopted the definition of *stress* offered by Salas, Driskell, and Hughes (1996) as "a process by which certain environmental demands . . . evoke an appraisal process in which perceived demand exceeds resources and results in undesirable physiological, psychological, behavioral, or social outcomes" (p. 6). Practically, our initial investigations involved extensive interviewing and observing of actual Navy crews. As a function of this effort, we found that the following characteristics, which can be defined as stressors, all appear to be present in the operational environment:

- multiple information sources
- incomplete, conflicting information
- rapidly changing, evolving scenarios
- requirement for team coordination
- adverse physical conditions
- performance pressure
- time pressure
- high work/information load
- auditory overload/interference
- threat

A second benefit of these interviews and observations was to determine which stressors are likely to have an impact on decision-making performance. Of these, it was necessary to select a subset of stressors that could be induced reliably and safely in experiments. It was decided that the following stressors were candidates for manipulation in experimental studies.

Task-related stressors

- workload/time pressure
- uncertainty/ambiguity
- auditory overload

Ambient stressors

- auditory interference
- performance pressure
- fatigue/sustained operations

Once potential stressors for manipulation were identified, a survey of the literature was conducted to determine the most common ways these stressors have been manipulated in past research. Manipulation techniques were assessed on the basis of reliability with which stress could be created, the face validity of the procedure, ethical concerns, safety, and feasibility. On the basis of this review and interaction with fleet personnel, the following initial stress manipulations were established.

Workload/time pressure. Workload and time pressure were manipulated using the tactical task (problem) scenario. Specifically, the number of tasks an operator performed simultaneously was increased as a manipulation of workload. Time pressure was defined as the time allowed to process a radar contact, and manipulated by changing the distance at which a potentially hostile contact was inserted into the scenario (i.e., the closer the contact was to own ship on initial detection, the less time the team had to prosecute it).

Uncertainty/ambiguity. Uncertainty was described as the amount of information available regarding a contact's intent. Under high uncertainty conditions less information was available, as, for example, when an identification mode was not available on a contact.

Auditory overload. Auditory overload refers to an overload of information being received through auditory channels. In actual CICs, operators are required to monitor several channels of auditory information simultaneously (both intraship and external communications) to receive information that is crucial to their task. Auditory overload was manipulated by creating scenario-specific tapes that could be fed into operator headsets during an exercise. Tapes included information that was pertinent to the task and scenario events.

Other Issues

There are a number of other decisions that we made early regarding the boundaries of the problem. The first is that a major portion of the effort would be devoted toward investigating and improving team performance. The importance of understanding team-level issues has been well documented (Guzzo & Salas, 1995; Hackman, 1990). In this particular case, it was clear that we were dealing with a *team* task according to the definition offered by Salas, Dickinson, Tannenbaum, and Converse (1992). That is, we were dealing with more than two individuals who held a shared, valued goal and who were interdependent (i.e., successful task completion depended on the coordination of multiple operators). Obviously, the decision to focus on team performance and training issues had a significant impact on the project; the extent of this impact is made apparent in the remainder of this chapter and others in this volume (see chaps. 8–10 and 12).

A second focus of the program was primarily on scenario or exercise-based training. By this we mean that we were interested in training interventions that were experiential—wherein trainees received some measure of hands-on practice on the task. This is not to say that other types of interventions (e.g., lecture based) were excluded from consideration, only that we emphasized scenario-based training. Our reasoning here was that first we were more interested in training advanced skills—teamwork and decision making—that would require a training strategy that allowed trainees to gain experience and receive feedback on task performance. In addition, we were most interested in developing training for application to deployed (shipboard) contexts; these were less conducive to "classroom-style" training than scenario-based training. Finally, although scenario-based training has application to a wide variety of tasks, we did not find much in the research literature to guide its development.

Selecting Theoretical Perspectives

As noted, we had a difficult challenge at the outset of the program—that is, to select particular training approaches that would yield the greatest gains in performance. To begin to cope with this challenge, we initially decided that there were three broad categories of interventions that we believed would be effective.

1. Increase overall performance readiness (i.e., knowledge and skill levels): We reasoned that individuals and teams highly competent in executing a task will be more resilient to the negative effects of stress on performance. Therefore, a goal of TADMUS was to improve methods to train tactical decision making in teams and individuals so that the likelihood of reaching task proficiency is enhanced and the amount of time required to reach proficiency is reduced.

2. Training stress coping skills: We reasoned that a class of training interventions would be geared toward training operators to cope with the stressors that confronted them in the task environment. For example, sev-

eral theorists have suggested that exposing trainees to stress in training can inoculate them from the impact of stress in the actual task performance situation (e.g., Novaco, Cook, & Sarason, 1983; Sarason, Johnson, Berberich, & Siegel, 1979). Methods to implement stress exposure techniques in a tactical decision-making environment were investigated in TADMUS (e.g., see Johnston & Cannon-Bowers, 1996).

3. Target individual and teamwork skills that are particularly vulnerable to the effects of stress: On the basis of results from baseline studies, we believed it possible to identify skills that are particularly vulnerable to the effects of stress. An effort was made to concentrate attention on these skills in training so as to reduce the impact of stress on tactical decision-making performance. For example, we know that the ability to communicate is affected by stress (e.g., Kleinman & Serfaty, 1989); hence we focused effort specifically on training communication skills.

With these three categories of intervention in mind, we were still left with the task of selecting theories that would guide the research. Initially, we looked toward the learning and instructional literature for possible answers. However, it quickly became apparent that this class of theories provided a very small portion of the solution. This is because instructional theories have a tendency to focus on what Salas and Cannon-Bowers (1997) refer to as the *methods of training*, that is, how to train. Some of these go as far as linking training task characteristics to specific training methods. Further, they describe the process of skill acquisition, that is, how people learn and generalize new knowledge and skills. What they do not do is provide information regarding the *content* of training. To understand this, more general theories of human performance are required.

Given what has been said thus far, we selected a number of theoretical perspectives that involve the manner in which individuals and teams perform and learn complex decision-making skills. We further divided these into what we consider "major" influences—those perspectives that pervaded our thinking across many training interventions, and less pervasive influences—those that had a more limited impact. Specifically, major theoretical perspectives include naturalistic decision making, team performance and effectiveness, and shared mental models.

All of these were adopted because they had something important to say about how individuals and teams cope with the demands and difficulties of a task. Essentially, each of these theoretical perspectives describe how people perform—how they gather and use information, how they perceive the task situation, how they interact with their teammates, and how they adjust their strategies in accordance with task demands. As such, they delineate "expert" performance and can provide a target for training. In particular, naturalistic decision making describes how people make complex, real-world decisions; team performance and effectiveness theories describe how expert teams perform and the requisite competencies needed for such performance—they begin to help us answer the question of how to turn a team of experts into an expert team; and shared mental model theory helps to explain how teams can be adaptable and flexible in response to stressors. Taken together, these perspectives cover the range of performance that we needed to train; each is described in more detail in later sections.

The more minor (but very important) theoretical perspectives we selected include automaticity, goal orientation, meta-cognition, and stress inoculation.

These perspectives all offer insight into how people perform and learn in the targeted environments. First, automaticity theory describes how overlearning of particular aspects of the task can benefit overall task performance. Meta-cogntion theory (which is associated with naturalistic decision making in some senses) explains how the meta-cognitive process can benefit both learning and performance of complex decision making. Goal orientation holds that attention can be focused on various outcomes in training and predicts that an emphasis on mastery of the task is superior to an emphasis on maximizing performance outcomes. Finally, borrowing heavily from the clinical psychology literature, stress inoculation suggests that people can be taught systematically to cope with the stressors likely to confront them in the operational environment. These perspectives are explained in more detail in the following sections.

It is clear that there are other important theoretical perspectives that we could have selected to guide our work. However, as a result of an extensive investigation of the operational problem and thorough review of the literature, we selected this subset because it covers the major aspects of learning and performance that we felt were most crucial. The next section provides more detail on these perspectives.

Major Theoretical Perspectives

At this point we can describe the theoretical perspectives associated with TADMUS. To do so, we first describe the theory itself and how it relates to the performance of interest. We then list briefly the implications of the theory for training—that is, what the theory might suggest by way of training design.

Naturalistic Decision Making

For many years, a set of theories commonly referred to now as classical decision-making theories were accepted by most decision researchers. Briefly, these theories tended to be prescriptive in nature, suggesting that decision makers should use a rational approach to decision making (see Klein, Orasanu, Calderwood, & Zsambok, 1993). However, in the mid- to late 1980s, a trend toward more "naturalistic" views of decision making began gaining popularity. According to Orasanu and Connolly (1993) this was because it was "not feasible to apply classical decision-making research analyses to many real-life situations" (p. 19) because they do not account for the expertise of the decision maker. Moreover, these authors and others (e.g., Brehmer, 1990) have argued that it is impossible to isolate decision making and study it apart from the task in which it resides. Hence, the naturalistic decision making movement can be broadly characterized as one involving the study of real-world task performance and action in which decisions happen to be embedded (for more explanation see Cannon-Bowers, Salas, & Pruitt, 1996; Cohen, 1993; Zsambok & Klein, 1997). It suggests that to conduct meaningful research, study participants, experimental tasks, and other details must be selected or constructed carefully to resemble closely how performance occurs in the real world.

Recognition-Primed Decision Making

One major theoretical approach to decision making that falls under the general heading of naturalistic decision making is recognition-primed decision making (RPD; see Klein, 1989). According to this theory, people interpret a situation in which they find themselves by comparing it with similar, previously experienced situations. When confronting a new situation, they use their memory of the old situation to create a tentative representation for the new situation. This representation accounts for the observed situation data, creates expectations about activities that should be observed in the future, and specifies constraints about the characteristics of the situation that may not be observed. The situation representation is continually tested with new data. Data that are consistent with expectations confirm the representation for predicting future events and for inferring characteristics about the event that may not have been observed. Disconfirming data can either refine the representation or indicate that it must be replaced altogether, but there is evidence that sometimes it is ignored or explained away.

In general, the naturalistic decision making family of theories has had a significant impact on TADMUS research. First of all, in general we decided at the outset of the program to conduct field investigations with actual Navy participants whenever possible. This was because we believed that our ability to generalize the results of our efforts to actual operators would be hurt drastically if we relied on laboratory tasks.

Second, naturalistic decision making suggests that expertise develops over time on the basis of the decision makers' experiences. As such, we concluded early on that the goal of decision-making training should be to accelerate the development of expertise (see Cannon-Bowers & Bell, 1997). According to Cannon-Bowers and Bell (1997) one way to accomplish this is through use of practice-based interventions such as simulation and training that is embedded into the task environment. Therefore, we adopted an event-based approach to training (Dwyer, Fowlkes, Oser, Salas, & Lane, 1997) that allowed us to craft scenarios from a set of events. These events are tied to training objectives and provide a basis for feedback. Over time, this process of providing "managed experiences" with appropriate performance feedback should lead decision makers to have knowledge structures and mental models necessary to confront novel situations.

Finally, the RPD theory itself has important implications as a basis on which to design training and decision support. In terms of training system design, the theory raises a number of questions regarding how to develop training so as to accelerate the acquisition of expertise. For example, how should scenarios be developed and structured so that they foster recognition; how should feedback be delivered so that maximum learning occurs; and what is the range of cues that decision makers must be exposed to in training? In terms of decision support, TADMUS research drew heavily on this theoretical perspective (see Morrison, Kelly, Moore, & Hutchins, chap. 16, this volume).

Team Performance and Effectiveness

Despite over 50 years of research in the team training area, little existed at the outset of the program to guide development of team training interventions. However, as noted, we were committed to studying team performance and training because it is obviously a crucial aspect of task performance. What follows is a very brief account of the current thinking in team performance and training when the program began (in the late 1980s), an overview of how things have progressed in the past 10 years, and a description of the implications of all of this for training. For more details on this topic, see Salas, Cannon-Bowers, and Blickensderfer (1997), and Salas, Bowers, and Cannon-Bowers (1995).

A Brief History of Team Performance and Effectiveness Research

In the mid- to late 1980s, researchers in the team area were concerned with several factors. First, a body of research on how teams evolve over time (e.g., Glickman et al., 1987; Morgan, Glickman, Woodard, Blaiwes, & Salas, 1986) was being developed. Briefly, this line of work sought to determine the phases of development that characterized team performance. Several important conclusions can be drawn from this work. First, teams appear to develop two related tracks of skill—those associated with the technical aspects of the task (labeled *taskwork*) and those associated with the team aspects of the task (labeled *teamwork*; see Morgan et al., 1986). In addition, it was found that teamwork skills are consistent across tasks and that they are related to the team's effectiveness (McIntyre & Salas, 1995).

In the early 1990s researchers turned their attention toward developing models of team effectiveness (e.g., Hackman, 1990; Salas et al., 1992; Tannenbaum, Beard, & Salas, 1992). These models generally attempted to relate characteristics both internal and external to the team with team effectiveness. For example, Tannenbaum et al. (1992) proposed an input–process–output model that included individual characteristics, team characteristics, task characteristics, and work structure as influences on team processes and outcomes. For a more complete description of this model see Tannenbaum et al., Salas et al. (1992), and Salas, Cannon-Bowers, and Johnston (1997).

Another trend in the 1990s was to begin specification of the knowledge, skills, and attitudes required for effective team performance. Early work in this area suggested that seven dimensions of team behavior seemed to best describe expert team performance. These include: mission analysis, assertiveness, adaptability/flexibility, situational awareness, decision making, leadership, and communication (see Prince, Brannick, Prince, & Salas, 1992; Prince & Salas, 1993). This work was an important beginning in determining the content of team training.

Subsequent effort in this area led Cannon-Bowers and her colleagues to propose that teams require several categories of competencies (i.e., knowledge, skills, and attitudes) to be effective (Cannon-Bowers, Tannenbaum, Salas, & Volpe, 1995). Briefly, these authors maintained that team competencies may be either generic or specific to the task, or generic or specific to the team. Combining these categories yields four classes of competencies required for teams—generic (i.e., those that apply across tasks and teams), task contingent (i.e., those that are specific to the task at hand but that apply across teams), team contingent (i.e., those that are specific to the team but not to the task), and context driven (i.e., those that are specific to both the task and the team).

It is beyond the scope of this chapter to list all of the competencies included in the Cannon-Bowers et al. (1995) framework. Some illustrations include the following:

1. Team knowledge requirements: cue/strategy associations, knowledge of teammate characteristics, shared task models, knowledge of team interaction patterns, and task sequencing
2. Team skill requirements: adaptability, shared situational awareness, mutual performance monitoring, communication, decision making, interpersonal skills, team leadership, assertiveness, and conflict resolution
3. Team attitude requirements: collective efficacy, shared vision, team cohesion, mutual trust, collective orientation, and importance of teamwork

The implications of these lines of inquiry for the development of TADMUS training interventions were vast. Essentially, these theoretical positions helped us to define the content of training—that is, which teamwork competencies needed to be trained. They also helped us to define the target of training by defining the characteristics of expert teams and team performance. In addition, they had an impact on the manner in which we developed performance measures in training (e.g., see Cannon-Bowers & Salas, 1997; Smith-Jentsch, Johnston, & Payne, chap. 4, this volume). Overall, the team performance and effectiveness literature provided a solid foundation on which to develop training interventions.

Shared Mental Model Theory

One of the important theories of team performance we relied on is shared mental model theory. The notion of mental models has been invoked as an explanatory mechanism by those studying skilled performance and system control for a number of years (Jagacinski & Miller, 1978; Rouse & Morris, 1986; Veldhuyzen & Stassen, 1977). According to Rouse and Morris (1986), a mental model can be defined as a "mechanism whereby humans generate descriptions of system purpose and form, explanations of system functioning and observed system states, and predictions of future system states" (p. 360).

In the area of cognitive psychology, researchers have suggested that mental models are important more generally to the understanding of how humans interact and cope with the world (Rouse & Morris, 1986). For example, Williams, Hollan, and Stevens (1983) maintain that mental models allow people to predict and explain system behavior, and help them to understand the relationship between system components and events. Wickens (1984) contended further that mental models provide a source of people's expectations. In an even more general view, Johnson-Laird (1983) suggested that people "understand the world by constructing working models of it in their mind" (p. 10). Mental models enable people to draw inferences and make predictions, to understand phenomena, to decide what actions to take, to control system execution, and to experience events vicariously (Johnson-Laird, 1983).

In reviewing the literature pertaining to mental models, Rouse and Morris (1986) concluded that a number of common themes can be drawn among theories that describe the purpose of mental models; namely that mental models serve to help people describe, explain, and predict system behavior. It must also be noted that most theorists conceptualize mental models as more than simple mental images. Instead, mental models are manipulable, enabling people to predict system states by mental manipulation of model parameters (see Johnson-Laird, 1983, for a detailed description of mental model functioning). Klein (1978) suggested, for example, that expert decision makers engage in a mental simulation that allows them to predict the ramifications of a potential decision before taking action.

With respect to training, a number of theorists have hypothesized that training that fosters development of accurate mental models of a system will improve performance. According to Rouse and Morris (1986), for example, one of the purposes of instruction is to develop mental models necessary to execute the task. Recent evidence suggests that the manner in which people cognitively structure information about a task has an impact on the way new information is assimilated and learned (cf. Eberts & Brock, 1987). Information that is compatible with existing mental models will be easier to learn (Wickens, 1984). Preexisting models of the task can also have an impact on training. Rouse and Morris (1986) contend in this regard that incorrect mental models can impede learning. Furthermore, evidence suggests that preexisting models may be difficult to eliminate; they appear to persist in the face of correct information (e.g., DiSessa, 1982). Finally, evidence suggests that to be most effective and generalizable, training must provide a conceptual model of the system, along with specific guidance or cueing in how general principles of system functioning are applied (Rouse & Morris, 1986).

Team Performance and Shared Mental Models

Research into team effectiveness has shown that effective teams can maintain performance even under conditions of high workload when communication opportunities are reduced (Kleinman & Serfaty, 1989). This ability has been labeled *implicit coordination*—it depends on the team's ability to draw on a common understanding of the task. Recently, several authors have hypothesized that the mechanisms that allow this type of performance are *shared mental models* (Cannon-Bowers, Salas, & Converse, 1993; Rouse, Cannon-Bowers, & Salas, 1992). Shared mental models allow team members to predict the needs of the task and anticipate the actions of other team members in order to adjust their behavior accordingly. In other words, team members appear to form expectations of what is likely to happen. Particularly when a novel situation arises, teams that cannot strategize overtly must exercise shared or common models of the task and team to maintain performance (Cannon-Bowers et al., 1993; Kleinman & Serfaty, 1989; McIntyre, Morgan, Salas, & Glickman, 1988; Orasanu, 1990). The role of mental models in explaining team behavior, then, stems from their ability to allow team members to generate predictions about task and team demands in absence of communication among team members (Cannon-Bowers et al., 1993).

The notion of shared mental models and how they relate to team effectiveness presented thus far has several implications for the understanding of team performance and training. As an explanatory mechanism, the team mental model construct is useful in understanding how teams are able to coordinate behavior and select task strategies in absence of explicit coordination activities. Under conditions of high workload, time pressure, and other kinds of stress, such implicit coordination appears to be critical (Kleinman & Serfaty, 1989).

With respect to training, the shared mental model idea suggests that training strategies designed to foster development of shared mental models has the potential to improve team performance. Such interventions must have as a goal the development in the trainees of a shared knowledge of the task and its demands, of the role of each team member in relation to all others, and of the characteristics of each of the other members of the team. Moreover, training for shared mental models must allow team members to build common models of the task and to develop and maintain accurate shared situation assessment.

Other Theoretical Perspectives

As noted, several other theoretical perspectives influenced TADMUS research. These are detailed next.

Automaticity

The theory of controlled versus automatic processing (see Fisk, Ackerman, & Schneider, 1987; Shriffin & Schneider, 1977) provides a basis on which to hypothesize that overtraining may be a viable means to help individuals be more resilient to stress effects. The theory postulates that certain task components can be "automatized"; that is, performed with little demand on cognitive resources. Automatized skills can be performed rapidly and effortlessly even under extremely exceptional conditions such as stress (Kirlik, Fisk, Walker, & Rothrock, chap. 5, this volume). The implication of this research for tactical decision making training is that it may be possible to train certain components of the decision-making task to automaticity free up critical resources for higher order cognitive functioning (Hodge et al., 1995). This is particularly critical in stressful or emergent situations. In terms of training system design, this research can lead to principles for overtraining decision makers, and for part-task training (i.e., training only selected aspects of the larger task).

Goal Orientation

Another theoretical perspective that has relevance to the training of tactical decision-making skills has to do with goal orientation in training (Kozlowski, chap. 6, this volume). Briefly, this is an intervention that can be implemented prior to practice sessions that focuses trainees' attention on their own mastery of the task rather than on their performance outcomes in training. According to Kozlowski (chap. 6, this voulme), a "mastery orientation" causes trainees to focus their attention on how well they are learning the task. As such, it helps to trigger meta-cognitive strategies in training (i.e., whereby trainees attend to their progress and learning state), which helps them to place emphasis on those aspects of the task that they may not understand or believe they have mastered. Mastery orientations also help to build self-efficacy—they focus the trainee's attention on successes so as to increase confidence. In contrast, a "performance" orientation places emphasis on how well the trainee is performing during training, stressing the performance outcome rather than the learning process.

In terms of training design, goal orientation has the potential to provide a relatively simple, low-cost mechanism to improve training effectiveness. In addition, because goal orientation drives trainees to consider different aspects of the task more deeply, it can also be expected that long-term retention and transfer will be enhanced.

Meta-Cognition

In conjunction with the naturalistic decision making movement, a number of researchers have argued that meta-cognitive skills can be crucial in stressful decision-making situations. It should be noted that in this case we are referring to meta-cognition during actual task performance, rather than meta-cognition during the learning process as discussed earlier. In terms of task performance, it has been argued that well-developed meta-cognitive skills can help decision makers assess the amount of time they have to make a decision, to avoid decision biases, and to reduce errors (Cohen, Freeman, & Wolf, 1996).

In terms of training, this line of thinking implies that training operators in task-specific meta-cognitive skills will improve performance. These skills involve the ability of the decision maker to assess his or her own decision-making process as the problem unfolds. In terms of the design of such training, it would seem that some type of scenario-based practice strategy could provide a vehicle to demonstrate and reinforce desired behaviors.

Stress Inoculation

As mentioned above, several researchers have suggested that exposing people to stress in training may inoculate them from the effects of stress in task performance (Ivancevich, Matteson, Freedman, & Phillips, 1990; Novaco, Cook, & Sarason, 1983; Zakay & Wooler, 1984). Although much of the research in this area has been conducted in clinical settings, it is clear that it had implications for task training. In fact, over the years, stress inoculation training has grown in popularity (Gebhardt & Crump, 1990). In fact, several cognitive–behavioral stress-coping training programs have been shown to be effective (Johnston & Cannon-Bowers, 1996; Meichenbaum, 1985; Smith, 1980).

According to Johnston and Cannon-Bowers (1996), *stress exposure training* (SET) refers to stress-coping training that extends beyond the clinical psychology domain. Further, SET has three objectives: to build skills that promote effective performance under stress, to build task-specific confidence or self-efficacy, and to enhance familiarity with the stress environment (Johnston & Cannon-Bowers, 1996). It rests on the notion that when people have accurate expectations regarding what to expect in the stress environment, have confidence in their ability to cope with such stressors, and have an opportunity to practice dealing with the stress so that appropriate skills can be developed, maximal performance can be expected.

The implications of this position for developing training are outlined in Johnston and Cannon-Bowers (1996). Briefly, they delineate a three-phased approach including (a) presentation of requisite knowledge regarding typical stress reactions, (b) practice building meta-cognitive skills and problem-solving skills, and (c) practice while being exposed to the actual stressors. Taken together, these activities should prepare trainees to cope with difficult environmental and task demands.

Hypotheses for Training

At this point it is possible to summarize the major hypotheses for training that we derived for empirical consideration. Table 1 provides an overview of the theoretical perspectives we considered under TADMUS along with the training interventions that flowed from them. Because the remaining chapters of this volume (as well as other work—see Appendix B at the end of this volume) describes these in detail and documents their effectiveness, we do not repeat them here. Instead, we briefly review each of these as they relate to the underlying theory or perspective.

In Table 1, the first column lists the theoretical perspective that we drew from. The second column provides a hypothesis for training that was generated from this perspective. The third column shows the training intervention that we developed on the basis of the hypothesis. The fourth column lists references pertinent to the training intervention (either theoretical or empirical). In the following sections we explain briefly the information in the table, organized by the training intervention; many of these are covered in more detail in other chapters.

1. Event-based training. This approach seeks to provide a systematic mechanism to insert crucial events into training scenarios. It is consistent with the naturalistic decision making perspective in that it allows scenario-based training to be structured in a manner that accelerates the development of expertise.

2. Team adaptation and coordination training. This intervention is designed to teach team members about the importance of teamwork and to introduce them to important teamwork skills. It grew out of work in the team performance area, using categories of teamwork skill that were derived from past effort. For more detail, see Serfaty, Entin, and Johnston (chap. 9, this volume).

3. Team leader training. The goal of this type of training is to instruct team leaders in sound observation and debriefing skills. It also stems from the literature into team effectiveness; greater detail regarding this intervention can be found in Tannenbaum, Smith-Jentsch, and Behson (chap. 10, this volume).

4. Cross-training. This intervention is consistent with both the team performance literature and shared mental model theory. The idea behind cross training is that team members can learn about the demands of the task from the perspective of their teammates. This should enable them to better anticipate the behavioral and information needs of their teammates. Cross training is reviewed by Blickensderfer, Cannon-Bowers, and Salas (chap. 12, this volume) and has been found to be effective in improving team performance (see Cannon-Bowers, Salas, Blickensderfer, & Bowers, 1998; Volpe, Cannon-Bowers, Salas, & Spector, 1996).

5. Interpositional knowledge training. Also consistent with shared mental model theory is the notion that training team members about the roles and responsibilities of their teammates will provide higher degrees of shared knowledge. This knowledge can then be used to generate predictions about teammate performance and reduce the need for overt communication (see Duncan

Table 1. Major Training Hypotheses and Interventions Tested Under Tactical Decision Making Under Stress (TADMUS)

Theoretical perspective	Hypothesis for training design	Training intervention	TADMUS Sources
NDM	Expertise can be accelerated by exposing trainees to carefully crafted scenarios.	Event-based approach to training	Dwyer, Oser, & Fowlkes, 1995; Johnston, Cannon-Bowers, & Smith-Jentsch, 1995; Dwyer et al., 1997
TP&E	Familiarizing trainees with teamwork skills can improve team performance.	Team adaptation and coordination training	Serfaty, Entin, & Johnston, chapter 9, this volume; Salas, Cannon-Bowers, & Johnston, 1997; Salas & Cannon-Bowers, 1997
TP&E	Training team leaders to properly observe and provide team feedback will improve team performance.	Team leader training	Cannon-Bowers, Tannenbaum, Salas, & Volpe, 1995; Tannenbaum, Smith-Jentsch, & Behson, chapter 10, this volume; Kozlowski, Gully, McHugh, Salas, & Cannon-Bowers, 1996
TP&E, SMMs	Cross training will allow team members to build accurate shared mental models of the task.	Cross training	Cannon-Bowers, Salas, & Converse, 1993; Rouse, Cannon-Bowers, & Salas, 1992; Volpe, Cannon-Bowers, Salas, & Spector, 1996; Blickensderfer, Cannon-Bowers, & Salas, chapter 12, this volume; Cannon-Bowers, Salas, Blickensderfer, & Bowers, 1998
SMMs	Training team members to better understand roles and responsibilities of their teammates will enhance shared mental models and improve team performance.	Interpositional knowledge training	Duncan et al., 1996; Salas, Cannon-Bowers, & Johnston, 1997

SMMs, TP&E	Self-correction training	Training teams to give themselves meaningful feedback will help to build shared mental models and improve performance.	Blickensderfer, Cannon-Bowers, & Salas, 1997; Smith-Jentsch, Payne, & Johnston, 1996; Blickensderfer, Cannon-Bowers, & Salas, 1994
Automaticity	Part-task training	Extracting and training consistent aspects of the task will free up resources for more difficult aspects of the problem.	Hodge et al., 1995; Kirlik, Fisk, Walker, & Rothrock, chapter 5, this volume
Meta-cognition	Meta-cognitive training	Helping trainees to be aware of their decision-making process during performance will improve decision-making performance.	Cohen, Freeman, & Wolf, 1996; Cohen, Freeman, & Thompson, chapter 7, this volume
Goal orientation	Mastery learning	Helping trainees to focus their attention on their own learning will improve their confidence and performance.	Kozlowski, chapter 6, this volume; Kozlowski, Gully, Smith, Nason, & Brown, 1995
Stress inoculation	Stress exposure training	Exposing trainees to the stressors likely to confront them and training coping skills will improve performance.	Driskell & Johnston, chapter 8, this volume; Inzana, Driskell, Salas, & Johnston, 1996; Johnston & Cannon-Bowers, 1996; Saunders, Driskell, Johnston, & Salas, 1996

Note. NDM = naturalistic decision making; TP&E = team performance and effectiveness; SMMs = shared mental models.

et al., 1996, for a more detailed description of the intervention and a test of its effectiveness).

6. Team self-correction training. Yet another intervention that stems from the team effectiveness area and shared mental model theory involves team self-correction. This intervention instructs team members in how to engage in posttraining feedback sessions that enable them to enhance their shared knowledge. A number of researchers have explored this type of intervention (e.g., Blickensderfer, Cannnon-Bowers, & Salas, 1997).

7. Part-task training. Consistent with theory of automaticity, a part-task training intervention was developed and tested. The notion here is that the consistent aspect of the task can be isolated and trained so that it does not require cognitive resources during task performance (i.e., because these resources are better applied to higher order aspects of the problem). For more detail, see Kirlik, Fisk, Walker, and Rothrock (chap. 5, this volume).

8. Meta-cognitive training. An intervention based on meta-cognition was developed. This intervention was designed to train decision makers to use meta-cognitive strategies as a means to cope with difficult decision-making situations (see Cohen, Freeman, & Thompson, chap. 7, for a description of the intervention).

9. Mastery learning. Consistent with the notion of goal orientation in training, an intervention was developed that uses a mastery learning orientation. The idea here is to focus the trainees' attention on their own state of learning. Kozlowski (chap. 6, this volume) provides more detail on this intervention.

10. Stress exposure training. Given what was said about stress exposure training, this intervention seeks to implement and test the three-phased approach described earlier. (For more detail on this intervention, see Johnston and Cannon-Bowers, 1996; Driskell and Johnston, chap. 8, this volume).

Summary and Conclusion

Our purpose in this chapter was to provide background information regarding the theoretical underpinnings of TADMUS. As noted, we attempted to cope with the problem of choosing training interventions to test (and predicting which would be most effective) by adopting and elaborating a set of theoretical perspectives regarding performance in the types of tasks we were interested in. Of course, there are probably many other training interventions that may have be effective—no single research program can exhaust all possibilities. However, we are confident that in adopting this theoretically based approach we were able to develop and test several meaningful and fruitful training solutions. The remainder of the chapters elaborate the major findings of the effort.

References

Blickensderfer, E. L., Cannon-Bowers, J. A., & Salas, E. (1994). Feedback and team training: Exploring the issues. *Proceedings of the Human Factor and Ergonomics Society 38th annual meeting* (pp. 1195–1199). Nashville, TN: Human Factors and Ergonomics Society.

Blickensderfer, E. L., Cannon-Bowers, J. A., & Salas, E. (1997, April). *Training teams to self-correct: An empirical investigation.* Paper presented at the 12th annual meeting of the Society for Industrial and Organizational Psychology, St. Louis, MO.

Brehmer, B. (1990). Strategies in real-time, dynamic decision making. In R. Hogarth (Ed.), *Insights in decision making: A tribute to Hillel J. Einhorn* (pp. 262–279). Chicago: University of Chicago Press.

Cannon-Bowers, J. A., & Bell, H. R. (1997). Training decision makers for complex environments: Implications of the naturalistic decision making perspective. In C. Zsambok & G. Klein (Eds.), *Naturalistic decision making* (pp. 99–110). Hillsdale, NJ: Erlbaum.

Cannon-Bowers, J. A., & Salas, E. (1997). A framework for developing team performance measures in training. In M. T. Brannick, E. Salas, & C. Prince (Eds.), *Assessment and measurement of team performance: Theory, research, and applications* (pp. 45–62). Hillsdale, NJ: Erlbaum.

Cannon-Bowers, J. A., Salas, E., Blickensderfer, E. L., & Bowers, C. A. (1998). The impact of cross-training and workload on team functioning: A replication and extension of initital findings. *Human Factors, 40,* 92–101.

Cannon-Bowers, J. A., Salas, E., & Converse, S. A. (1993). Shared mental models in expert team decision making. In N. J. Castellan, Jr. (Ed.), *Individual and group decision making: Current issues* (pp. 221–246). Hillsdale, NJ: Erlbaum.

Cannon-Bowers, J. A., Salas, E., & Pruitt, J. S. (1996). Establishing the boundaries for decision-making research. *Human Factors, 38,* 193–205.

Cannon-Bowers, J. A., Tannenbaum, S. I., Salas, E., & Volpe, C. E. (1995). Defining team competencies and establishing team training requirements. In R. Guzzo & E. Salas (Eds.), *Team effectiveness and decision making in organizations* (pp. 330–380). San Francisco: Jossey-Bass.

Cohen, M. (1993). The bottom line: Naturalistic decision making. In G. Klein, J. Orasanu, & R. Calderwood (Eds.), *Decision making in action: Models and methods* (pp. 265–269). Norwood, NJ: Ablex.

Cohen, M. S., Freeman, J. T., & Wolf, S. (1996). Metacogntition in time-stressed decision making: Recognizing, critiquing, and correcting. *Human Factors, 38,* 206–219.

DiSessa, A. A. (1982). Unlearning aristotelian physics: A study of knowledge-based learning. *Cognitive Science, 6,* 37–75.

Duncan, P. C., Rouse, W. B., Johnston, J. H., Cannon-Bowers, J. A., Salas, E., & Burns, J. J. (1996). Training teams working in complex systems: A mental model-based approach. In W. B. Rouse (Ed.), *Human/Technology Interaction in Complex Systems, 8,* 173–231.

Dwyer, D. J., Fowlkes, J. E., Oser, R. L., Salas, E., & Lane, N. E. (1997). Team performance measurement in distributed environments: The TARGETs methodology. In M. T. Brannick, E. Salas, & C. Prince (Eds.), *Assessment and management of team performance: Theory, research, and applications* (pp. 137–154). Hillsdale, NJ: Erlbaum.

Dwyer, D. J., Oser, R. L., & Fowlkes, J. E. (1995, October). *Symposium on distributed simulation for military training of team/groups: A case study of distributed training and training performance.* Paper presented at the 39th annual Human Factors and Ergonomics Society Meeting, Santa Monica, CA.

Eberts, R. E., & Brock, J. F. (1987). Computer-assisted and computer managed instruction. In G. Salvendy (Ed.), *Handbook of human factors* (pp. 976–1011). New York: Wiley.

Fisk, A. D., Ackerman, P. L., & Schneider, W. (1987). Automatic and controlled processing theory and its application to human factors problems. In P. A. Hancock (Ed.), *Human factors psychology* (pp. 159–197). New York: North-Holland.

Gebhardt, D. L., & Crump, C. E. (1990). Employee fitness and wellness programs in the workplace. *American Psychologist, 45,* 262–272.

Glickman, A. S., Zimmer, S., Montero, R. C., Guerette, P. J., Campbell, W. J., Morgan, B.

B., Jr., & Salas, E. (1987). *The evolution of teamwork skills: An empirical assessment with implications for training* (Report No. TR87-016). Orlando, FL: Naval Training Systems Center.

Guzzo, R., & Salas, E. (Eds.). (1995). *Team effectiveness and decision making in organizations.* San Francisco: Jossey Bass.

Hackman, J. R. (Ed.). (1990). *Groups that work (and those that don't): Creating conditions for effective teamwork.* San Francisco: Jossey-Bass.

Hodge, K. A., Rothrock, L., Kurlik, A. C., Walker, N., Fisk, A., Phipps, D. A., & Gay, P. E. (1995). Training for tactical decision making under stress: Towards automatization of component skills (Tech. Rep. No. HAPL-9501). Atlanta, GA: Human Attention and Performance Laboratory, Georgia Institute of Technology.

Hogan, R., & Hogan, J. C. (1982). Subjective correlates of stress and human performance. In E. A. Alluisi, & E. A. Fleishman (Eds.), *Human performance and productivity: Stress and performance effectiveness.* Hillsdale, NJ: Erlbaum.

Inzana, C. M., Driskell, J. E., Salas, E., & Johnston, J. H. (1996). Effects of preparatory information on enhancing performance under stress. *Journal of Applied Psychology, 81,* 429–435.

Ivancevich, J. M., & Matteson, M. T. (1980). *Stress and work: A managerial perspective* (pp. 141–163). Glenview, IL: Scott, Foresman.

Ivancevich, J. M., Matteson, M. T., Freedman, S. M., & Phillips, J. S. (1990). Worksite stress management interventions. *American Psychologist, 45,* 252–261.

Jagacinski, R. J., & Miller, R. A. (1978). Describing the human operator's internal model of a dynamic system. *Human Factors, 20,* 425–433.

Janis, I., & Mann, L. (Eds.). (1977). *Decision making.* New York: Free Press.

Johnson-Laird, P. (1983). *Mental models.* Cambridge, MA: Harvard University Press.

Johnston, J. H., & Cannon-Bowers, J. A. (1996). Training for stress exposure. In J. E. Driskell & E. Salas (Eds.), *Stress and human performance* (pp. 223–256). Mahwah, NJ: Erlbaum.

Johnston, J. H., Cannon-Bowers, J. A., & Smith-Jentsch, K. A. (1995). Event-based performance measurement system for shipboard command teams. *Proceedings of the First International Symposium on Command and Control Research and Technology* (pp. 274–276). Washington, DC: Institute for National Strategic Studies.

Klein, G. A. (1978). Phenomenological vs. behavioral objectives for training skilled performance. *Journal of Phenomenological Psychology, 9,* 139–156.

Klein, G. A. (1989). Recognition-primed decisions. In W. B. Rouse (Ed.), *Advances in man–machine systems research* (pp. 47–92). Greenwich, CT: JAI Press.

Klein, G. A., Orasanu, J., Calderwood, R., & Zsambok, C. E. (1993). *Decision making in action: Models and methods.* Norwood, NJ: Ablex.

Kleinman, D. L., & Serfaty, D. (1989). Team performance assessment in distributed decision making. In R. Gilson, J. P. Kincaid, & B. Goldiez (Eds.), *Proceedings: Interactive Networked Simulation for Training Conference* (pp. 22–27). Orlando, FL: Naval Training Systems Center.

Kozlowski, S. W. J., Gully, S. M., McHugh, P. P., Salas, E., & Cannon-Bowers, C. A. (1996). A dynamic theory of leadership and team effectiveness: Developmental and task contingent leader roles. *Research in Personnel and Human Resources Management, 14,* 253–305.

Kozlowski, S. W. J., Gully, S. M., Smith, E. M., Nason, E. R., & Brown, K. G. (1995, April). *Learning orientation and learning objectives: The effects of sequenced mastery goals and advance organizers on performance, knowledge, meta-cognitive structure, self-efficacy, and skill generalization for complex tasks.* Paper presented at the 10th Annual Conference of the Society for Industrial and Organizational Psychology, Orlando, FL.

Marrow, A. J. (1969). *The practical theorist: The life and work of Kurt Lewin.* New York: Basic Books.

McIntyre, R. M., Morgan, B. B., Jr., Salas, E., & Glickman, A. S. (1988). Teamwork from team training: New evidence for the development of teamwork skills during operational training. *Proceedings of the 10th Interservice / Industry Training Systems Conference* (pp. 21–27). Orlando, FL: National Security Industry Association.

McIntyre, R. M., & Salas, E. (1995). Measuring and managing for team performance: Emerging principles from complex environments. In R. Guzzo & E. Salas (Eds.), *Team effectiveness and decision making in organizations* (pp. 149–203). San Francisco: Jossey-Bass.

Meichenbaum, D. (1985). *Stress inoculation training*. New York: Pergamon.

Morgan, B. B., Jr., Glickman, A. S., Woodard, E. A., Blaiwes, A. S., & Salas, E. (1986). *Measurement of team behaviors in a Navy training environment* (Report No. TR-86-014). Norfolk, VA: Old Dominion University, Center for Applied Psychological Studies.

Novaco, R., Cook, T., & Sarason, I. (1983). Military recruit training: An arena for stress-coping skills. In D. Meichenbaum & M. Jaremko (Eds.), *Stress reduction and prevention* (pp. 377–418). New York: Plenum.

Orasanu, J. (1990). *Shared mental models and crew performance* (Tech. Rep. No. 46). Princeton, NJ: Princeton University.

Orasanu, J., & Connolly, T. (1993). The reinvention of decision making. In G. Klein, J. Orasanu, R. Calderwood, & C. E. Zsambok (Eds.), *Decision making in action: Models and methods* (pp. 3–20). Norwood, NJ: Ablex.

Prince, A., Brannick, M. T., Prince, C., & Salas, E. (1992). Team process measurement and implications for training. *Proceedings of the 36th Annual Meeting of the Human Factors Society* (pp. 1351–1355). Santa Monica, CA: Human Factors Society.

Prince, C., & Salas, E. (1993). Training and research for teamwork in the military aircrew. In E. L. Wiener, B. G. Kanki, & R. L. Helmreich (Eds.), *Cockpit resource management* (pp. 337–366). Orlando, FL: Academic Press.

Rouse, W. B., Cannon-Bowers, J. A., & Salas, E. (1992). The role of mental models in team performance in complex systems. *IEEE Transactions on Systems, Man, and Cybernetics, 22,* 1296–1308.

Rouse, W. B., & Morris, N. M. (1986). On looking into the black box: Prospects and limits in the search for mental models. *Psychological Bulletin, 100,* 349–363.

Salas, E., Bowers, C. A., & Cannon-Bowers, J. A. (1995). Military team research: Ten years of progress. *Military Psychology, 7,* 55–75.

Salas, E., & Cannon-Bowers, J. A. (1997). Methods, tools, and strategies for team training. In M. A. Quinones & A. Ehrenstein (Eds.), *Training for a rapidly changing workplace: Applications of psychological research* (pp. 249–279). Washington, DC: American Psychological Association.

Salas, E., Cannon-Bowers, J. A., & Blickensderfer, E. L. (1997). Enhancing reciprocity between training theory and practice: Principles, guidelines, and specifications. In J. K. Ford & Associates (Eds.), *Improving training effectiveness in work organizations* (pp. 291–322). Hillsdale, NJ: Erlbaum.

Salas, E., Cannon-Bowers, J. A., & Johnston, J. H. (1997). How can you turn a team of experts into an expert team? Emerging training strategies. In C. Zsambok & G. Klein (Eds.), *Naturalistic decision making* (pp. 359–370). Hillsdale, NJ: Erlbaum.

Salas, E., Cannon-Bowers, J. A., & Kozlowski, S. W. J. (1997). The science and practice of training: Current trends and emerging themes. In C. Zsambok & G. Klein (Eds.), *Naturalistic decision making* (pp. 359–370). Hillsdale, NJ: Erlbaum.

Salas, E., Dickinson, T. L., Tannenbaum, S. I., & Converse, S. A. (1992). Toward an understanding of team performance and training. In R. W. Swezey & E. Salas (Eds.), *Teams: Their training and performance* (pp. 3–29). Norwood, NJ: Ablex.

Salas, E., Driskell, J. E., & Hughes, S. (1996). Introduction: The study of stress and human performance. In J. E. Driskell & E. Salas (Eds.), *Stress and human performance* (pp. 1–45). Mahwah, NJ: Erlbaum.

Sarason, I. G., Johnston, J. H., Berberich, J. P., & Siegel, J. M. (1979). Helping police officers to cope with stress: A cognitive–behavioral approach. *American Journal of Community Psychology, 7,* 593–603.

Saunders, T., Driskell, J. E., Johnston, J. H., & Salas, E. (1996). The effect of stress inoculation training on anxiety and performance. *Journal of Occupational Health Psychology, 1,* 170–186.

Shiffrin, R. M., & Schneider, W. (1977). Controlled and automatic human information processing: II. Perceptual learning, automatic attending, and a general theory. *Psychological Review, 84,* 127–190.

Smith, R. E. (1980). A cognitive/affective approach to stress management training for athletes. In C. H. Nadeau, W. R. Halliwell, K. M. Newell, & G. C. Roberts (Eds.), *Psychology of motor behavior and sport—1979* (pp. 54–73). Champaign, IL: Human Kinetics.

Smith-Jentsch, K. A., Payne, S. C., & Johnston, J. H. (1996, April). Guided team self-correction: A methodology for enhancing experiential team training. In K. A. Smith-Jentsch (chair), *When, how and why does practice make perfect?* Paper presented at the 11th annual conference of the Society for Industrial and Organizational Psychology, San Diego, CA.

Tannenbaum, S. I., Beard, R. L., & Salas, E. (1992). Team building and its influence on team effectiveness: An examination of conceptual and empirical developments. In K. Kelley (Ed.), *Issue, theory, and research in industrial/organizational psychology* (pp. 117–153). Amsterdam: Elsevier.

Veldhuyzen, W., & Stassen, H. G. (1977). The internal model concept: An application to modeling human control of large ships. *Human Factors, 19,* 367–380.

Volpe, C. E., Cannon-Bowers, J. A., Salas, E., & Spector, P. (1996). The impact of cross training on team functioning. *Human Factors, 28,* 87–100.

Wickens, C. D. (1984). *Engineering psychology and human performance.* Columbus, OH: Charles E. Merrill.

Williams, M. D., Hollan, J. D., & Stevens, A. L. (1983). Human reasoning about simple physical systems. In D. Gentner, & A. L. Stevens (Eds.), *Mental models* (pp. 131–153). Hillsdale, NJ: Erlbaum.

Zakay, D., & Wooler, S. (1984). Time pressure, training, and decision effectiveness. *Ergonomics, 27,* 273–284.

Zsambok, C., & Klein, G. (1997). *Naturalistic decision making—Where are we now?* Mahwah, NJ: Erlbaum.

3

Decision Making Under Stress: Creating a Research Methodology

Joan H. Johnston, John Poirier, and Kimberly A. Smith-Jentsch

A major Tactical Decision Making Under Stress (TADMUS) program requirement was to develop a theoretically based research methodology that would allow for consistent experimental control over collecting baseline data on tactical decision making (TDM) under stress and testing hypotheses on individual and team training interventions (Cannon-Bowers, Salas, & Grossman, 1991). Creating a viable methodology was challenging because the focus of the research—air warfare in a ship combat information center (CIC)—is a fast-paced, stressful, and complex team task that is difficult to replicate in a laboratory setting. Conducting research on board ships and in high-fidelity team training facilities was not possible at the beginning of the program because few teams were available to draw conclusions about the viability of training interventions. Therefore, conducting research meant developing laboratories with adequate levels of cognitive fidelity so that the research products could be generalized to high-fidelity environments. Furthermore, a research methodology was required that would stimulate a team to perform a highly complex task; previously unheard of in team and group research. Nevertheless, the research methodology, including the development of two laboratory test beds, was completed and implemented.

The main purpose of this chapter is to describe important lessons learned from our experience in developing the TADMUS research methodology. Those lessons provide the foundation for the chapters in this book. This chapter is organized into four sections: a description of the CIC air warfare task environment, an explanation of how the research methodology was developed, a discussion on validation of the methodology, and, finally, a list of important lessons learned.

The Task Environment

An initial TADMUS task was to conduct traditional, cognitive, and team task analyses to specify CIC operator cognitive and psychomotor tasks (including equipment requirements), team task interdependencies, and stressors (e.g., Department of the Navy, 1989; Duncan et al., 1996; Zachary et al., 1991). This information was crucial in establishing a theoretically based rationale for the design and development of research test beds, training materials, performance measures, identification of manipulable stressors, and identification of appropriate training interventions. Following is a brief description of the CIC environment including air warfare team level tasks, air warfare team member tasks, and stressors.

The CIC is the central information processing and decision-making area for a surface combatant. It is where sensor data are displayed and analyzed, weapons control systems are staffed and operated, command and control functions are performed, and most tactical decisions are made. The Navy's composite warfare command doctrine specifies that CIC teams be partitioned into subteams that specialize in functionally different warfare areas: air, surface, and subsurface (Department of the Navy, 1985). Each warfare area is composed of team members who provide surveillance and tactical information through a chain of command to the primary CIC decision makers, the tactical action officer (TAO) and the commanding officer (CO).

The main objective of the air warfare team is to provide surveillance for ownship defense against such potential air threats as missiles and aircraft (Department of the Navy, 1989). The team performs a series of tasks: detecting, tracking, identifying radar contacts, taking action on these contacts, and performing battle damage assessment in the event of weapons release. Specifically, the TADMUS air warfare team members of interest were the electronic warfare supervisor, tactical information coordinator, identification supervisor, the air warfare coordinator, the TAO, and the CO. The electronic warfare supervisor reports detection and identification of potentially hostile air, surface, and subsurface contacts. The identification supervisor is responsible for resolving identification conflicts and assigning identification to tracks. The tactical information coordinator manages tracking and identification of aircraft and surface contacts. The air warfare coordinator manages continuous review and evaluation of the air tactical situation. And, finally, the TAO and CO are responsible for tactical employment and defense of the ship.

The team hierarchy is such that the CO is the primary decision maker, with the TAO as second in command. The air warfare coordinator and electronic warfare supervisor report directly to the TAO, the identification supervisor reports directly to the tactical information coordinator, who reports to the air warfare coordinator (Department of the Navy, 1989). The CIC team makes decisions according to "command by negation." That is, a team member will report his or her intentions to the superior and the superior will only respond if he or she does not want the team member to follow through. A strong interdependency exists among team members because full task execution requires extensive knowledge and skill with respect to coordination, exchanging, and validating critical ship sensor information and relating it to the tactical situation, or the current status of potential threats (Serfaty, Entin, & Volpe, 1993; Volpe, Cannon-Bowers, Salas, & Spector, 1996).

Each air warfare team member operates a complex set of computer equipment that shows current status of ship sensor information in the form of text and geographical displays. Figure 1 shows a bird's eye view of a typical CIC with each air warfare operator watchstation chair noted in black (Department of the Navy, 1989). The team's communications are linked through these displays, through multiple internal and external verbal communications nets operated with headset gear, and through physical cues.

The requirement for rapid, brief, and accurate team communications combined with accurate and timely manipulation of complex watchstation consoles displays creates a considerable amount of pressure to perform. The critical factors affecting performance, according to subject matter experts (SMEs), include time pressure, sustained operations (fatigue), performance pressure, workload, auditory overload, and information ambiguity. The latter three are considered to be top stressors due to the fast pace characterizing of the air warfare environment. For example, accurate team-based decisions may be required within about 1 minute after detection of aircraft by ship sensors. As a result of such pressures, team members may exhibit a narrowing of attention, inaccurate data assessment, slowed responses, and erroneous and biased conclusions (Driskell & Salas, 1991; Perrin, Barnett, & Walrath, 1993).

The task analyses revealed that TDM differs from the more traditional or formalized types of decision making commonly studied in the research literature. Therefore, alternative theories to explain and predict TDM processes were needed to support TADMUS research methodology development (see Cannon-Bowers & Salas, chap. 2, this volume). Such a theoretical framework was found in a relatively new movement or paradigm in decision-making research called naturalistic decision making (Zsambok & Klein, 1997). The next section discusses the theoretical underpinnings of naturalistic decision making and how they apply to TDM.

Research Methodology Development

Means, Salas, Crandall, and Jacobs (1993) argued that decision making in such settings as the CIC must be highly adaptive to extreme time pressure, uncertainty and information ambiguity, high workload, team coordination demands and task complexity. Indeed, it is just such an environment to which naturalistic decision making applies. In fact, Orasanu and Connolly (1993) identified eight factors that typify decision making in naturalistic environments: ill-structured problems; uncertain, dynamic environments; shifting, ill-defined or competing goals; multiple event-feedback loops; time constraints; high stakes; multiple players; and organizational norms and goals that must be balanced against the decision maker's personal choice. Such environments stand in sharp contrast to those in which the more traditional type of decision making occurs (e.g., decision makers

Typical Combat Information Center Watchstation Configuration

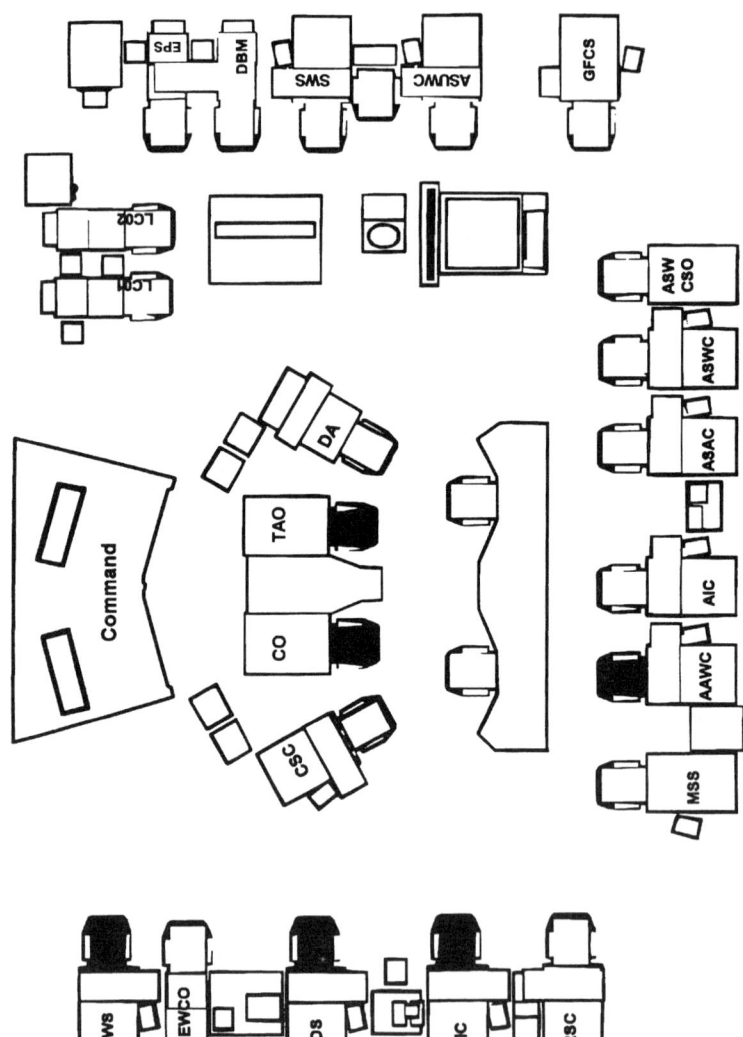

Figure 1. Typical Ship combat information center. EWS = electronic warfare supervisor; EWCO = electronic warfare commanding officer; IDS = identification supervisor; TIC = tactical information coordinator; RSC = radar system coordinator; CSC = combat system coordinator; MSS = missile system supervisor; AAWC = anti-air warfare coordinator; AIC = air intercept controller; ASAC = anti-submarine air coordinator; ASW = anti-submarine warfare; ASWC = anti-submarine warfare commander; GFCS = gun fire control supervisor; ASUWC = anti-surface warfare commander; SWS = surface warfare supervisor; DBM = database manager; EPS = engagement planning supervisor; CO = commanding officer; TAO = tactical action officer; DA = display assistant; LS = launch supervisor.

have sufficient time to generate options, conduct option assessment, and select a course of action; the consequences of an incorrect response are not severe; decisions are reached by consensus of a group of decision makers; and workload is manageable).

What, then, is naturalistic decision making? Zsambok (1997) defined it as "the way people use their experience to make decisions in field settings" (p. 4). Decision makers recognize familiar elements of a situation on the basis of their past experience and existing memory structures ("schemata") then use that information to evaluate the current situation (i.e., to create "mental models" or make "situation assessments"; Lipshitz & Orit, 1997). In other words naturalistic decision making involves, among other things, a form of recognition-primed decision making.

Cannon-Bowers and Bell (1997) and others have asserted that the acquisition of naturalistic decision making skills is contingent on the context in which it is applied (Zsambok & Klein, 1997). They have recommended that as much of the knowledge-rich task environment as possible be provided to the trainee in order to develop the appropriate cue–strategy relationships embedded in the task. First, we applied this guidance to methodology development to create a dynamic environment without compromising experimental control. To maintain a realistic task environment we operationally defined the independent variable of stress as consisting of the three most common CIC operator stressors: workload, information ambiguity, and auditory overload. Second, we chose to use a team-level CIC simulation test bed to include five CIC team members. Indeed, Navy CIC team training had adopted this approach for some time, whereby multiple aircraft, missiles, and conflicting problems were incorporated into training scenarios so that team members could practice multiple tactical strategies (Duncan et al., 1996).

The next step in methodology development was to create specific realistic, but stressful air warfare scenarios for a team of CIC operators. Because the primary goal was to develop a baseline of TDM performance and test the impact of various training strategies on mitigating stressors, we determined it would be best to compare current Navy team training strategies to our proposed training strategies. We decided that our methodology would adapt, with some specific and important changes, current air warfare team training exercises that were conducted in high-fidelity CIC simulations at various shore-based training facilities (Duncan et al., 1996). The CIC simulation is capable of running tactical scenarios that present the same information that operators see on their shipboard consoles. A typical training session involves a team of instructors leading a prebrief (summary of the tactical situation and ownship capabilities for handling it) on training scenario details, then team members working a scenario displayed on the simulated combat console. The scenarios are designed to contain many types of problems converging on the team, simulating high workload and ambiguity. Instructor-led debriefing covers a variety of tactical issues and problems the team may encounter.

Specific changes and refinements to this approach involved developing the air warfare scenarios so that stressors could be manipulated and individual and team performance processes and outcomes could be tracked and assessed. To this end, the next three sections briefly describe the approach we developed to create an effective research protocol for conducting experiments on tactical decision making under stress. It included development of event-based scenarios, embedded scenario stressors, and the research test bed.

Event-Based Scenarios

Early on in the TADMUS program, we determined that assessment of individual and team performance *processes and outcomes* would enable us to develop a baseline of TDM and to assess the impact of the training interventions on mitigating stressors (Johnston, Smith-Jentsch, & Cannon-Bowers, 1977; Smith-Jentsch, Payne, & Johnston, 1996). Consequently, it was determined that identification of multiple, prespecified training events would allow for meaningful comparison of performance under different levels of stress, and for comparison of individual and team measures of processes and outcomes.

The event-based approach to scenario development guided efforts in developing two pairs of scenarios. Navy SMEs with extensive operational experience from a number of training commands assisted in the development of one pair of scenarios situated in the northern Arabian (Persian) Gulf, and the second pair situated in the southern part of the Gulf near the Straits of Hormuz. Each scenario pair shared three "priority threat" events that SMEs agreed would require the full attention of the air warfare team. In accordance with naturalistic decision making theory, SMEs were encouraged to develop the scenarios to represent the essential characteristics and demands of TDM on a CIC air warfare team in littoral warfare operations. They include operations in shallow/confined waters, neutral and hostile countries in close proximity, ambiguous rules of engagement, combined force interactions, modern military weapons systems among neutral and friendly forces, heavy concentration of neutral/friendly traffic, and requirements to coordinate and communicate with the chain of command located external to the ship. Hall, Dwyer, Cannon-Bowers, Salas, and Volpe (1993) present a detailed description of scenario events for one of the pairs.

Included in scenario development were the typical briefing materials Navy trainees receive before conducting a team training simulation exercise: geopolitical background, rules of engagement, ship's mission, ship weapons and firing status, battle force organization, identification matrix, identification/warning call procedures, communication plan, status board information, situation update, maneuvering board, call signs and identification friend or foe (IFF) codes, and threat summary. A role player script was created to enhance the realism of the scenario, add auditory overload, and control responses to team member requests for resources or information from sources external to the five-person team.

Embedded Scenario Stressors

High-stress scenarios in each pair were created by adding visual and auditory workload and information ambiguity. Greater workload was created by increasing the number of aircraft, ships, false contacts, and the number and complexity of requests for information from the team members using a role player script. Hall et al. (1993) proposed that by increasing the number of contacts with known information, team member mental and communications activity would increase with respect to recognizing, reporting, identifying, monitoring, and taking action. Information ambiguity was created by increasing the number of potentially hostile relationships among aircraft and ships that had to be "deconflicted" by the team members (Hall et al., 1993). Hall et al. proposed that adding contacts with very little available information would also increase cognitive workload and team communications activities among team members. To determine whether the scenarios would have the expected impact on performance and perceptions, an initial assessment of the effect of added stressors was made; the results of that assessment are described in later sections.

Test Bed Development

As the information on air warfare tasks and stressors was being gathered and scenarios were being developed, a parallel effort ensued to identify a test bed simulation for air warfare teams. To maintain experimental control, we determined that choosing a low physical fidelity simulation was acceptable as long as cognitive fidelity in a team simulation was maintained and naturalistic decision making activities would not be compromised. Later research literature supported our assumptions by concluding that physical fidelity could be minimized to conduct team coordination research (Bowers, Salas, Prince, & Brannick, 1992; Driskell & Salas, 1992; Travillian, Volpe, Cannon-Bowers, & Salas, 1993; Volpe et al., 1996; Weaver, Bowers, Salas, & Cannon-Bowers, 1995).

We determined that two team simulation test beds would be developed: (a) the Decision Making Evaluation Facility for Tactical Teams (DEFTT), including a stand-alone version called the Georgia Institute of Technology Aegis Simulation Platform (GT-ASP), and (b) the Tactical Navy Decision Making (TANDEM) system. DEFTT was designed for use with experienced Navy trainees, whereas TANDEM was designed for use with novice Navy trainees and naive participants (Dwyer, Hall, Volpe, Cannon-Bowers, & Salas, 1992). The following section provides brief descriptions of DEFTT, GT-ASP, and TANDEM.

DEFTT. To develop a test bed for use with experienced Navy trainees, a systematic review of Navy training and research devices was conducted. DEFTT was chosen because it met the following requirements for such a facility. **Cognitive fidelity** was adequate in that it provided all necessary tactical cues to simulate a multithreat battle problem (Holl & Cooke,

1989). The system software had been designed, developed, and configured to support a variety of tactical training requirements at three Navy shore-based training facilities. It could be configured using a local area network to allow **air warfare team activities** to take place (e.g., detecting, identifying, and acting). DEFTT incorporated an experimental control station that contained a **scenario exercise generation and execution system** for controlling and updating a network of at least five operator consoles, and it had an **automated time-tagged performance** data capturing system for some of the consoles. In addition, **development, portability, and maintenance** costs were acceptable because it had been developed from low cost commercial-off-the-shelf hardware and software (Holl & Cooke, 1989). **Ease of use** was a consideration in choosing the DEFTT system because preparation for a research experiment required just 30 minutes.

Figure 2 shows the layout of the DEFTT system configuration, including an experimenter control station (ECS; a Hewlett Packard 9000 Series) that we used to generate and control the experimental air warfare scenarios following their development on paper. The ECS was connected through a local area network with six IBM-compatible 486 personal computers that served as operator workstations. A large-screen monitor was situated next to the TAO's station to simulate the AEGIS CIC large-screen display. The TAO was provided with a simulated AEGIS display status (ADS); the air warfare coordinator, tactical information coordinator, and identification supervisor each had simulated command and decision displays. The electronic warfare supervisor had both a command and decision display and a simulated early warning radar station (SLQ-32) display.

Figure 3 shows an example of the DEFTT command and decision display. The radar screen shows ownship (designated with a plus sign) in the center of the display surrounded by unknown, hostile, and friendly "tracks." Unknown tracks are designated as boxes (surface craft) and half boxes (air or subsurface craft). Hostile tracks are illustrated as diamonds and half diamonds. Friendly tracks are represented by circles and half circles. An operator "hooks" a track with a track ball mouse so that information about it (e.g., identification, evaluation, track number, bearing, range, depth, speed, and course) is displayed to the left of the radar screen on the "character readout." An operator at a DEFTT console uses the keyboard or trackball mouse to execute operations listed at the bottom of the display shown in Figure 3 (i.e., ownship engine rudder, aircraft control, weapon control, track manager, and display control).

Each scenario was designed to run on DEFTT for approximately 30 to 35 minutes. A multichannel communications system provided two (internal and external) channels for each team member. All verbal communications among team members were recorded, and a videocamera trained on the TAO's ADS display recorded actions in manipulating the display. At least two people were required for conducting the experiments: one for operating the ECS, conducting DEFTT familiarization training, managing data recording devices, and acting as the role player, and a second for conducting research interventions and questionnaire distribution.

To conduct DEFTT research to include teams of Navy students the

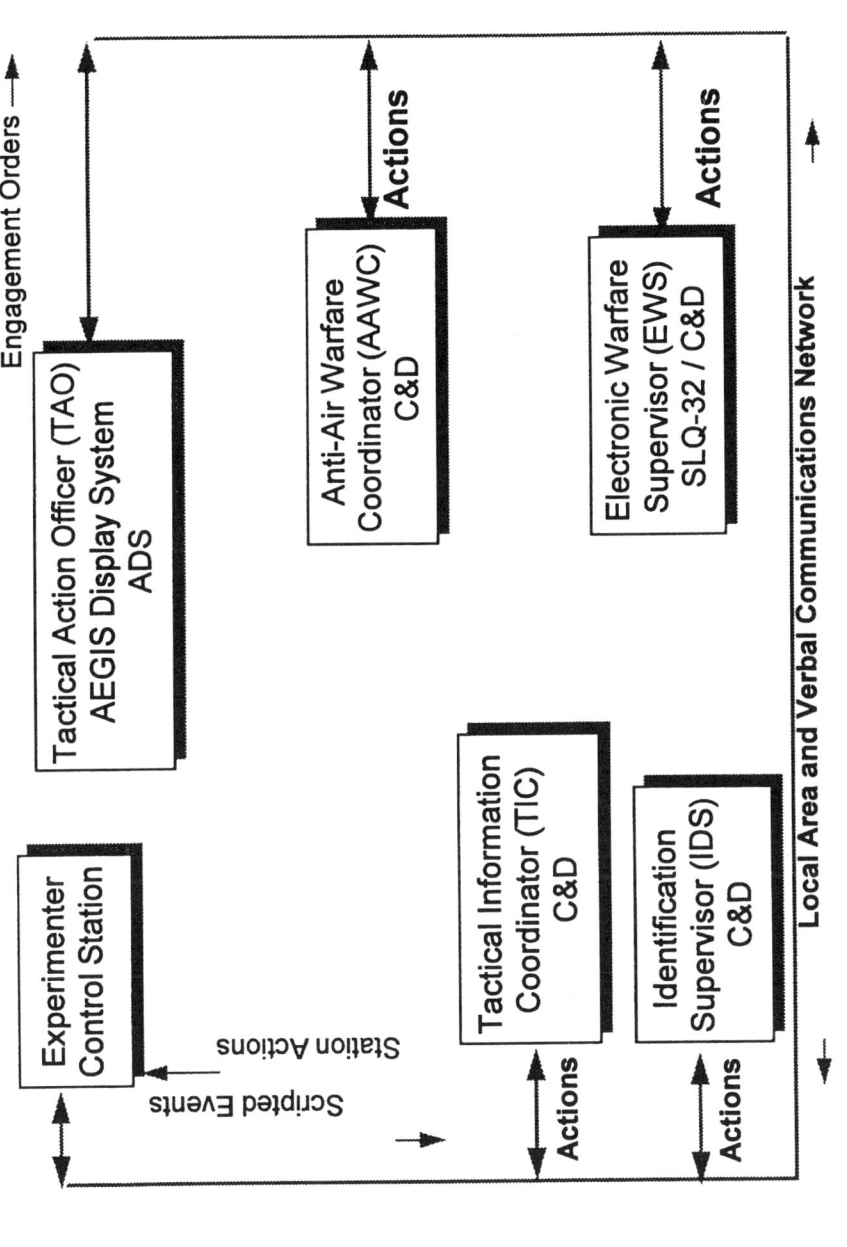

Figure 2. Decision Making Evaluation Facility for Tactical Teams (DEFTT) system configuration. C&D = command and decision.

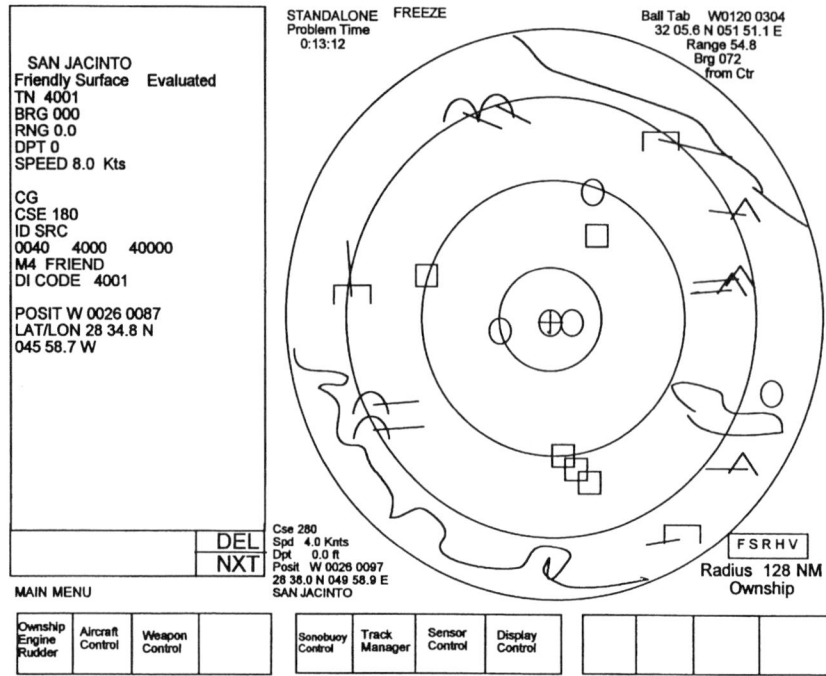

STANDALONE   FREEZE
Problem Time
0:13:12

Ball Tab   W0120 0304
32 05.6 N 051 51.1 E
Range 54.8
Brg 072
from Ctr

SAN JACINTO
Friendly Surface   Evaluated
TN 4001
BRG 000
RNG 0.0
DPT 0
SPEED 8.0 Kts

CG
CSE 180
ID SRC
0040   4000   40000
M4 FRIEND
DI CODE 4001

POSIT W 0026 0087
LAT/LON 28 34.8 N
045 58.7 W

DEL
NXT

Cse 280
Spd 4.0 Knts
Dpt 0.0 ft
Posit W 0026 0097
28 38.0 N 049 58.9 E
SAN JACINTO

F S R H V
Radius 128 NM
Ownship

MAIN MENU

| Ownship Engine Rudder | Aircraft Control | Weapon Control | | Sonobuoy Control | Track Manager | Sensor Control | Display Control | | | | | |

Figure 3.   Decision Making Evaluation Facility for Tactical Teams (DEFTT) command and decision display.

DEFTT system was installed at two Navy shore-based training facilities. In addition, in 1993 the primary Navy shore-based training facility for TAOs had installed the DEFTT system as a permanent training device. Consequently, most of the research data were obtained on ad hoc teams at this facility.

GT-ASP. Following development of the DEFTT system, a stand-alone version of the command and decision display was modified to conduct individual part-task training experiments (similar to Figure 3) at the Georgia Institute of Technology. They labeled the system the Georgia Institute of Technology-AEGIS Simulation Platform. A number of advanced tools were developed to support scenario scripting and generation, automated performance data capture, automated feedback, and an on-line performance support system during training. Details can be found in Kirlik, Rothrock, Walker, and Fisk (1996).

TANDEM. A main reason for developing TANDEM was to conduct basic individual and team training experiments that were not ready to be introduced directly to experienced Navy trainees. Consequently, TANDEM was developed to serve as a stand-alone or a three-person team simulation with operator radar displays and pull-down menus that require tactical decision-making activities much less complicated to master than DEFTT, but include gathering, integration, and interpretation of tactical information. Like DEFTT, TANDEM has scenario generation, scenario execution, and performance data capture modules, as well as performance measurement and feedback tools (Dwyer et al., 1992). Figure 4 shows a bird's eye view of the TANDEM system configured as three IBM compatible computers connected through a local area network and updated by an ECS (Delta). Team members Alpha, Bravo, and Charlie communicated using headsets.

Figure 5 shows an example of a TANDEM display screen. The primary objective of the team task is to correctly identify and process multiple radar targets. Similar to DEFTT, scenario workload and information ambiguity can be manipulated. Each unknown target may be classified as one of six types of contacts or tracks (submarine civilian, submarine military, surface civilian, surface military, aircraft civilian, and aircraft military). Accessing the information menus (similar to the DEFTT character readout) listed to the right of the radar display, the team must decide whether the contact is hostile or friendly and destroy or clear that target on the basis of the target's intentions. A target correctly identified and processed (i.e., destroyed or cleared) will provide the team with a set number of points that accumulate to a total team score at the end of a scenario. The team loses points if they make an identification error. To achieve optimal performance, the team members must share information and work together to arrive at a correct decision with each contact.

A number of individual and team experiments have been conducted with this system. (See Blickensderfer, Cannon-Bowers, & Salas, 1997; Dwyer, 1995; Inzana, Driskell, Salas, & Johnston, 1996; Johnston, Driskell & Salas, 1997; Kozlowski, chap. 6, this volume, for descriptions of research methodologies. More details on the TANDEM capabilities can be found in Dwyer et al., 1992.)

Validation Experiments

To assess the viability of the research methodology developed for DEFTT, three questions were posed based on the requirements set forth in previous sections: (a) Can experimental procedures be conducted in a reliable and economical manner? (b) Are relevant and realistic task elements identified and included in the methodology? (c) Do stressors significantly affect preliminary measures of performance and perceptions? DEFTT experiments that were conducted to establish a reliable and valid methodology are described below.

Background

In 1993, a DEFTT laboratory was set up at a Navy training command on the east coast in order to begin testing the research methodology. Data were collected on four intact Aegis ship air-warfare teams. Each person represented one of the air warfare positions (TAO, air warfare coordinator,

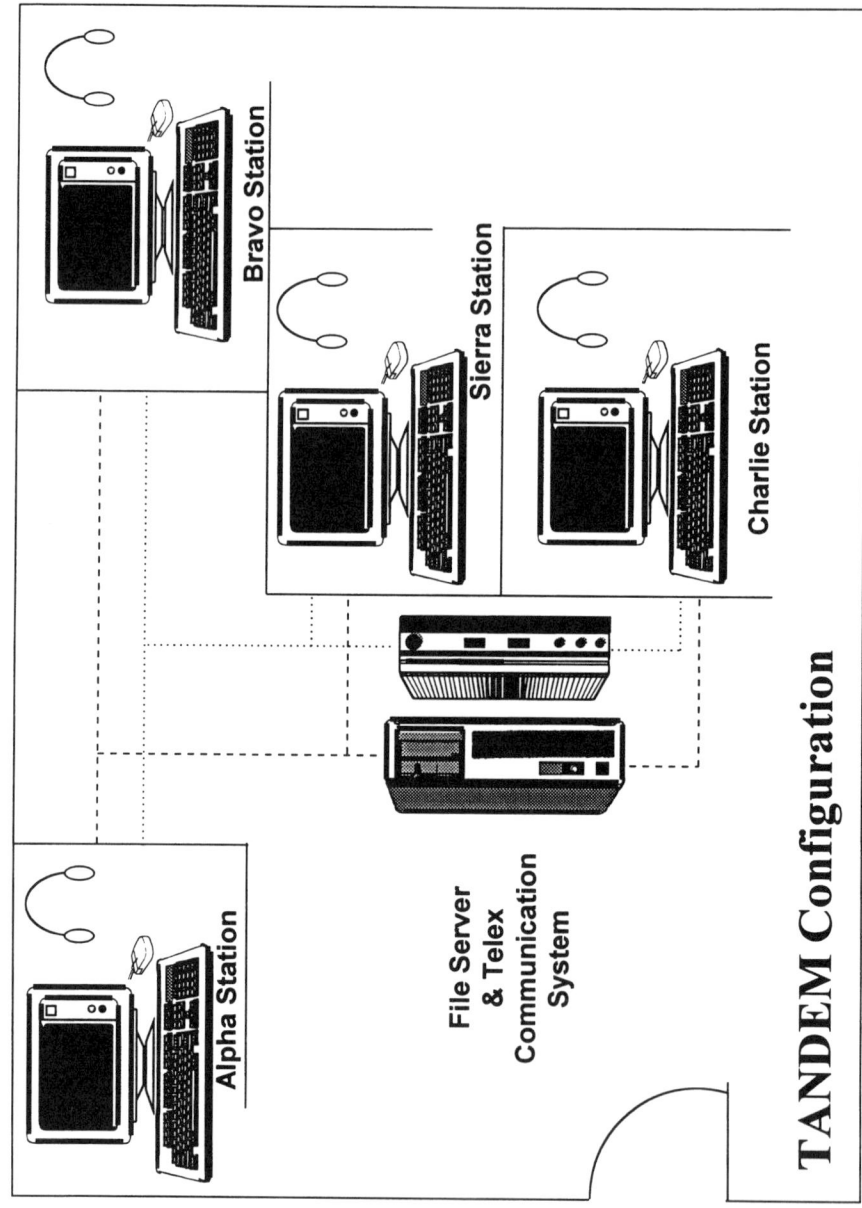

Figure 4. Tactical Naval Decision Making (TANDEM) system configuration.

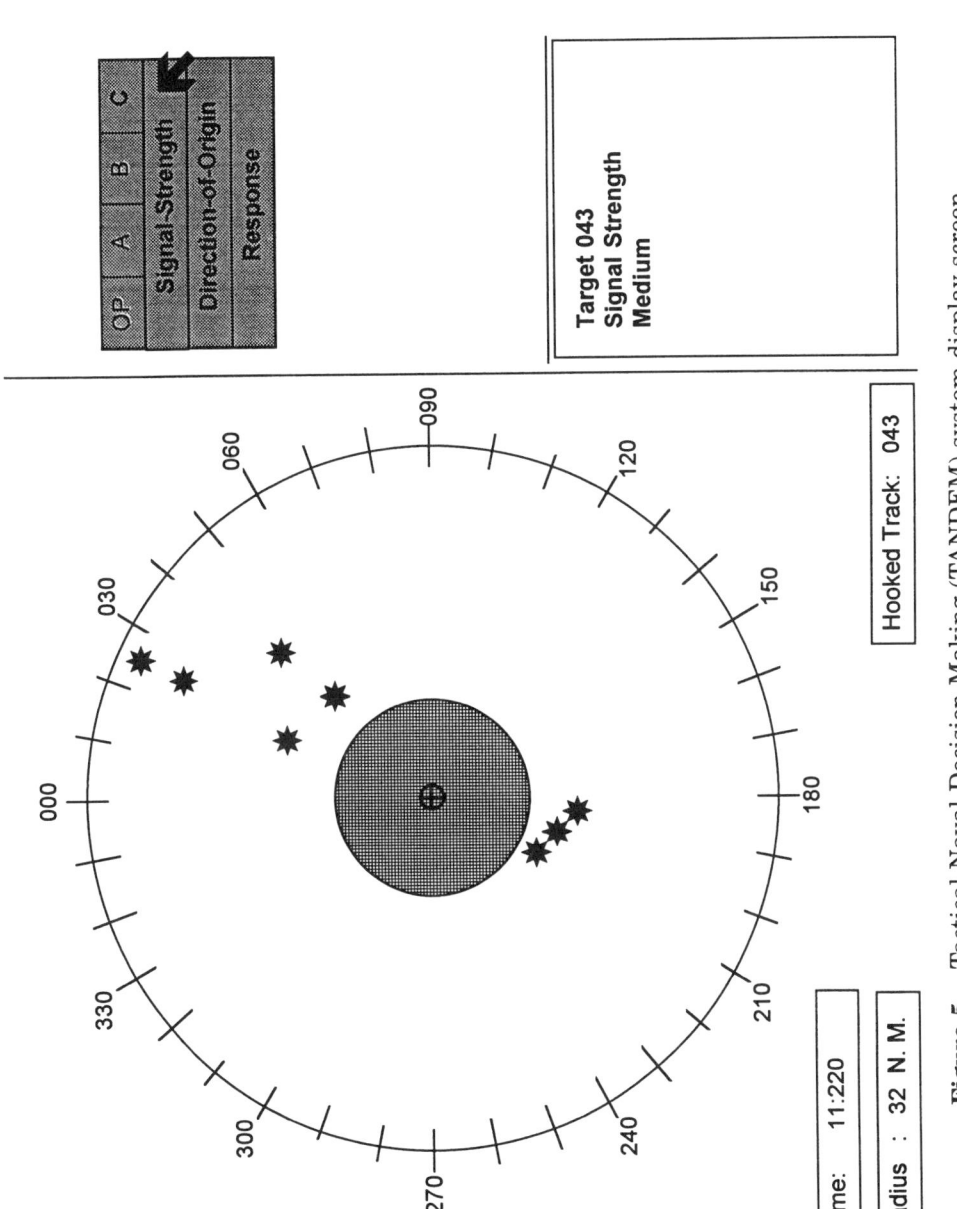

Figure 5.   Tactical Naval Decision Making (TANDEM) system display screen.

tactical information coordinator, identification supervisor, and electronic warfare supervisor). In addition, a second DEFTT system being used as a training device at another Navy training facility allowed us to collect data on moderately experienced ad hoc teams. Therefore, data were available to compare performance at two levels of team experience (moderate and high). Scenario stress level (moderate and high) represented another independent variable. All of the teams participated in at least one pair of the moderate and high stress scenarios. Whenever possible, attempts were made to partially control for training effects by counterbalancing scenarios across teams.

The dependent variables included assessment of individual performance, team performance, and perceived stress. The Behavior Observation Booklet (BOB) was developed to assess, on a 5-point scale, the adequacy of individual task performance with respect to the priority threat events in each of the four scenarios (see Hall et al., 1993, for greater detail on its development). The Air Warfare Team Performance Index (ATPI) served as a measure of overall adequacy of team taskwork performance on the air warfare detect-to-engage sequence and is described in greater detail by Johnston, Smith-Jentsch, and Cannon-Bowers (1997). To assess stress perceptions, a Likert-type scale was developed to determine perceived workload, ambiguity, and general stress with respect to each scenario.

Procedure

Because of constraints placed on the research by participant availability, the experimental procedure had to be designed to last 1 day per team. In many respects, the procedure was designed to be similar to typical simulation training evolutions that we observed at a number of Navy training commands (Duncan et al., 1996). It began with a 90-minute DEFTT familiarization training that included team role familiarization, individualized buttonology practice on operator consoles, team practice on two 15-minute DEFTT training scenarios, and a paper-and-pencil test to establish DEFTT knowledge. Included in the familiarization training was a list of procedural artificialities. For example, DEFTT did not simulate operator station configuration and equipment layout, which actually plays a significant role in how the team accomplishes the air warfare task. However, operator station functionality was sufficiently simulated. The air warfare "team" was composed of five members for research purposes, whereas on board ship the air warfare team is at least twice that number. The ad hoc teams consisted of officers only, whereas a ship CIC team includes both officers and enlisted personnel. The communications network consisted of all internal communications feeding over one channel and all external communications feeding over a second channel, whereas shipboard communications are fed over multiple auditory channels. The teams were requested to perform the simulation exercise under those constraints (e.g., verbally communicating actions that may have been passed by electronic messages).

Next, teams were provided a general briefing on the Arabian Gulf scenarios that included rules of engagement, geopolitical background, ship's mission, weapons and firing status, battle force organization, identification matrix, identification/warning call procedures, communication plan, and status board information. Just prior to each scenario a "situation update" prebrief was provided that included specific details on ship's mission, a maneuvering board with significant aircraft and ships at the start of the problem, call signs and IFF codes, and a threat summary. Following the situation update, team members were allowed to organize themselves according to the designated TAO's requirements for handling warning call procedures and assignment of aircraft and surface ship identification responsibilities. When scenarios were initialized on operator displays at the start of the problem team members were told to use specific communications they would normally use in a CIC so that the role player could handle the responses. Following scenario completion, teams were allowed to debrief themselves with respect to how the scenario unfolded.

Reliable and Economical Procedures

On average, each scenario run lasted 60 minutes, including the prebriefing and debriefing sessions. We found that all four scenarios could be run over the course of a day with little variation in procedures. It was determined that training interventions could be tested using the four scenarios with a pre–post test design if the intervention was limited to about 2 hours. Our conclusion was that a reasonably reproducible and economical procedure had been established on the basis of trial runs. Although post hoc, surveys and interviews with research participants and experiment support staff provided valuable input to make changes to increase experimental control. For example, to keep the scenarios true to their original design, a detailed role-player script was developed to control participants' and support staffs' responses to the team's requests for information and to prevent the team and the role player from changing the prescribed scenario.

Relevant and Realistic Task Elements Identified

In addition to testing the reliability of experimental procedure, anecdotal responses from intact ship teams and ad hoc teams reported that, procedural artificialities aside, the task elements were relevant and realistic compared with actual shipboard CIC experiences. Many of the ad hoc teams complained that the role-player communications were overwhelming the team members' ability to communicate among themselves. In a number of these instances, Navy instructors supporting the research reminded trainees that information overload was a common occurrence in a CIC and that exposure to it was a good preparatory training experience. These reactions were a fairly consistent finding in subsequent research experiments. In general, most teams did not recognize similarities in scenario pairs because the distracting stressors in the high-stress scenarios masked them.

Stressors Significantly Affecting Performance and Perceptions

Individual performance. It was expected that individuals in the high-stress condition would perform more poorly on their tasks than in the moderate stress condition, and that the individuals with more experience would perform better than those with moderate experience. Before evaluating individual performance, two SMEs practiced with the *Behavior Observation Booklet* (BOB) to evaluate a single experienced ship team's members that had been videotaped performing on the scenarios. The SMEs practiced with the BOB until they achieved consensus on member performance. Then, each SME evaluated each experimental team member's performance across eight events for each scenario. Westra (1994) reported an acceptable index of rater agreement ($T = .74$). The higher stress scenario caused poorer team member performance on three out of the eight events. Nevertheless, a majority of the experienced team members were not significantly affected; the more experienced air warfare coordinator performed significantly better than the inexperienced air warfare coordinator in both the moderate- and high-stress scenarios. This was to be expected because the scenarios focused very heavily on air warfare.

Team performance. It was expected that teams in the low-stress condition would perform significantly better than teams in the high-stress condition and that the more experienced teams would perform significantly better than the less experienced teams. Prior to evaluating team performance, two SMEs practiced applying the ATPI to an experienced team that had performed on the scenarios. Then, the SMEs practiced with the ATPI until they achieved consensus on team performance. Next, each SME evaluated the four experimental experienced ship teams with the ATPI. Westra reported interrater agreement was weak ($T = .40$). Only one of these raters evaluated the four moderately experienced teams. Results showed that both teams performed more poorly in the high-stress scenarios on five of the six air warfare tasks. However, the more experienced teams performed better than the less experienced teams on five of the six tasks (Westra, 1994).

Stressor perceptions. The stressor perception scale was developed for the specific purpose of estimating reactions to the scenario stressors. Following each scenario run, team members were requested to rate on a 7-point scale their reactions to scenario workload, time pressure, ambiguity, and overall stress level. Westra (1994) reported weak trends in perceived stress differences within scenario pairs, but it was found that participants ($n = 69$) perceived significantly higher levels of stress when comparing reactions between the moderate-stress scenario from one pair and the high-stress scenario from the second pair.

Conclusion

These findings appear to support the validity of the research methodology. However, we do not want to overstate the reliability and validity of the performance measures and the stressor perceptions measure. They were experimental tools in and of themselves and had inherent deficiencies that had to be improved (e.g., labor intensive, lacked behavioral specificity, low rater agreement, lack of internal reliability assessments). In addition, there were limitations with respect to the application of the tools in that only one and two raters had participated in assessing performance. The stressor perceptions measure had individual items for each type of stressor, which limits internal consistency reliability. In summary, initial efforts in validating the methodology were just that—initial efforts.

Although tradeoffs were made to maintain experimental control by continuing experiments with DEFTT, training research was eventually conducted at high-fidelity training sites and in ship CICs using the same experimental procedures (Bailey, Johnston, Smith-Jentsch, Gonos, & Cannon-Bowers, 1995; Smith-Jentsch, Payne, & Johnston, 1996). Efforts to compare data collected from all three sites indicate the procedure and at least one training intervention can generalize to the real-task environment (Smith-Jentsch et al., 1996).

Lessons Learned for Conducting Team Research in Complex Environments

Lesson 1: Multiple Kinds of Task Analyses Enable Establishing a Theoretically Based Methodology for Studying Decision Making Under Stress

The cognitive, team, and traditional task analyses performed helped to establish the theoretical underpinnings of the methodology. For example, the cognitive task analysis led to identifying naturalistic decision making theory to guide maintaining multiple stressors in our scenarios. Our results enabled stronger generalization of TDM to high-fidelity training environments.

Lesson 2: Teams Can Get Engaged and Perceive Scenarios to be Realistic in a Test Bed that Elicits the Cognitive Requirements of the Task

Physical fidelity may be impossible and unnecessary to achieve, but as long as cognitive fidelity is maintained, team decision-making research in complex environments can be addressed. However, high-fidelity team research environments should be identified prior to developing a low-fidelity simulation to ensure that transfer of research methods does not require costly changes.

Lesson 3: Scenarios for Studying Decision Making in Complex Environments Need to Be Carefully Crafted

Our findings showed that it was necessary to include as much realism in the scenarios as possible. Therefore, we enlisted the help of SMEs as much as we could, and we conducted iterative tests of the scenarios early in the research to establish realism. However, there will always be artificialities, and these should be identified early to ensure research participant expectations are met and to plan for role players and other supporting experimental procedures.

Lesson 4: The Event-Based Approach to Assessing Complex Performance Works

Scenario development should consist of designing "events" that elicit the behaviors of interest. For example, to assess the impact of complex stressors on TDM, scenario "pairs" should be developed that share events so that comparisons of performance can be made under similar but more stressful conditions. Care should be taken in adding stressors that will not distract teams from performing the shared events.

Lesson 5: Conduct a Pre-Test of the Methodology with Experienced Participants to Ensure Future Research is Cost Effective

Experienced teams will tell you where your procedure and simulation are faulty and what you should be able to supply in terms of needed information. Early in the research, experienced participants were invaluable in providing us the information we required to enhance realism without increasing costs.

Lesson 6: A Variety of Test Beds Helps Develop a Better Understanding of Decision Making Under Stress

Although we were able to collect a considerable amount of data on DEFTT, data collected on TANDEM, GT-ASP, and during shipboard experiments enabled us to conduct more research, draw much stronger conclusions, and develop more robust recommendations than had we limited ourselves to just one test bed.

Lesson 7: Methodology Development to Study Complex Behavioral Phenomena is Labor Intensive and Expensive at First, But the Payoffs Are Worthwhile

It took 2 years of intensive work to develop the methodology and the multiple test beds. However, the opportunity to conduct team research in the various test bed facilities has greatly broadened our capability and confidence in designing training for high-performance teams.

Development of the TADMUS research methodology resulted in the ability to (a) create acceptable levels of fidelity (e.g., cognitive requirements, team interactions, and task stressors) for stimulating realistic team interactions, (b) enable opportunities for assessing individual and team performance processes and outcomes, (c) support research designs for testing the impact of training interventions on performance under stress, and (d) include Navy trainees and ship teams as research participants. Developing these capabilities was a crucial key to the success of the TADMUS program.

References

Bailey, S. S., Johnston, J. H., Smith-Jentsch, K. A., Gonos, G., & Cannon-Bowers, J. A. (1995). Guidelines for facilitating shipboard team training. *Proceedings of the 17th Annual Interservice/Industry Training Systems and Education Conference* (pp. 360–369). Washington, DC: American Defense Preparedness Association.

Blickensderfer, E., Cannon-Bowers, J. A., & Salas, E. (1997). Fostering shared mental models through team self-correction: Theoretical bases and propositions. In M. Beyerlein, D. Johnson, & S. Beyerlein (Eds.), *Advances in interdisciplinary studies in work teams series* (Vol. 4, pp. 249–279). Greenwich, CT: JAI Press.

Bowers, C. A., Salas, E., Prince, C., & Brannick, M. (1992). Games teams play: A methodology for investigating team coordination and performance. *Behavior Methods, Instruments and Computers, 24,* 503–506.

Cannon-Bowers, J. A., & Bell, H. H. (1997). Training decision makers for complex environments: Implications of the naturalistic decision making perspective. In C. E. Zsambok & G. K. Klein (Eds.), *Naturalistic decision making* (pp. 99–110). Mahwah, NJ: Erlbaum.

Cannon-Bowers, J. A., Salas, E., & Grossman, J. D. (1991, June). *Improving tactical decision making under stress: Research directions and applied implications.* Paper presented at the International Applied Military Psychology Symposium, Stockholm, Sweden.

Department of the Navy. (1985, June). *Composite warfare commander's manual* (Naval Warfare Publication 10-1). Washington, DC: Chief of Naval Operations.

Department of the Navy. (1989, March). *CG 52 class combat system doctrine* (COMNAVSURFLANT Instruction C3516.39). Norfolk, VA: Naval Surface Force.

Driskell, J. E., & Salas, E. (1991). Overcoming the effects of stress on military performance: Human factors, training, and selection strategies. In R. Gal and A. D. Mangelsdorff (Eds.), *Handbook of military psychology* (pp. 183–193). Chichester, NY: Wiley.

Driskell, J. E., & Salas, E. (1992). Can you study real teams in contrived settings? The value of small group research to understanding teams. In R. W. Swezey & E. Salas (Eds.), *Teams: Their training and performance* (pp. 101–124). Norwood, NJ: Ablex.

Duncan, P. C., Rouse, W. B., Johnston, J. H., Cannon-Bowers, J. A., Salas, E., & Burns, J. J. (1996). Training teams working in complex systems: A mental model-based approach. In W. B. Rouse (Ed.), *Human/technology interaction in complex systems* (Vol. 8, pp. 173–231). Greenwich, CT: JAI Press.

Dwyer, D. J. (1995). Training for performance under stress: Performance degradation, recovery, and transfer (CD-ROM). *Proceedings of the 17th Annual Interservice/Industry Training Systems Conference* (pp. 154–162). Washington, DC: American Defense Preparedness Association.

Dwyer, D. J., Hall, J. K., Volpe, C. E., Cannon-Bowers, J. A., & Salas, E. (1992). *A performance assessment task for examining tactical decision making under stress* (Special Report No. 92-002). Orlando, FL: Naval Training Systems Center.

Hall, J. K., Dwyer, D. J., Cannon-Bowers, J. A., Salas, E., & Volpe, C. E. (1993). Toward assessing team tactical decision making under stress: The development of a methodology for structuring team training scenarios. *Proceedings of the 15th Annual Interservice/Industry Training Systems Conference* (pp. 357–363). Washington, DC: American Defense Preparedness Association.

Holl, R. E., & Cooke, J. R. (1989). Rapid software development: A generic tactical simulator/trainer. *Proceedings of the 11th Annual Interservice/Industry Training Systems Conference* (pp. 337–342). Washington, DC: American Defense Preparedness Association.

Inzana, C. M., Driskell, J. E., Salas, E., & Johnston, J. H. (1996). The effects of preparatory information on enhancing performance under stress. *Journal of Applied Psychology, 81,* 429–435.

Johnston, J. H., Driskell, J. E., & Salas, E. (1997). Vigilant and hypervigilant decision making. *Journal of Applied Psychology, 82.*

Johnston, J. H., Smith-Jentsch, K. A., & Cannon-Bowers, J. A. (1997). Performance measurement tools for enhancing team decision making. In M. T. Brannick, E. Salas, & C. Prince (Eds.), *Team performance assessment and measurement: Theory, method, and application* (pp. 311–327). Hillsdale, NJ: Erlbaum.

Kirlik, A., Rothrock, L., Walker, N., & Fisk, A. D. (1996). Simple strategies or simple tasks? Dynamic decision making in "complex" worlds. *Proceedings of the Human Factors and Ergonomics Society 40th Annual Meeting* (pp. 184–188). Santa Monica, CA: Human Factors and Ergonomics Society.

Lipshitz, R., & Orit, B. S. (1997). Schemata and mental models in recognition primed decision making. In C. E. Zsambok & G. K. Klein (Eds.), *Naturalistic decision making* (pp. 293–303). Mahwah, NJ: Erlbaum.

Means, B., Salas, E., Crandall, B., & Jacobs, T. O. (1993). Training decision makers for the real world. In G. Klein, J. Orasanu, R. Calderwood, & C. E. Zsambok (Eds.), *Decision making in action: Models and methods* (pp. 306–326). Norwood, NJ: Ablex.

Orasanu, J., & Connolly, T. (1993). The reinvention of decision making. In G. A. Klein, J. Oransanu, R. Calderwood, & C. E. Zsambok (Eds.), *Decision making in action: Models and methods* (pp. 3–20). Norwood, NJ: Ablex.

Perrin, B. M., Barnett, B. J., & Walrath, L. D. (1993). Decision making bias in complex task environments. *Proceedings of the 37th Annual Human Factors and Ergonomics Society Annual Meeting* (pp. 1107–1111). Santa Monica, CA: Human Factors and Ergonomics Society.

Serfaty, D., Entin, E. E., & Volpe, C. (1993). Adaptation to stress in the team decision-making and coordination environment. *Proceedings of the 37th Annual Human Factors and Ergonomics Society Annual Meeting* (pp. 1228–1232). Santa Monica, CA: Human Factors and Ergonomics Society.

Smith-Jentsch, K. A., Payne, S. C., & Johnston, J. H. (1996, April). Guided team self-correction: A methodology for enhancing experiential team training. In K. A. Smith-Jentsch (chair), *When, how, and why does practice make perfect?* Paper presented at the 11th annual conference of the Society for Industrial and Organizational Psychology, San Diego, CA.

Travillian, K. K., Volpe, C. E., Cannon-Bowers, J. A., & Salas, E. (1993). Cross-training highly interdependent teams: Effects on team process and team performance. *Proceedings of the 37th Annual Human Factors and Ergonomics Society Conference* (pp. 1243–1247). Santa Monica, CA: Human Factors and Ergonomics Society.

Volpe, C. E., Cannon-Bowers, J. A., Salas, E., & Spector, P. (1996). The impact of cross training on team functioning. *Human Factors, 38,* 87–100.

Weaver, J. L., Bowers, C. A., Salas, E., & Cannon-Bowers, J. A. (1995). Networked simulations: New paradigms for team performance research. *Behavior Research Methods, Instruments, and Computers, 27,* 12–24.

Westra, D. (1994). *Analysis of TADMUS pre-experimental data: Considerations for experimental design.* Orlando, FL: Enzian Corp.

Zachary, W. W., Zaklad, A. L., Hicinbothom, J. H., Ryder, J. M., Purcell, J. A., & Wherry, R. J. (1991). *COGNET representation of tactical decision-making in ship-based Anti-Air Warfare*. Spring House, PA: CHI Systems.

Zsambok, C. E. (1997). Naturalistic decision making: Where are we now? In C. E. Zsambok & G. K. Klein (Eds.), *Naturalistic decision making* (pp. 3–16). Mahwah, NJ: Erlbaum.

Zsambok, C. E., & Klein, G. K. (Eds.). (1997). *Naturalistic decision making*. Mahwah, NJ: Erlbaum.

4

Measuring Team-Related Expertise in Complex Environments

Kimberly A. Smith-Jentsch, Joan H. Johnston,
and Stephanie C. Payne

This chapter describes an approach to developing and evaluating a human-performance measurement system designed to assess training needs. We and others have argued previously that such a system must describe, diagnose, and evaluate processes that lead to effective outcomes (Cannon-Bowers & Salas, 1997a; Johnston, Smith-Jentsch, & Cannon-Bowers, 1997). A series of analyses was performed to evaluate whether the performance-measurement tools developed as part of the Tactical Decision Making Under Stress (TADMUS) program possessed these properties. Results from these analyses are summarized here to aid both scientists and practitioners in obtaining reliable and diagnostic ratings of critical performance processes.

The chapter begins with an outline of the principles that guided the development of TADMUS performance measures. Next, the approach taken to assess the psychometric properties of one of these measures, the Anti-Air Teamwork Observation Measure (ATOM), is described. Lessons learned regarding how to obtain reliable descriptions of teamwork, diagnose specific teamwork deficiencies, and link these deficiencies to team outcomes are presented. Finally, directions for future research are offered.

Guiding Principles

It has been argued previously (Cannon-Bowers & Salas, 1997a; Johnston et al., 1997) that a performance-measurement system designed to support team training should have components that target both processes and outcomes at the team and individual levels of analysis. Figure 1 illustrates this taxonomy of human-performance indicators (see Cannon-Bowers & Salas, 1997a). The need to differentiate among these types of performance measures is driven by two guiding principles.

PERFORMANCE MEASUREMENT SCHEME

	INDIVIDUAL	TEAM
P R O C E S S	• COGNITIVE PROCESSES • POSITION - SPECIFIC TASKWORK SKILLS	• INFORMATION EXCHANGE • COMMUNICATION • SUPPORTING BEHAVIOR • TEAM LEADERSHIP
O U T C O M E	• ACCURACY • LATENCY	• MISSION EFFECTIVENESS • AGGREGATE LATENCY & ACCURACY

Figure 1. Taxonomy of human performance measures. From "A Framework for Developing Team Performance Measures in Training," by J. A. Cannon-Bowers and E. Salas, in M. T. Brannick, E. Salas, and C. Prince (Eds.), *Team Performance Assessment and Measurement: Theory, Methods, and Applications* (p. 56), 1997, Hillsdale, NJ: Erlbaum. Copyright 1997 by Lawrence Erlbaum Associates. Adapted with permission.

Principle 1: Remediation Should Emphasize Processes That Are Linked to Outcomes

Process measures describe the strategies, steps, or procedures used to accomplish a task. These can be distinguished from outcome measures, which assess the quantity and quality of the end result. Process measures, often referred to as *measures of performance*, or MOPs, evaluate the human factor involved in complex systems. In contrast, outcome measures, often referred to as *measures of effectiveness*, or MOEs, are influenced by much more than human performance. These measures also contain variance accounted for by equipment, the surrounding environment, and luck. However, outcome measures tend to be favored both in the military and in industry, because they appear to be more objective and are easier to obtain than process measures.

Whereas performance outcomes are important to measure, outcome-based data alone are not particularly useful for determining training needs and providing trainees with feedback that supports the instructional process (Johnston, Cannon-Bowers, & Smith-Jentsch, 1995; Johnston et al., 1997). This is because outcome measures do not specify which aspects of human performance are deficient. Furthermore, because outcome measures are affected by more than human performance, feedback to trainees on the basis of outcomes alone may be misleading and detrimental to learning. For example, teams sometimes stumble on the correct decision in a training scenario despite the use of flawed processes. When the only feedback provided to such a team is outcome based, these flawed processes are not corrected; worse yet, these processes may be inadvertently reinforced. Conversely, there are situations when even the best-laid plans result in a suboptimal outcome. Overemphasis on the outcome in these cases may negatively reinforce effective processes.

For trainees to generalize appropriately what is learned in training, specific processes contributing to effective and ineffective outcomes must be pulled apart by the instructor and made explicit to the trainees. In other words, the goal in decision-making training should not be to train teams to "make the right decision" in a given scenario, but rather to "make the decision right" using processes that on average will lead to more favorable outcomes (Cannon-Bowers & Salas, 1997a). Recently researchers have begun to compile empirical evidence that supports this type of learning orientation (Johnson, Perlow, & Pieper, 1993; Kozlowski, Gully, Smith, Nason, & Brown, 1995).

Whereas process measures are critical for providing feedback to trainees, outcome measures are needed to identify which processes are, in fact, more effective. Thus, measures of both outcome and process are necessary to assess training needs.

Principle 2: Individual- and Team-Level Deficiencies Must Be Distinguished to Support the Instructional Process

Processes and outcomes can and should be evaluated at both the individual and team levels of analysis. This is important for identifying the appropriate type of remediation. For example, a team may make an incorrect decision because information was not circulated effectively among team members (i.e., a team-level problem). The same incorrect decision may result if information is circulated appropriately but was incorrect to begin with as a result of a technical error made by an individual. In the first case, team training focused on information exchange may be appropriate remediation, whereas in the second case, a single team member may require additional technical training.

Another way of labeling the distinction between individual and team processes is to talk about "teamwork" versus "taskwork" (Cannon-Bowers & Salas, 1997b; Cannon-Bowers, Tannenbaum, Salas, & Volpe, 1995). *Taskwork* can be thought of as the requirements of a job that are position specific and usually technical in nature (e.g., the ability to interpret a radar display). *Teamwork*, on the other hand, includes processes that individuals use to coordinate their activities (e.g., communication).

Development of Event-Based Performance Measures

We advocate an event-based approach to obtaining diagnostic measures of individual and team processes that can be linked empirically to important outcomes. This approach involves, first, identifying multiple independent, prespecified situations, or "events," that will occur within a performance exercise. These events are designed to elicit behaviors that are consistent with scenario-learning objectives. Second, measures of both processes and outcomes are taken at each of these events. Finally, consistent relationships among processes and outcomes across a large pool of events can be explored.

In support of TADMUS research, event-based measures were developed to evaluate processes and outcomes in the Combat Information Center (CIC) environment at both the individual and team levels of analysis. These measures are depicted in Figures 2 through 5. Previous descriptions may vary because these measures evolved considerably over the course of the TADMUS program (e.g., Bailey, Johnston, Smith-Jentsch, Gonos, & Cannon-Bowers, 1995; Dwyer, 1992; Johnston et al., 1995; Westra, 1994).

Individual Outcomes

Figure 2 illustrates a measure that was developed to evaluate individual-level outcomes for specific positions within the CIC: the Sequenced Actions and Latencies Index (SALI). Using this measure, raters can assess the timeliness of actions taken by individual team members. These required actions differ across team positions.

EVENT: 1	AT SCENARIO START	AT SCENARIO START	AT SCENARIO START	AT 10 MINUTES 30 SECONDS
TRACK NUMBER: CRAFT TYPE:	7026 COMMERCIAL HELICOPTER	7027 COMMERCIAL HELICOPTER	7030 MILITARY AIRCRAFT	7037 MILITARY AIRCRAFT
THREAT EVALUATION	0 1 2 3	0 1 2 3	0 1 2 3	0 1 2 3

0	Did not report evaluation
1	Reported incorrect/delayed evaluation
2	Reported incorrect evaluation, but corrected it within 60 seconds
3	Reported correct evaluation within 60 seconds

Figure 2.   Individual outcome measure: Sequenced Actions and Latencies Index for the anti-air warfare coordinator position. From "Performance Measurement Tools for Enhancing Team Decision Making," by J. H. Johnston, K. A. Smith-Jentsch, and J. A. Cannon-Bowers, in M. T. Brannick, E. Salas, and C. Prince (Eds.), *Team Performance Assessment and Measurement: Theory, Methods and Applications* (p. 319), 1997, Hillsdale, NJ: Erlbaum. Copyright 1997 by Lawrence Erlbaum Associates. Adapted with permission.

Figure 2 depicts latencies assessed for the anti-air warfare coordinator (AAWC). One of the AAWC's responsibilities is to make threat evaluations. Thus, the SALI for the AAWC asks raters to evaluate the timeliness of threat evaluations made for prespecified aircraft at key scenario events. For each scenario event, the top row indicates the time that a particular aircraft could have been detected. The second row lists the type of aircraft and track number to aid raters in making their observations. The third row provides a 4-point scale on which raters evaluate the timeliness of a particular action, which, in this case, is threat evaluation.

Individual Processes

The behavioral observation booklet (BOB) was developed to evaluate individual processes used by team members to accomplish position-specific tasks. The BOB for a tactical action officer is depicted in Figure 3. Previous descriptions of the BOB vary (Hall, Dwyer, Cannon-Bowers, Salas, & Volpe, 1993; Johnston et al., 1995). As shown in Figure 3, scenario events and expected outcomes are listed in a critical incidents format. The first column provides a brief description of significant objective-based events in chronological order. The second column lists the appropriate actions to be taken in response to those events by a particular team position. Finally, in the third column, assessors check whether or not a team member performed the expected action in response to a key event.

		ACTION TAKEN	
EVENT	**EXPECTED ACTION**	**YES**	**NO**
SCENARIO TIME 13:48 COMAIR TN 7031 leaves Khark Island bearing 026, range 38 nm, no ESM squawking mode III 4132	Order designation as UAE; order Level III warning	✓	
SCENARIO TIME 16:00 COMAIR TN 7031 closes to range 33 nm bearing 018, speed 350 kt, altitude 5000	Order standby for illumination with SM fire control radar if TN 7031 closes to 20 nm		✓

Figure 3. Individual process measure: Behavioral Observation Booklet for the tactical action officer.

Team Outcomes

Given the nature of the CIC team task, performance outcomes include the timeliness of track identifications, assessments of intent, and decision to employ weapons (Zachary, Zaklad, Hicinbothom, Ryder, & Purcell, 1993). Figure 4 illustrates a measure developed to assess these team-level outcomes: the Anti-Air Warfare Team Performance Index (ATPI). Raters evaluate team outcomes using an anchored rating scale (0–3). The anchors are designed to make the rating criteria consistent across instructors. Examination of performance outcomes across events allows the instructor to pinpoint the type of situations with which a team consistently has problems.

Team Processes

The Anti-Air Teamwork Observation Measure (ATOM) was developed to evaluate team-level processes that contribute to performance outcomes at key scenario events (Smith-Jentsch, Payne, & Johnston, 1996). The ATOM requires raters to evaluate teams on 11 specific teamwork behaviors that are categorized under four dimensions. These ratings are based on notes taken during a performance exercise using the ATOM worksheet (see Figure 5). This worksheet provides a scenario timeline that lists prescribed event information in the left column. Assessors record detailed descrip-

EVENT:	AT SCENARIO START	AT SCENARIO START	AT SCENARIO START	AT 10 MINUTES 30 SECONDS
TRACK NUMBER: CRAFT TYPE:	7026 COMMERCIAL HELICOPTER	7027 COMMERCIAL HELICOPTER	7030 MILITARY AIRCRAFT	7037 MILITARY AIRCRAFT
SHOOT / NO SHOOT	0 1 2 3	0 1 2 3	0 1 2 3	0 1 2 3

0	Fired missile at target; fired other weapons
1	Shot missile, but self-destructed it before contact
2	Initiated entire firing sequence and stopped just short of release
3	Did not shoot at targets

Figure 4. Team outcome measure: Anti-Air Warfare Team Performance Index. From "Performance Measurement Tools for Enhancing Team Decision Making," by J. H. Johnston, K. A. Smith-Jentsch, and J. A. Cannon-Bowers, in M. T. Brannick, E. Salas, and C. Prince (Eds.), *Team Performance Assessment and Measurement: Theory, Methods and Applications* (p. 319), 1997, Hillsdale, NJ: Erlbaum. Copyright 1997 by Lawrence Erlbaum Associates. Reprinted with permission.

GROUND TRUTH

BEGIN EVENT ONE	
00-00 DETECTED TACAIR "A" AT KHARK 044/38	
00-30 LOST CONTACT ON TACAIR 7030	
01-00	
01-30 REGAIN CONTACT ON TACAIR 7030 VIC KHARK ISLAND	
02-00	
02-30	
03-00	

Figure 5.   Worksheet for Anti-Air Teamwork Observation Measure.

tions of teamwork behaviors as they occur in the right column across from the associated point in the scenario timeline. Finally, the incidence of each targeted behavior is estimated (see Figure 6), and overall ratings are assigned for each dimension using a 5-point Likert scale (see Figure 7).

Approach to Assessing Properties of the Anti-Air Teamwork Observation Measure

Although both individual- and team-level processes are important for identifying appropriate remediation, we have found that measures of individual taskwork performance are much more common than sound measures of teamwork processes. This is likely due to the fact that teamwork processes are less tangible and objective to evaluate. Thus, the remainder of this chapter focuses on our efforts to develop a reliable, diagnostic, and valid measure of processes that "transform a team of experts into an expert team" in the CIC (Salas, Cannon-Bowers, & Johnston, 1997, p. 359).

The analyses summarized in this chapter were performed on event-

Communication Behavior Ratings

Improper Phraseology	0	1-3	>3
Inaudible Communication	0	1-3	>3
Excess Chatter	0	1-3	>3
Incomplete Reports	0	1-3	>3

Figure 6. Anti-Air Teamwork Observation Measure target behavior ratings: Communication.

based performance ratings assigned to teams that had participated in one or more of the TADMUS exercises over the course of the program. The methods and scenarios that were used are described in detail in chapter 3, this volume. This method was consistently employed for each of the teams included in our sample. However, these teams participated in different TADMUS experiments. Some of the teams received training prior to performing, whereas others did not. In addition, teams did not all perform the same number of scenarios (ranging from one to four), nor did they perform the scenarios in the same order. Finally, teams varied on a number of experience-related variables. This diverse database provided us with the opportunity to examine team processes that consistently emerge as correlates of effective decision making across various types of teams despite event-related, practice, and training-related variability.

Although the methods for conducting scenario exercises were consistently employed across teams, the measures originally used to evaluate performance differed to suit the unique focus of each study. Therefore, to examine empirically the processes that were linked to effective team decision making across the data set, 100 videotaped performance exercises were reevaluated using the ATOM.

Condition-blind subject matter experts assigned ratings of teamwork processes and team outcomes for three preplanned events within each of these exercises. A few of the videotapes had audio problems for one or more events, so that a total of 283 event-level ratings were obtained for each metric. The following sections summarize lessons learned on the basis of analyses of these ratings.

Describing Team Processes

Several years into the TADMUS program, a group of subject matter experts was assembled to discuss the ATOM dimensions and the rating scales used to assess them. The goal of this focus group was to develop descriptions of teamwork dimensions that would allow raters to provide reliable and distinct ratings. The original version of the ATOM included 5-point Likert scales that targeted seven teamwork dimensions: commu-

Overall Communication Rating

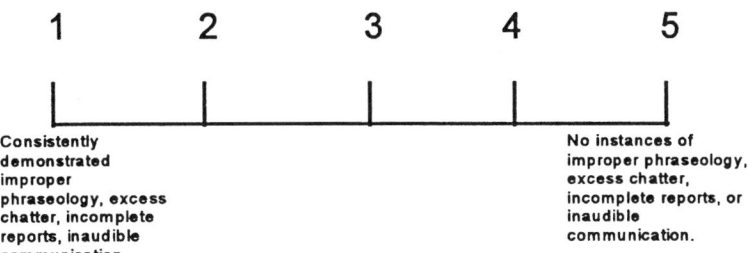

Figure 7. Anti-Air Teamwork Observation Measure overall dimension rating: Communication.

nication, feedback, monitoring, coordination, team initiative/leadership, backup behavior, and situation awareness (Dickinson & McIntyre, 1997; McIntyre & Salas, 1995). Some of these dimension ratings had demonstrated poor interrater reliability. Moreover, many of the dimension ratings were highly correlated, indicating that they did not describe distinct performance processes. By examining definitions of the original seven dimensions, discussing these with subject matter experts, and comparing interrater reliabilities and correlations among ratings, we identified a number of potential reasons for poor reliability and high intercorrelations among dimensions.

Definitions of Teamwork Dimensions to Be Rated

It appeared that dimensions that required raters to infer a state of mind (e.g., team monitoring, situation awareness) tended to be less reliable than those that were based on the observation of overt behaviors (e.g., provided backup). In addition, many of the high correlations among dimensions seemed to stem from the fact that a rater could infer one only from observing the other. For example, raters found it virtually impossible to evaluate whether team members were monitoring one another's activities unless someone on the team provided feedback or backup. Thus, separate ratings of monitoring, backup, and feedback did not provide additional information.

Another example involved the situation awareness dimension. As shown in Table 1, the definition provided for *situation awareness* included "avoid tunnel vision and maintain awareness of all relevant contacts." On the basis of discussions among the focus group members, it was decided that situation awareness is an individual-level cognitive component of performance. Although an individual team member's situation awareness clearly can influence the rest of the team, situation awareness itself is not a "team process." On the other hand, when an individual passes along something he or she is aware of to other team members, that is a team process.

Table 1.   Original Seven ATOM Teamwork Dimensions

Dimension	Definition
Communication	Report all contacts or detections as prescribed.
	Use proper terminology and communications procedure.
	Avoid excess chatter or nets.
	Pass complete information to correct members.
	Acknowledge requests from others.
	Acknowledge receipt of information.
Feedback	Respond to others' requests for information.
	Accept reasonable suggestions.
	Avoid nonconstructive, sarcastic comments.
	Provide specific, constructive suggestions to others.
	Ask for advice when needed.
	Ask for input regarding performance.
Monitoring	Observe and keep track of performance of other team members.
	Recognize when a team member makes a mistake.
	Recognize when a team member performs exceptionally well.
	Listen on the nets to other team member communications.
Coordination	Pass relevant information to others in a timely manner.
	Pass relevant information to others in an efficient manner.
	Be familiar with relevant parts of other members' jobs.
	Facilitate performance of other team members.
	Carry out individual tasks in a synchronized manner.
	Avoid distractions during critical operations.
Team initiative/ leadership	Encourage others to make decisions.
	Provide direction and support to other members.
	Explain exactly what is needed during exercise.
	Quickly take action when members become overwhelmed.
	Provide needed information to other members.
	Encourage others to take on extra duties.
Backup behavior	Provide assistance to others who need it.
	Ensure member who was assisted is aware of what was done.
	Complete own duties while helping others.
	Ask for help when needed rather than struggling.
Situation awareness	Note deviations from steady state.
	Identify potential or anticipated problems.
	Avoid tunnel vision and maintain awareness of all relevant contacts.
	Recognize other members' need for some of one's own information.
	Recognize need for action.
	Be proactive versus reactive.

Note. ATOM = Anti-Air Teamwork Observation Measure.

Subject matter experts lobbied for a dimension labeled *team situation assessment* as a replacement for situation awareness. Team situation assessment, which included the accuracy and timeliness of team decisions regarding aircraft identifications and evaluations of hostile versus friendly intent, was temporarily added as one of the ATOM dimensions (Johnston et al., 1997). However, it was later deemed to be a team-level outcome. We found that subject matter experts were reluctant to remove situation assessment from the list of team processes. After many hours of debate, it became clear that this reluctance did not stem from a disagreement over the fact that situation assessment was an outcome but from the belief that as an outcome, situation assessment was most important to measure and to debrief after an exercise. In the end, it was agreed that ratings of team situation assessment at key scenario events would serve as the outcome against which team and individual processes would be examined. In other words, we were interested in identifying team processes that lead to accurate and timely situation assessments.

Number of Performance Dimensions to Be Rated

A second difficulty raters expressed regarding the original ATOM ratings involved redundancy across the definitions of the seven dimensions. For example, the definition for *situation awareness* had included "recognizing other members' need for information," whereas the *coordination* dimension included "passing relevant information to others in a timely manner." Additionally, the coordination dimension included "facilitating performance of other team members," whereas the team initiative/leadership dimension included "providing direction and support to other team members."

We asked subject matter experts to develop a list of teamwork behaviors that they considered distinct, objectively observable, important for CIC performance, and variable across teams. Eleven such behaviors were identified. We asked these experts individually to sort the 11 behaviors into meaningful categories and then to reach consensus. The resulting four dimensions—information exchange, communication, supporting behavior, and team initiative/leadership—are shown in Table 2. Data collected sub-

Table 2.   Final Four ATOM Teamwork Dimensions

Dimension	Definition
Information exchange	Seeking information from all available sources
	Passing information to the appropriate persons before being asked
	Providing "big picture" situation updates
Communication	Using proper phraseology
	Providing complete internal and external reports
	Avoiding excess chatter
	Ensuring communications are audible and ungarbled
Supporting behavior	Correcting team errors
	Providing and requesting backup or assistance when needed
Team initiative/ leadership	Providing guidance or suggestions to team members
	Stating clear team and individual priorities

Note. ATOM = Anti-Air Teamwork Observation Measure.

sequently from an independent group of shipboard instructors indicated that examples of CIC team performance could be reliably placed into these categories.

A new ATOM was developed that required ratings of the 11 targeted behaviors first, followed by overall ratings of the four dimensions on a 5-point Likert scale. Rater training provided assessors with a common frame of reference for evaluating each of the targeted behaviors. This training included information, demonstration of effective and ineffective behaviors, practice categorizing those behaviors, and feedback from instructors and their peers (for more detail, see chapter 11, this volume).

A factor analysis performed on 283 ratings of the 11 targeted behaviors supported the proposed four-factor structure (Smith-Jentsch, 1995). In addition, we found that raters' descriptions of CIC teamwork were more consistent using the new ATOM dimensions. Interrater reliabilities ranged from a low of .82 for team initiative/leadership to a high of .91 for communication. These findings are consistent with previous research in which it has been demonstrated that three to five dimensions generally account for subject variability, regardless of the number of dimension ratings collected (Hobson, Mendel, & Gibson, 1981; Sackett & Hakel, 1979; Slovic & Lichtenstein, 1971).

In summary, each of the four dimensions used was defined by behaviors that could be objectively observed during a training scenario. These behavioral definitions of teamwork allowed raters to provide detailed and reliable descriptions of CIC teamwork. Although such descriptions are necessary, they are not sufficient for the diagnosis of performance deficiencies and prescription of appropriate remediation. Diagnosticity goes beyond describing what took place in a given situation by determining the causes for observed performance breakdowns. The following section describes lessons learned regarding the diagnosticity of ATOM ratings.

Diagnosing Stable and Unique Performance Deficiencies

If a measure is to be used to provide developmental feedback or to prescribe remediation for a team, ratings must be shown to reflect relatively stable and distinct behavioral patterns. For instance, after viewing a performance exercise, multiple raters may agree that a key piece of information was not passed among team members. However, it may be unclear why it was not passed. If information was passed effectively among team members in all other cases, this observation alone would not necessarily indicate that the team has a deficiency in the area of information exchange. It may be that in this instance a team member failed to notice the piece of information or misinterpreted his or her radar scope. However, if multiple observations indicate that team members consistently failed to pass key pieces of information, one might infer that the team requires additional training in information exchange.

High correlations among team ratings of the same dimension across multiple events provide evidence of a relatively stable tendency. Such evidence of stability, often referred to as *convergent validity*, is one indication that a measure is diagnostic. Conversely, relationships among different dimensions within a scenario event should be low. If such correlations are as high as or higher than correlations among ratings of the same dimension, one may conclude that the ratings are tapping a single construct. Therefore, evidence that multiple dimension ratings are relatively independent of one another must be shown for a measure to be considered truly diagnostic. Such evidence is referred to as *discriminant validity*.

Demonstrating that dimension ratings measure relatively stable and distinct performance constructs has proved to be difficult at both the individual and the team levels. Correlations among ratings of the same dimension assigned to individuals in multiple exercises are generally weak, whereas different dimensions assessed in the same exercise are often highly correlated (McEvoy, Beatty, & Bernardin, 1987; Sackett & Dreher, 1982). Several potential reasons for this have been offered.

First, it has been suggested that high correlations among different dimension ratings often result from rater errors (Bernardin & Beatty, 1984). Raters are believed to have a limited cognitive ability to discriminate among multiple performance dimensions (McEvoy et al., 1987; Sackett & Dreher, 1982). Moreover, raters are often prone to decision biases, such as halo error, in which judgments of one dimension are colored by impressions formed on other dimensions. Second, it has been suggested that correlations among different dimensions within an exercise may represent true "exercise effects" that mirror stable dispositions toward behaving in particular situations (Sackett & Dreher, 1982). Similarly, low correlations among the same dimension ratings across exercises may represent situation-specific tendencies to employ a particular performance strategy. For example, it has been demonstrated that interpersonal context (i.e., work, personal, stranger interaction) has a significant impact on the stability of assertive behavior (Smith-Jentsch, Salas, & Baker, 1996).

It is not surprising, then, that few studies have demonstrated strong evidence of convergent validity for team-level dimension ratings because these ratings include the situation-related variability of multiple team members. For example, Brannick, Prince, Prince, and Salas (1995) examined the construct validity of teamwork dimension ratings given to two-person flight crews. Multiple raters were able to provide reliable ratings of the various performance dimensions. However, correlations among ratings of the same dimension across multiple scenarios were generally nonsignificant.

To examine the diagnosticity of ATOM dimension ratings, we compared correlations among ratings of the four teamwork dimensions within and across events in the TADMUS scenarios. First, these analyses were performed on the overall dimension ratings assigned by raters using a 5-point Likert scale. Second, we calculated composite dimension ratings by summing the scores assigned to specific behavioral categories within each dimension. It was predicted that these composite dimension ratings would be more diagnostic than overall dimension ratings. Specifically, we expected raters to be more prone to rating biases, such as halo error, when assigning the overall dimension ratings. Ratings of the individual behaviors that were included in composite dimension ratings were considered to be more objective. A summary of the results of these analyses follows.

Global Versus Composite Dimension Ratings

Results indicated that composite ratings of the same dimension were significantly correlated across the three events within a scenario. These correlations ranged from .32 for information exchange to .67 for communication. Correlations among overall ratings of the same dimension across the three events in a scenario were also significant. These correlations (.67–.76) tended to be somewhat higher than those found for the composite dimension ratings. Therefore, both composite dimension ratings and overall dimension ratings demonstrated convergent validity across events within a scenario.

Next, we compared discriminant validity for the two types of dimension ratings. This was done by computing correlations among ratings of different dimensions within scenario events. Results indicated that the average correlation between composite ratings of different dimensions within an event was .15, which suggested that these ratings discriminated between unique performance deficiencies. A team that scored low on communication could have scored high on supporting behavior.

As we did with the composite dimension ratings, computed correlations among overall ratings of different dimensions within an event. In contrast to the composite dimension ratings, overall dimension ratings were not found to be diagnostic. Correlations among overall ratings of different dimensions within an event (.59–.73) were each as high as or higher than correlations among overall ratings of the same dimension ratings across events.

Impact of Computer-Generated Performance Data

Payne (1996) investigated one potential source of bias that may have contributed to the lack of discriminant validity found for the overall dimension ratings. For 80 of the teams, one rater had computer-generated performance data available. It was hypothesized that these data may have biased ratings by coloring overall impressions of a team. A number of previous studies have demonstrated that raters weight quantitative data differently from qualitative data, especially if computer generated, because they trust such data more than they trust subjective judgments (Mosier, 1997). If computer-generated data are reliable and relevant to the construct of interest, one would expect these data to improve the quality of ratings. However, if raters are given access to computer-generated data that are unrelated to some of the dimensions they are to evaluate, such data may introduce bias and exhibit lower discriminant validity.

The computer-generated performance data that were available to the rater allowed him to determine what information team members had on their screens and when. By marrying these data with team communications, a rater could determine with greater confidence whether team members failed to pass a relevant piece of information to which they had access. Therefore, it would have been appropriate for these data to influence the ratings assigned for information exchange. However, the computer-generated data should not have been useful for determining the quality of behaviors within the other dimensions (e.g., phraseology).

A single rater's policy for arriving at overall ratings was compared for two groups of teams in our data set: those for which the rater had access to computer-generated data and those for which such data were unavailable. Results revealed that the rater's policy for combining cues did, in fact, change significantly when computer-generated data were available. When such data were not available to the rater, ratings of information exchange behaviors contributed only to overall ratings of information exchange. However, specific information exchange behaviors accounted for variance in the overall ratings of communication, supporting behavior, and team initiative/leadership, as well as of information exchange when the rater had access to the computer-generated data. As a result, overall ratings of the four dimensions were less distinct for this group of teams. Composite dimension ratings, which were more diagnostic than overall ratings for both groups of teams, did not appear to be influenced by the computer-generated data.

Differential Impact of Training Interventions

The evidence presented thus far suggests that composite dimension ratings can be used to diagnose distinct performance deficiencies. To assess further the diagnosticity of these ratings, we examined whether they were sensitive to differential training-related improvement. Average ratings assigned to teams that had participated in three TADMUS training strategies were compared to an equivalent no-treatment control group within the data set.

The three training strategies compared were (a) team model training (TMT; see Duncan, Rouse, Johnston, Cannon-Bowers, Salas, & Burns, 1996), (b) team adaptation and coordination training (TACT; see chapter 9, this volume), and (c) team leader training (see chapter 10, this volume). Each of these strategies was previously shown to be effective when assessed using different measures that focused on the specific behaviors targeted in each study. The analyses described here were not performed to validate the training strategies themselves, therefore, but to assess whether ATOM composite ratings were sufficiently diagnostic to differentiate their unique effects.

Each strategy emphasized different aspects of teamwork in the attempt to improve performance. Therefore, one would expect that whereas each training strategy may improve overall team performance, the impact on the four ATOM dimensions should differ in predictable ways. Although the three training strategies are explained in greater detail elsewhere, the following brief descriptions allow one to speculate on which of the four ATOM dimensions should be most affected.

Team model training. A portion of the teams that participated in TMT training were included in the sample that was reevaluated using the ATOM. The TMT training focused on improving knowledge regarding communication procedures within a combat information center. This training was designed to make trainees familiar with the location of various team members in the CIC, their roles, and the types of communications that pass among them. Trainees were instructed to navigate through computer-based modules, which allowed them to listen to incoming and outgoing communications from the perspective of any team position they wished. These communications introduced trainees to appropriate and realistic terminology to be used when reporting information. Thus, one might predict that teams that received TMT would demonstrate improved communication skills.

Team adaptation and coordination training. Teams that participated in TACT were taught, in a group setting, how to anticipate one another's need for information. In particular, team members were instructed to provide periodic situation updates so that their team would develop a shared understanding, specifically of the tactical action officer's (TAO) priorities. This was expected to help team members determine which pieces of information were most critical to pass during periods of high stress. Given the focus of TACT, it was predicted that teams that received this type of training would demonstrate improved information exchange and team initiative/leadership skills.

Team leader training. Team leader training focused on improving the feedback skills of those who led their team's prebriefs and debriefs between each performance exercise. This training included lecture, demonstration of effective and ineffective briefings, practice in role-play exercises, and feedback.

Results indicated that trained leaders were more likely to critique themselves and ask the team for feedback on their own performance during team debriefs. Additionally, trained leaders were more likely to guide their teams toward a critique of the four teamwork dimensions. In turn, team members admitted more mistakes and offered more suggestions on how to improve their performance (see chapter 10, this volume). Therefore, it was expected that teams briefed by trained leaders would demonstrate improved performance on each of the four teamwork dimensions.

Comparison of training effects. Table 3 illustrates the significant training effects found when comparing the dimension ratings received by a group of equivalent no-treatment control teams with those received by teams that had participated in TMT, TACT, and team leader training. These results indicated that the three training strategies did not uni-

Table 3. Differential Training Effects on ATOM Composite Dimension Ratings

TADMUS training interventions	Significant training effects on ATOM ratings			
	Information exchange	Communication	Supporting behavior	Team initiative/ leadership
Team model training		*		
Team adaptation and coordination training	*			*
Leadership training	*		*	*

Note. ATOM = Anti-Air Teamwork Observation Measure.

formly influence teamwork behavior. In fact, the pattern of effects appears to support the diagnosticity of the ATOM ratings.

Teams in the sample that participated in TMT received higher ratings on communication than did the control teams. This was expected given that TMT allowed trainees to listen to standard reports of various kinds being stated correctly by those at different watchstations. TACT teams, on the other hand, did not demonstrate improved communication behaviors. Because TACT training did not specifically address these behaviors (i.e., phraseology, completeness, brevity, clarity), these results were also expected. TACT teams did, however, receive significantly higher ratings than did control teams on team initiative/leadership, as well as on information exchange. These effects were predicted because several of the behaviors within these dimensions (i.e., providing situation updates, passing information before having to be asked, stating priorities) were directly emphasized in TACT.

The debriefs given as part of team leader training encouraged team members to critique themselves on all four dimensions. These teams received higher ratings for supporting behavior, team initiative/leadership, and information exchange, but no significant improvement was evident for communication behaviors. Although these teams critiqued their communication skills as part of each debrief, an instructor was not present to ensure that team members identified appropriate communication procedures. Because team members did not necessarily have recent shipboard experience, they may not have had the requisite knowledge to determine on their own what the correct procedures should be. On the other hand, these teams seem to have been able to solve coordination problems stemming from information exchange, supporting behavior, and team initiative/ leadership without the intervention of an instructor.

In summary, the differential training-related improvement reflected by ATOM ratings was consistent with the emphasis of the training strategies examined. These results provided additional evidence in support of the diagnosticity of ATOM ratings. When scores on specific behaviors were summed within a dimension, the resulting composite ratings demonstrated both convergent and discriminant validity. This conclusion was based on patterns of correlations among dimension ratings within and across scenario events. In addition, composite ratings were found to be sensitive to unique training-related effects from three TADMUS training strategies.

In contrast, subjective overall ratings of the four teamwork dimensions did not differentiate unique performance deficiencies. Moreover, these ratings were shown to be biased by computer-generated performance data that were available for a subset of the teams examined. Together, the findings discussed in this chapter provide strong evidence that the diagnosis of performance deficiencies and associated developmental feedback provided to trainees should be based on specific targeted behaviors. This is consistent with research on goal setting that has repeatedly demonstrated that specific goals are more effective than general goals (Locke & Latham, 1990; Locke, Shaw, Saari, & Latham, 1981).

Validity of Evaluations

The previous sections summarized evidence suggesting that composite dimension ratings reliably discriminated among four unique aspects of teamwork. This section summarizes analyses conducted to investigate the validity of those composite dimension ratings. We were interested in determining whether the four teamwork dimension ratings each represented superior performance strategies.

Because most of the teams included in the TADMUS data set did not have a future together as a CIC team, we were unable to evaluate the predictive validity of dimension ratings. To explore whether these ratings represented valid evaluations of a team's performance, we took a concurrent validation approach. Such an approach involves examining linkages between predictor variables and criterion variables that are collected at the same point in time. To examine concurrent validity for ATOM dimension ratings, we explored (a) whether more experienced teams received higher ratings than less experienced teams and (b) whether teams that made correct decisions at key events received higher dimension ratings than those that did not.

Experience-Related Performance Differences

If composite dimension ratings represented superior performance strategies, one would expect more experienced teams to receive higher scores than less experienced teams. As one strategy for examining the validity of ATOM ratings, therefore, we compared scores assigned to three groups of teams in our sample that differed in experience as a CIC team.

Participants consisted of watchteams from Navy ships, ad hoc teams from the Surface Warfare Officer School (SWOS) and from the Naval Post Graduate School (NPGS). As shown in Table 4, these teams were known to differ on a number of experience-related variables. First, only the

Table 4. Known Group Differences Across Experience-Related Variables

Sample poulations	Experience-related variables		
	Past and future as a team	Team member service	Experience on watchstation assigned for exercise
Naval ships	Yes	Naval officers and enlisted	• All team members high and recent
Surface Warfare Officer School (SWOS)	No	All Naval officers	• Mixed • Few recent
Navy Post Graduate School (NPGS)	No	Officers from Navy, Army, Marines, and Air Force	• Mixed • Few had high or recent

watchteams from Navy ships had a past and a future together as a CIC team. Second, teams from SWOS and NPGS were made up of people who did not have recent experience in the particular team position they were asked to perform. In fact, some of these team members had never worked the particular watchstation assigned to them. Third, although the ship teams and SWOS teams consisted solely of Navy personnel, teams from NPGS were made up of a mix of members from each of the services.

On the basis of these experience-related differences between the three known groups of teams in the TADMUS data set, we expected ship teams to outperform both SWOS and NPGS teams. Furthermore, SWOS teams were expected to outperform NPGS teams. A comparison of dimension ratings across the three known groups revealed mean differences that were generally consistent with these hypotheses. Table 5 illustrates the pattern of means and comparisons that were statistically significant.

Ship teams received higher ratings on average than did SWOS and NPGS teams on all four dimensions. Average ratings assigned to SWOS teams were higher than those of NPGS teams in all cases but one. NPGS teams received higher average ratings on team initiative/leadership than did SWOS teams. In retrospect, it appears that this may have been due to the fact that these teams were made up of more senior individuals who were currently in graduate school. Although team members from NPGS had less technical experience on the task than those from SWOS, they were likely to have had more experience in a leadership role.

Linkages Between Processes and Outcomes

If higher ratings on the four teamwork dimensions represent superior performance strategies, one would expect that each of these would be positively related to important team outcomes. We have argued earlier that an event-based approach to performance measurement allows one to investigate process–outcome linkages systematically. Using this approach, we examined whether teams that made accurate decisions regarding key

Table 5. Summary of Known Group Comparisons

Dimension	Relative performance
Information exchange	Ships > SWOS and NPGS
Communication	Ships and SWOS > NPGS
Supporting behavior	Ships and SWOS > NPGS
Team initiative–leadership	Ships and NPGS > SWOS

Note. > denotes significant differences, $p < .05$. SWOS = Surface Warfare Officer School; NPGS = Navy Post Graduate School.

scenario events were more likely to have demonstrated teamwork processes consistent with the four ATOM dimensions.

A multiple regression analysis indicated that composite ratings of information exchange, communication, and team initiative/leadership each contributed uniquely to the accuracy of team decisions. Together, these three dimensions accounted for 16% of the variability across events. Supporting behavior ratings assigned to teams did not significantly add to the prediction of correct decisions.

Supporting behavior is the mechanism by which team members compensate for one another's weaknesses by correcting errors and shifting workload. Because these behaviors are compensatory, they are not observed unless a need for such behavior exists. If the need to correct an error or to provide backup to another teammate who is overloaded never arises during a given scenario, it is impossible for a rater to assign a meaningful rating.

The need to provide supporting behavior was not controlled for in the TADMUS scenarios. Teams that frequently made mistakes had many opportunities to demonstrate supporting behavior, whereas the more technically competent teams had fewer opportunities to demonstrate supporting behavior. It is not surprising, therefore, that supporting behavior ratings did not consistently predict event outcomes in this data set. However, as we have reported earlier, more experienced CIC teams exhibited more supporting behaviors on average than did inexperienced teams. This finding suggested that supporting behavior is an important dimension of teamwork that mature teams use to regulate and maintain effective performance.

The preceding results provided evidence that composite ratings of the four ATOM dimensions represented superior teamwork strategies. First, more experienced teams in this data set tended to receive higher ratings on these dimensions than did less experienced teams. Second, information exchange, communication, and team initiative/leadership were each uniquely related to the accuracy of team decisions across different scenario events. Although supporting behavior ratings did not account for unique variance in team decisions, we have suggested that this was because the need to demonstrate supporting behavior was not controlled for in the TADMUS scenarios.

Performance-Measurement Lessons Learned

On the basis of the research summarized in this chapter, we offer a number of lessons learned for scientists and practitioners interested in developing and evaluating a human-performance measurement system.

Lesson 1: Define Team Performance Ratings Behaviorally

Raters provided more reliable ratings of dimensions that were defined by overt behaviors such as communications than of those that required them to draw inferences regarding team members' states of mind (e.g., situation awareness, team monitoring). Therefore, it is recommended that teamwork dimension ratings be defined in terms of specific behaviors that are objectively observable or audible to raters.

In addition to team skills, team-related attitudes and cognition play a significant role in effective teamwork (Cannon-Bowers & Salas, 1997a). However, we suggest that these types of team competencies can be more reliably measured through direct means than through rater inferences during a training exercise. For example, a number of researchers have successfully evaluated situation awareness by pausing a scenario at various points and having participants describe their mental representations of the environment (Endsley, 1995; Stout, Cannon-Bowers, & Salas, 1996). In addition, think-aloud protocols and structured postscenario interviewing techniques (e.g., probed protocol analysis) have been used to evaluate metacognitive processes (Cohen, Freeman, Wolf, & Militello, 1995; Gill, Gordon, Moore, & Barbera, 1988; Klein, Calderwood, & MacGregor, 1989; Means & Gott, 1988).

Lesson 2: Strive for the Simplest Factor Structure That Can Adequately Describe Teamwork

Researchers have repeatedly demonstrated that raters have a limited cognitive ability to provide distinct performance ratings on multiple dimensions. In fact, it has been shown that regardless of the number of dimensions for which ratings are requested, three to five dimensions generally are sufficient to account for subject variability (Hobson et al., 1981; Sackett & Hakel, 1979; Slovic & Lichtenstein, 1971). Our results indicated that a four-factor model of CIC teamwork was more meaningful than the previous seven-factor model.

On the basis of our experience and that of previous researchers, it is recommended that scale developers strive for simplicity in defining components of teamwork. We believe that the four teamwork dimensions identified here are generalizable to a wide variety of settings. For example, a task force of instructors from the areas of seamanship, engineering, and damage control has concluded that the four ATOM dimensions are equally applicable in their domains.

Alternative sets of dimension definitions should be carefully examined for redundancy. Card sorting techniques can be useful for identifying appropriate groupings of teamwork behaviors. Ultimately, ratings collected at the level of these individual behaviors can be factor analyzed to derive teamwork performance dimensions empirically.

Lesson 3: Collect Assessments of Teamwork From Raters at the Level of Specific Behaviors Rather Than Overall Dimensions

We found that overall dimension ratings assigned using 5-point Likert scales were highly correlated with one another within an event, indicating that they were not useful for diagnosing unique performance deficiencies. On the other hand, ratings that were calculated by summing scores assigned to specific behaviors within a dimension were found to be relatively independent from one another within an event. In addition, these composite dimension ratings were not biased by computer-generated data as were overall ratings. Finally, composite ratings were sensitive to unique effects produced by different TADMUS training strategies.

Therefore, to diagnose teamwork deficiencies, it is recommended that raters be asked to evaluate instances of specific behaviors within a dimension, which can later be combined mathematically. Moreover, raters' behavioral observations rather than overall ratings should be used to provide developmental feedback, because overall ratings are prone to bias and do not provide trainees with specific detail about what they did wrong or how to make improvements.

Lesson 4: Do Not Allow Raters Access to Computer-Generated Performance Data While Rating Dimensions That Are Conceptually Unrelated to That Data

It has been demonstrated that raters weight objective data differently from subjective data, particularly if the data are generated by computer. We found, for example, that the rater's policy for arriving at overall ratings changed when computer-generated data were made available. Although the data were conceptually relevant to only one of the teamwork dimensions, overall ratings of all four were affected. This bias significantly reduced the diagnosticity of overall dimension ratings.

Therefore, caution should be taken when raters are asked to combine system-generated data with their own subjective observations. If it is determined that such data are likely to introduce bias, the following recommendations are offered: Either (a) evaluation responsibilities may be divided up such that individuals who have access to computer-generated performance data only rate dimensions that are conceptually relevant to that data or (b) raters may evaluate all dimensions that are conceptually irrelevant to computer-generated performance data prior to obtaining access to that data.

Lesson 5: Identify Process–Outcome Linkages Using an Event-Based Approach

We found that it was possible to uncover consistent relationships between processes and outcomes by systematically evaluating both at multiple events within a scenario. Using such an approach, accurate decisions at key events were reliably linked to effective information exchange, communication, and team initiative/leadership. These relationships held up despite contextual differences across the scenario events in the data set.

Lesson 6: Provide Developmental Feedback That Emphasizes Processes Rather Than Outcomes

The relationship between effective decisions and effective teamwork was not 1:1. As we have argued earlier, poor team decisions result from individual- as well as team-level deficiencies. This point was illustrated by the fact that ratings of the four teamwork processes together accounted for only 16% of the variability in decision-making outcomes.

Unfortunately, data on the tactical experience and technical skill of individual team members were not consistently collected for all teams in the current sample. However, it was clear that these variables had an impact on team outcomes. For example, some team leaders who were proactive about making their priorities clear (i.e., demonstrating effective team initiative/leadership) made bad decisions because they lacked the tactical knowledge of which priorities were appropriate. In other cases, team members sought information from available sources (i.e., effective information exchange) but misinterpreted that information once they got it. Therefore, it is clear that a comprehensive assessment of training needs should involve the measurement of both individual- and team-level processes at multiple scenario events. Furthermore, these examples illustrate how providing outcome-based feedback alone has the potential to reinforce effective processes negatively. We reiterate that the instructional process is enhanced by an emphasis on processes rather than outcomes when developmental feedback is provided.

Lesson 7: Introduce a Number of Controlled Opportunities to Demonstrate Supporting Behavior Skill to Obtain Valid Comparisons Across Teams

To obtain valid comparisons of supporting behavior across teams, we recommend the use of several scenario events to introduce controlled opportunities to demonstrate such behavior. This can be accomplished in a number of ways. For example, previous research has demonstrated linkages between performance outcomes and supporting behaviors through the use of a confederate teammate (Smith, 1994). A confederate teammate can systematically commit errors that any team would need to correct. In addition, at predetermined points in the scenario, a confederate can demonstrate signs of stress that indicate that he or she requires backup.

The need for supporting behavior can be controlled without the use of a confederate as well by providing discrepant information to team members and examining whether they note and resolve the discrepancy. Similarly, one could script the need for workload redistribution. This can be accomplished by intentionally placing unusually high workload demands on a particular team position at predetermined points in a scenario.

Lesson 8: Target Information Exchange, Communication, Supporting Behavior, and Team Initiative/Leadership for Team Training

The event-based assessment of CIC team training needs described in this chapter uncovered four teamwork dimensions that are important for effective performance as well as trainable: information exchange, communication, supporting behavior, and team initiative/leadership. Preliminary data suggest that these dimensions are equally applicable to damage control, engineering, and seamanship teams.

We found that each of the four teamwork dimensions could be improved by one or more of the training interventions developed and tested under TADMUS. Given that each of these interventions lasted only 2 hours, it appears that such training would be well worth the investment. In addition, we recommend that the four ATOM dimensions be evaluated and debriefed as a part of scenario-based team training. A method for conducting this type of training, team dimensional training, is described in chapter 11, this volume.

Conclusion

Human performance measurement in complex environments is a multifaceted problem. The TADMUS program afforded a unique opportunity to advance the knowledge in this area. However, much remains to be learned. Directions for future research include (a) examining the appropriate use of advanced technologies such as eye tracking and voice recognition for gathering human performance data, (b) developing "expert" performance models against which student performance may be judged, (c) identifying appropriate means of integrating system-generated data with rater observations, (d) developing a greater understanding of the relationships among individual-level and team-level performance variables, and (e) developing principles for the measurement of multiteam processes and outcomes.

References

Bailey, S. S., Johnston, J. H., Smith-Jentsch, K. A., Gonos, G., & Cannon-Bowers, J. A. (1995). Guidelines for facilitating shipboard training. In *Proceedings of the 17th annual Interservice/Industry Training Systems and Education Conference* (pp. 360–369). Washington, DC: IITSEC.

Bernardin, H. J., & Beatty, R. W. (1984). *Performance appraisal: Assessing human performance at work*. Boston: Kent.

Brannick, M. T., Prince, A., Prince, C., & Salas, E. (1995). The measurement of team process. *Human Factors, 37*, 641–651.

Cannon-Bowers, J. A., & Salas, E. (1997a). A framework for developing team performance measures in training. In M. T. Brannick, E. Salas, & C. Prince (Eds.), *Team performance assessment and measurement: Theory, methods and applications* (pp. 45–62). Hillsdale, NJ: Erlbaum.

Cannon-Bowers, J. A., & Salas, E. (1997b). Teamwork competencies: The intersection of team member knowledge, skills, and attitudes. In H. F. O'Neil (Ed.), *Workforce readiness: Competencies and assessment* (pp. 151–174). Hillsdale, NJ: Erlbaum.

Cannon-Bowers, J. A., Tannenbaum, S. I., Salas, E., & Volpe, C. E. (1995). Defining team competencies and establishing team training requirements. In R. Guzzo & E. Salas (Eds.), *Team effectiveness and decision making in organizations* (pp. 330–380). San Francisco: Jossey-Bass.

Cohen, M. S., Freeman, J., Wolf, S., & Militello, L. (1995). *Training metacognitive skills in naval combat decision making* (Tech. Rep. No. N61339-92-C-0092). Arlington, VA: Cognitive Laboratories.

Dickinson, T. L., & McIntyre, R. M. (1997). A conceptual framework for teamwork measurement. In M. T. Brannick, E. Salas, & C. Prince (Eds.), *Team performance assessment and measurement: Theory, methods and applications* (pp. 19–44). Hillsdale, NJ: Erlbaum.

Duncan, P. C., Rouse, W. B., Johnston, J. H., Cannon-Bowers, J. A., Salas, E., & Burns, J. J. (1996). Training teams working in complex systems: A mental model-based approach. In W. B. Rouse (Ed.), *Human/technology interaction in complex systems* (Vol. 8, pp. 173–231). Greenwich, CT: JAI Press.

Dwyer, D. J. (1992). An index for measuring naval team performance. In *Proceedings of the Human Factors Society 36th annual meeting* (pp. 1356–1360). Santa Monica, CA: Human Factors and Ergonomics Society.

Endsley, M. R. (1995). Measurement of situation awareness in dynamic systems. *Human Factors, 37*, 65–84.

Gill, R., Gordon, S., Moore, J., & Barbera, C. (1988). The role of conceptual structures in problem-solving. In *Proceedings of the annual meeting of the American Society of Engineering Education* (pp. 583–590). Washington, DC: American Society of Engineering Education.

Hall, J. K., Dwyer, D. J., Cannon-Bowers, J. A., Salas, E., & Volpe, C. E. (1993). Toward assessing team tactical decision making under stress: The development of a methodology for structuring team training scenarios. In *Proceedings of the 15th annual Interservice/Industry Training Systems Conference* (pp. 87–98). Washington, DC: IITSEC.

Hobson, C. J., Mendel, R. M., & Gibson, F. W. (1981). Clarifying performance appraisal criteria. *Organizational Behavior and Human Performance, 28*, 164–188.

Johnson, D. S., Perlow, R., & Pieper, K. F. (1993). Differences in team performance as a function of type of feedback: Learning oriented versus performance oriented feedback. *Journal of Applied Social Psychology, 23*, 303–320.

Johnston, J. H., Cannon-Bowers, J. A., & Smith-Jentsch, K. A. (1995). Event-based performance measurement system for shipboard command teams. In *Proceedings of the First*

International Symposium on Command and Control Research and Technology (pp. 274–276). Washington, DC: Institute for National Strategic Studies.

Johnston, J. H., Smith-Jentsch, K. A., & Cannon-Bowers, J. A. (1997). Performance measurement tools for enhancing team decision making. In M. T. Brannick, E. Salas, & C. Prince (Eds.), *Team performance assessment and measurement: Theory, methods and applications* (pp. 311–330). Hillsdale, NJ: Erlbaum.

Klein, G. A., Calderwood, R., & MacGregor, D. (1989). Critical decision method for eliciting knowledge. *IEEE Transactions on Systems, Man, and Cybernetics, 19,* 462–472.

Kozlowski, S. W. J., Gully, S. M., Smith, E. M., Nason, E. R., & Brown, K. G. (1995). *Learning orientation and learning objectives: The effects of sequenced mastery goals and advanced organizers on performance, knowledge, metacognitive structure, self-efficacy and skill generalization of complex tasks.* Paper presented at the 10th annual conference of the Society for Industrial and Organizational Psychology, Orlando, FL.

Locke, E. A., & Latham, G. P. (1990). *A theory of goal setting and task performance.* Englewood Cliffs, NJ: Prentice-Hall.

Locke, E. A., Shaw, K. N., Saari, L. M., & Latham, G. P. (1981). Goal setting and task performance: 1969–1980. *Psychological Bulletin, 90,* 125–152.

McEvoy, G., Beatty, R., & Bernardin, J. (1987). Unanswered questions in assessment center research. *Journal of Business and Psychology, 2,* 97–111.

McIntyre, R. M., & Salas, E. (1995). Measuring and managing for team performance: Emerging principles from complex environments. In R. Guzzo & E. Salas (Eds.), *Team effectiveness and decision making in organizations* (pp. 149–203). San Francisco: Jossey-Bass.

Means, B., & Gott, S. P. (1988). Cognitive task analysis as a basis for tutor development: Articulating abstract knowledge representations. In J. Psotka, L. D. Massey, & S. A. Mutter (Eds.), *Intelligent tutoring systems: Lessons learned* (pp. 35–57). Hillsdale, NJ: Erlbaum.

Mosier, K. L. (1997). Myths of expert decision making and automated decision making aids. In C. Zsambok & G. Klein (Eds.), *Naturalistic decision making* (pp. 319–330). Hillsdale, NJ: Erlbaum.

Sackett, P. R., & Dreher, G. F. (1982). Constructs and assessment center dimensions: Some troubling empirical findings. *Journal of Applied Psychology, 67,* 401–410.

Sackett, P. R., & Hakel, M. D. (1979). Temporal stability and individual differences using assessment information to form overall ratings. *Organizational Behavior and Human Performance, 23,* 120–137.

Salas, E. Cannon-Bowers, J. A., & Johnston, J. H. (1997). How can you turn a team of experts into an expert team? Emerging training strategies. In C. Zsambok & G. Klein (Eds.), *Naturalistic decision making* (pp. 359–370). Hillsdale, NJ: Erlbaum.

Slovic, P., & Lichtenstein, S. (1971). Comparison of Bayesian and regression approaches to the study of information processing in judgment. *Organizational Behavior and Human Performance, 6,* 649–744.

Smith, K. A. (1994). *Narrowing the gap between performance and potential: The effects of team climate on the transfer of assertiveness training.* Unpublished doctoral dissertation, University of South Florida, Tampa.

Smith-Jentsch, K. A. (1995, May). *Measurement and debriefing tools refined and validated at SWOS.* Presentation at the meeting of the TADMUS Technical Advisory Board, Moorestown, NJ.

Smith-Jentsch, K. A., Payne, S. C., & Johnston, J. H. (1996). Guided team self-correction: A methodology for enhancing experiential team training. In K. A. Smith-Jentsch (Chair), *When, how, and why does practice make perfect?* Paper presented at the 11th annual conference of the Society for Industrial and Organizational Psychology, San Diego, CA.

Smith-Jentsch, K. A., Salas, E., & Baker, D. (1996). Training team performance-related assertiveness. *Personnel Psychology, 49,* 909–936.

Stout, R. J., Cannon-Bowers, J. A., & Salas, E. (1996). The role of shared mental models in developing team situational awareness: Implications for training. *Training Research Journal, 2*, 85–116.

Westra, D. (1994). *Analysis of TADMUS pre-experimental data: Considerations for experimental design.* Unpublished manuscript.

Zachary, W. W., Zaklad, A. L., Hicinbothom, J. H., Ryder, J. M., & Purcell, J. A. (1993). COGNET representation of tactical decision making in the Anti-Air Warfare. In *Proceedings of the 37th annual meeting of the Human Factors and Ergonomics Society* (pp. 1112–1116). Santa Monica, CA: HFES.

Part II _____

Individual Level Training Strategies and Research

5

Feedback Augmentation and Part-Task Practice in Training Dynamic Decision-Making Skills

*Alex Kirlik, Arthur D. Fisk, Neff Walker,
and Ling Rothrock*

Training High-Performance Skills

Current U.S. Navy command and control (C^2) systems in general, and in particular a shipboard combat information center (CIC), require performers to process large volumes of information at very rapid rates. These systems can impose extensive mental workload and task-coordination requirements in a variety of ways, including the need for rapid and accurate performance, the demand that several cognitive functions such as information retrieval and decision making be performed concurrently, and the requirement to obtain and integrate information from a number of perceptual modalities. From a cognitive perspective, the information-handling and coordination requirements, at both the individual and team levels, can become overwhelming, especially during times of high stress.

Successful operation of such systems, therefore, requires highly skilled operators who can perform rapidly, accurately, and in a coordinated manner even under high workload. The high-performance skills demanded of such operators typically require extensive practice to develop and are often characterized by striking qualitative differences between novices and experts. Such skill characteristics constitute the essence of an individual expert performer and clearly differentiate expert, highly coordinated teams from teams that are less skillful and whose performance degrades under high workload or stressful situations.

Because the development of high-performance skills requires extended practice, traditional training approaches (e.g., on-the-job training, limited

We extend our sincere thanks to Kevin Hodge, Dabby Phipps, Paul Gay, and Amy Bisantz for their many valuable contributions to the research reported here. The authors are grateful for the encouragement and support of our contract monitors, Eduardo Salas, Janis A. Cannon-Bowers, and Joan Johnston.

systems exercises) often result in minimally proficient operators who either never develop all required skills or do so only during the terminal phase of an assignment. Such problems can compromise the effective operation of C^2 systems. Difficulties in the transfer of component skills to new or modified systems and failure of personnel to retain required skills due to disuse further compound the potential cost and operational problems. Retention of high-performance skill components is not only of central importance to transfer of training but also is highly critical in emergency situations requiring the use of previously established skills that have not been used during a period of inactivity or reassignment to other duties.

In an environment requiring high-performance skills, the ability to rapidly train to a required proficiency level and ensure maintenance of that proficiency strongly influences operational effectiveness. The present definition of high-performance skills is taken from Schneider (1985). *High-performance skills* are defined as those activities that (a) require more than 100 hours of training (combined formal and on-the-job training); (b) result in easily identifiable differences among novice, intermediate, and expert skill levels; and (c) generally result in a high "washout" rate in the progression from entry level trainees to final, "mission ready" individuals. For any given high-performance skills training program, it is critical to minimize the training time necessary while maximizing the learning and retention rate of trainees.

Our goal in the present chapter is to discuss a variety of approaches for training high-performance skills in complex environments, present empirical evidence evaluating these approaches in a dynamic decision-making task, and finally to provide training principles and guidelines that are based on our findings. The task context for our research was based on the U.S. Navy CIC and, specifically, the duties of the AAWC within this system. Our initial efforts to understand specific training needs in this task context consisted of studying how the AAWC is currently trained at an existing Navy training center.

Field Study

We began our investigations of training to support naval CIC operations more than 4 years ago with a visit to a Navy precommissioning team training center. This land-based facility contains a full-scale hardware and software simulation of a ship-based CIC. Entire teams of CIC personnel, from the captain on down, receive team-based, crew-coordination training at this center just prior to boarding ship and taking to sea. Experts from the military and industry act as trainers and coaches, looking over the shoulders of trainees as the team conducts each training exercise. Each exercise, or scenario, simulates a plausible tactical situation unfolding over time. A typical scenario might begin as relatively unthreatening and calm. Soon thereafter, anomalous and ambiguous events begin to occur, and the team has to resolve these ambiguities by collecting, communicating, and integrating information. Finally, the team must decide what actions, if any, are necessary to carry out the goals of the mission. At the conclusion of each scenario, the team members move to a classroom, where they are debriefed by the training staff as well as by their own superiors within the team.

We focused our observations on the anti-air warfare coordinator (AAWC) station within the training center. The major duties of the AAWC are as follows (Zachary et al., 1991):

1. Supervise subordinates
2. Enter airspace-relevant information into the computer system
3. Direct and monitor the air situation
4. Assess the current air situation
5. Order and monitor air attacks
6. Communicate to superior officers
7. Coordinate with other members of the CIC team

The AAWC sits at a console with multiple display screens and control panels, wearing headphones and a microphone for communicating with other CIC team members. Even in this simulated training context, the AAWC's stress level as the scenario heats up can be readily inferred from the tone of his verbal communication and from his or her often frantic attempts to keep pace with events by scribbling updated information on the glass of his status displays. A trainer stands over his shoulder pointing out errors, giving advice, and occasionally reaching around the AAWC to operate a console control himself.

Observations on Training Opportunities

As researchers interested in skills training, we quickly became aware of three interesting features of this training context. First was the enormous cost of this type of training in terms of time, hardware, software, and human resources. We realized that even modest improvements in either the efficiency or the effectiveness of training could have huge benefits. Second, we observed that although the expressed goal of these training exercises was to enhance tactical decision making and team coordination, the AAWC and most other team members spent a considerable amount of time and effort merely trying to understand how to operate the console equipment. The hierarchical menu systems, programmable buttons, and trackball on the console panel required the trainee to learn a considerably sophisticated action grammar and to meet stringent motor control demands. In many cases, a series of events was rapidly unfolding on console displays while the AAWC had his or her head down and hands busy attempting to get the console equipment to obey his or her wishes.

The field study thus revealed that the need for the trainee to attend continuously to console manipulation activities severely compromised his or her ability to attend to, and benefit from, the tactical decision-making and team-coordination experiences that were the stated focus of training.

This observation directly motivated the portions of our research dealing with automaticity theory and part-task training on console manipulation skills (see the section A Part-Task Training Approach).

Third, we observed in our fieldwork what we believed to be an imbalance between the large amount of time and resources devoted toward creating a high-fidelity environment for training and the relatively modest amount of time and resources devoted toward providing trainees with beneficial feedback on their performance. Whereas the human trainers and postscenario debriefings provided important feedback to trainees, there was considerable room for improvement in terms of the nature of feedback provided and the manner in which it was presented. Some feedback information was presented so late, so inconsistently, or so abstractly that it appeared to be of limited value in efficiently moving trainees to higher levels of skill.

The field study thus revealed a need to improve feedback timeliness, specificity, and presentation mode. This observation directly motivated the portions of our research dealing with augmenting "over-the-shoulder" coaching with real-time automated feedback embedded within the training context (see the section A Feedback Augmentation Approach). In addition, the need to improve feedback quality, along with the results of our automated feedback experiments, motivated our approach to modeling individual decision-making strategies to target feedback toward a particular trainee's misconceptions and oversimplifications of the task environment (see the section A Diagnostic Approach to Feedback Augmentation). The remainder of the chapter discusses our part-task and feedback augmentation approaches in greater detail and empirical evaluations of these training interventions; it concludes with a discussion of lessons learned related to principles and guidelines for training high-performance skills in complex environments.

A Part-Task Training Approach

Literature Review on Part-Task Training

Our approach to part-task training was guided by an initial literature review (Fisk, Kirlik, Walker, & Hodge, 1992) and ultimately by automatic–controlled processing theory. Our review suggested that determining what kind of part-task training is most appropriate for a given situation, and indeed whether part-task training should be used, is not a simple matter. Various methods are available for decomposing an overall task into components to be separately trained (i.e., part-task training), and the choice of method must be driven by the type of task to be trained. *Segmentation* involves breaking up a task into a temporal series of stages and designing part-task training around these individual stages. *Simplification* involves making the task initially easier to perform in some manner and designing part-task training for the simplified task prior to having the trainee move to the full version of the task (e.g., training wheels). *Fractionation* involves decomposing a complex task into a set of individual activities that are performed in parallel in the overall task (e.g., walking and chewing gum) and designing part-task training around the individual activities.

General guidelines (see Fisk et al., 1992) are as follows: (a) The most successful method of segmentation has been *backward chaining*, in which the final segment of a task is trained prior to the sequential addition of all of the preceding tasks, and (b) the simplification technique is most successful for tasks that are initially extremely difficult to learn. By altering the task in such a way that it is easier to perform initially, overall performance is improved on the subsequent performance of the whole task. Although there is evidence that simplification may not be better than whole-task training, it is often cheaper and less frustrating for trainees trying to master a seemingly impossible task. (c) Fractionation is the least supported method in terms of the empirical studies reported to date. The lack of support for fractionation as a training procedure that is preferable to whole-task training is due to the separation of components that ultimately must be performed simultaneously. However, the fractionation method is beneficial if it is paired with some amount of whole- or dual-task practice.

One of the recurring themes that became evident from the review of the literature was that procedural items or psychomotor tasks benefit greatly from part-task training. It was also clear that, in general, the relative effectiveness of part- versus whole-task training depends on the type of task. We concluded that beginning a training program with part-task training and then proceeding to dual-task or whole-task training may be the most efficient training method. In addition, we concluded that training the AAWC's console manipulation skills was best done with a combination of both segmentation and simplification for task decomposition. Segmentation focuses the design of a part-task training system on the goal of preserving temporal sequences of activity as they occur in the full-task context. Simplification focuses the design of part-task training on the goal of eliminating elements of the full task that could interfere with acquiring the target skills. In the following section, we briefly describe the theory of skill acquisition underlying our selection of components included in the design of a part-task training system.

Automatic and Controlled Processing Theory

The distinction between the way people typically perform a novel task (i.e., in a slow, effortful, and error-prone manner) and how they perform the task after lengthy training or practice (i.e., in a fast, accurate, and low-effort manner) is not new to psychology (e.g., Norman, 1976, 1981; Posner & Snyder, 1975; Schneider & Shiffrin, 1977; Shiffrin & Schneider, 1977). William James (1890) discussed the transition from novice to skilled performance in terms of habit development. In fact, such a distinction is clear to anyone who reflects on the differences between his or her own initial and practiced performance in situations such as learning how to play a musical instrument, move the game pieces in chess, drive a car, or operate a computer. The changes that take place, both quantitatively (e.g., in terms of task accuracy, speed, amount of cognitive fatigue) and qualitatively (e.g., resource insensitivity, parallel rather than serial processing), characterize the essence of learning (especially from a phenomenological perspective).

Schneider and Shiffrin (1977; Shiffrin & Schneider, 1977) presented a theory that distinguishes these common learning situations from those in which a transition from effortful, slow, and generally poor task performance to relatively effortless, fast, and accurate performance fails to occur. Several laboratory paradigms were found, and their requisite conditions described, in which, regardless of the amount of task practice, no differences between initial, novel performance and final, practiced performance could be found. By analyzing and manipulating experimental conditions of various perceptual-motor tasks, these investigators found that two general types of processes appear to underlie a wide range of tasks in the experimental literature. Manipulation of these conditions accounted for the presence or absence of common learning effects through practice. These characteristics were denoted as tapping either "automatic" or "controlled" types of information processing (see Fisk, Ackerman, & Schneider, 1987, for a more detailed overview of the characteristics of automatic and controlled processing theory).

Automatic processes are characterized as fast, effortless, and unitized such that they allow for parallel operation with other information-processing task components. Furthermore, automatic processes are not easily altered by a person's conscious control and, once initiated, tend to run through to completion. Automatic processing is implied in the operations of various types of skilled behavior, such as playing a piano, reading, and some aspects of driving a car. These processes may be developed only through extensive practice under "consistent" conditions, which are typical of many skill acquisition situations. Although consistency is operationalized somewhat differently for diverse tasks, it generally involves the use of invariant rules, invariant components of processing, or invariant sequences of information-processing components to be used in successful task performance.

The use of controlled processes, on the other hand, is implied when consistency, such as of rules or sequences of information-processing components, is not present in the task. In addition, controlled processing is necessary when task requirements are novel or when consistencies of the task situation are not internalized by the person. Controlled processing is characterized as slow, requiring attentional effort, and amenable to quick alterations (e.g., switches of attention) under the learner's conscious control; this type of processing limits the availability of the attentional system, allowing mostly serial processing. Although researchers examining automatic and controlled processing have concentrated primarily on visual search, it is clear that the distinction between consistent-mapping and varied-mapping training has been intended to reflect general principles important throughout many aspects of human information processing (e.g., see Fisk & Schneider, 1983).

On the basis of this theory, we concluded that the design of part-task training systems must be guided by the identification of consistent components in the AAWC's task. We therefore analyzed the design of the AAWC's console interface and the activities needed to use the interface to identify the consistent components of the interface manipulation task. On the basis of this analysis, we decided to focus part-task training on two important aspects of the AAWC's task: manipulating console controls and interpreting the symbols present on console displays.

Knowledge of Keyboard Manipulation Skills

Keyboard skills of the AAWC require entering appropriate sequences of switch–button actions in a timely manner to obtain, display, and communicate information. Interviews indicated that for expert AAWC personnel, consistent patterns of control actions became automatized such that these actions could be performed accurately and rapidly without interference with (or from) other activities. This level of automaticity occurs in much the same manner that a skilled typist can time-share motor performance with other activities. Repeated references were made to the necessity for these manipulation skills to be performed rapidly and accurately without visually attending to console control interface. Consistent relationships exist between the cue to develop, send, receive, or review a given type of information on the display and the required control or button actions (e.g., move trackball to superimpose cursor on appropriate points or activate proper buttons). The speed and accuracy of the motor sequences required are similar to those that characterize entering a highly practiced telephone number sequence on a push-button telephone. Such a level of skill is crucial for effective AAWC performance. It may be recalled that the trainees we observed in our field study of team training frequently struggled with interface manipulation. Lack of proficiency in these component skills was reported by trainees as interfering with maximum gain from decision making and team coordination training.

Knowledge of Display Symbols

Our field observations of team training clearly pointed to a need for trainees to identify symbols (and their meaning) quickly and accurately on the display. Trainees frequently "lost" displayed symbols (although they were clearly presented on the screen) or incorrectly identified symbols. We expected that part-task training on the consistent interpretation of display symbols would be relatively simple to implement but would provide a large return in terms of freeing up attentional resources for the high-level cognitive aspects of the AAWC's task.

The capability of the AAWC to process various symbols rapidly and accurately is essential to effective performance. Because of the heavy time-sharing requirements imposed on the AAWC, and because of the need for rapid responses, the AAWC must be able to recognize appropriate symbols rapidly and assess the impact of displayed information in terms of situation assessment and evaluation. Most symbols used in the system, within a given context, are consistently mapped with a specific meaning. Therefore, this component skill represented a viable candidate for part-task training within the automatic processing framework. The following sections describe an experiment performed to evaluate our part-task training approach on the basis of this theory of automaticity.

Part-Task Training Experiment

Method. We evaluated our part-task training approach using a laboratory task requiring a representative set of AAWC activities and a training system designed to provide the trainee with experience at keyboard manipulation and symbol recognition tasks. The performance of participants using the part-task training system was compared to the performance of participants without such training, who instead were trained in a manner similar to that observed in our field study of current AAWC team training.

APPARATUS. We constructed a laboratory simulation of the AAWC task environment for the purpose of examining a variety of training interventions. Our simulation was closely modeled after the Decision Making Evaluation Facility for Tactical Teams (DEFTT) training system used in previous Navy training (see chapter 3, this volume). DEFTT, however, is a team training system, whereas our focus was on training individual performance. Our own simulation, the Georgia Tech AAWC Simulation Platform (GT-ASP), therefore, does not preserve the two explicitly team-oriented AAWC duties as previously described: supervise subordinates, coordinate with other members of the CIC team. GT-ASP does, however, use sound-generation technology to send vocal messages to the experimental participant to simulate what the AAWC hears over his or her headphones. Aside from these differences, GT-ASP is highly similar to DEFTT and preserves many of the psychologically important details of the training system we observed in our field study. The GT-ASP interface is shown in Figure 1.

To evaluate our part-task training approach, we also designed and constructed a separate, personal computer–based part-task training system. This system used an interface nearly identical to that used in GT-ASP. However, rather than containing a complete, dynamic scenario simulation capability like GT-ASP, the part-task training system presented the experimental participant with a series of discrete exercises, each of which required responding appropriately to symbols displayed on the display screen. We identified 24 consistent control action sequences that were used to manipulate the full-task GT-ASP interface, and we designed the

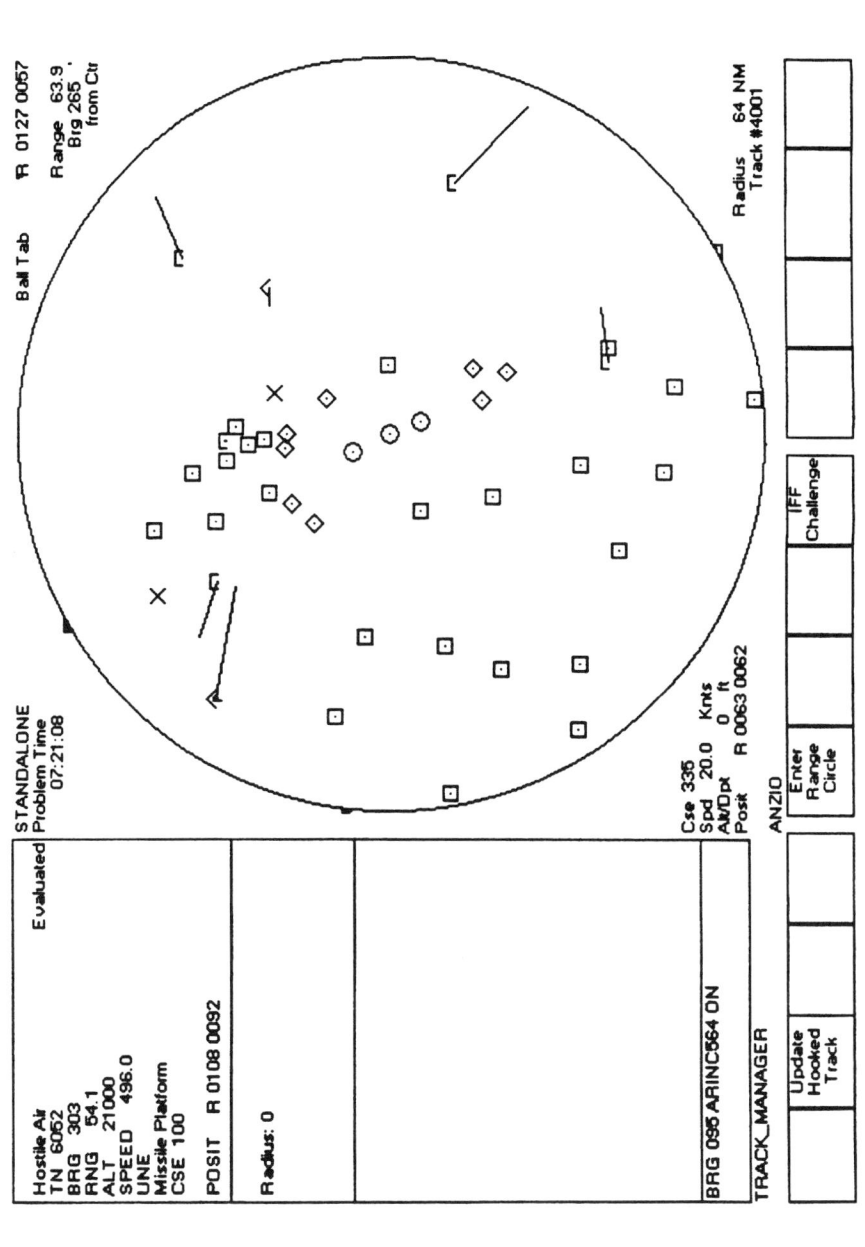

Figure 1. The GT-ASP interface. Vehicles, or "tracks," in the task environment are indicated by symbols on the round radar display. The left text window provides detailed state and identity information on any selected (by mouse) vehicle.

part-task training exercises around these 24 sequences. These exercises (e.g., change the identity of an item on the display, change the radius of the rings on the radar display) also required participants to interpret correctly the symbols presented on the part-task system display in order to perform correctly. In addition, the part-task system contained a "tutor," which consisted of instructions for performing each of the training exercises, as can be seen in Figure 2.

EXPERIMENTAL DESIGN. The experiment consisted of two groups, each containing 12 participants who were male undergraduate students at Georgia Tech. All participants attended an introductory session on the evening prior to the first day of data collection, in which they were briefed on the purpose of the experiment and provided with information on the contextual foundations of the task. The full-task group received training on the full GT-ASP task in a manner modeled after the "over-the-shoulder" training and postscenario debriefing used in the system observed in our field study. This group received fifteen 30-minute sessions of full-task training and then received three final sessions in which there was no interaction between the participants and the human trainers. The part-task group, on the other hand, began the experiment with 10 blocks of part-task training, each block containing 72 training exercises. In the first 5 blocks of training the tutor was available, and for the last 5 blocks the tutor was not available. This group then received 8 full-task training scenarios that were identical to the first 8 scenarios performed by the full-task group. The part-task group then received the same 3 final sessions performed by the full-task group, also without any interaction between participant and trainer. All scenarios were verified for representiveness and internal coherence by a highly experienced former naval officer who served as a subject-matter expert for this purpose. For a more detailed description of this experiment, see work by Hodge (1997).

PERFORMANCE MEASURES. Performance in GT-ASP is considered to be successful to the extent that a participant takes all and only those actions prescribed by the rules of engagement (ROE). The ROE prescribe the appropriate response to be taken in various task conditions. We used nine rules in our ROE modeled after those used in the Navy's DEFTT training system (e.g., if a hostile aircraft moves to within 50 miles of one's own ship, a particular warning should be issued to that aircraft; if a hostile aircraft moves to within 30 miles, the fire control illuminator should be turned on, and so on). We then developed a method for assessing the timeliness and accuracy of performance in GT-ASP on the basis of a *window-of-opportunity* construct.

As can be seen in Figure 3, the idea of measuring performance with time windows is modeled after signal detection theory. In signal detection theory, the world is in a state of either signal or noise, and the response is to indicate that a signal is either present or absent, yielding four performance outcomes. As shown in the figure, these outcomes are called a *hit*, a *miss*, a *false alarm*, and a *correct rejection*. Also as shown in the

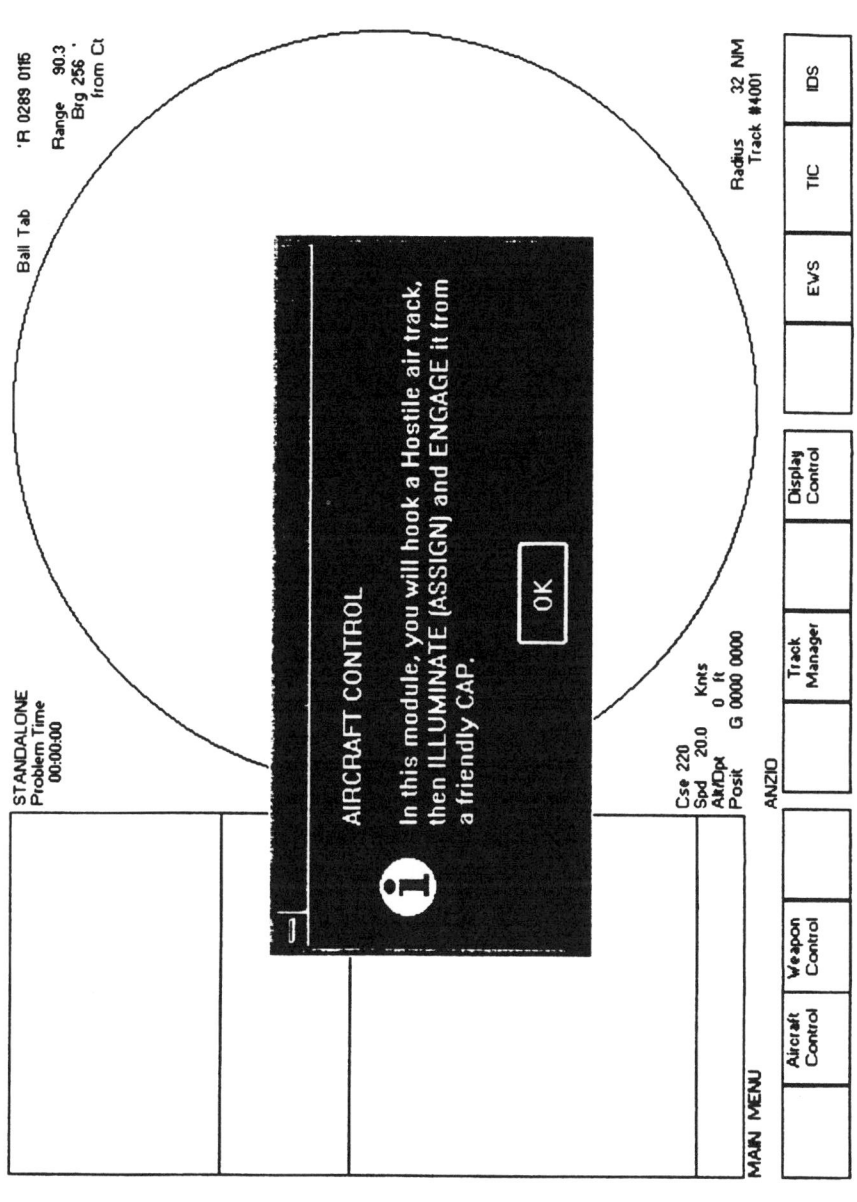

Figure 2. The part-task training system interface. In initial training sessions, a tutor leads the trainee through the sequence of actions necessary to accomplish each task.

Signal Detection Outcomes

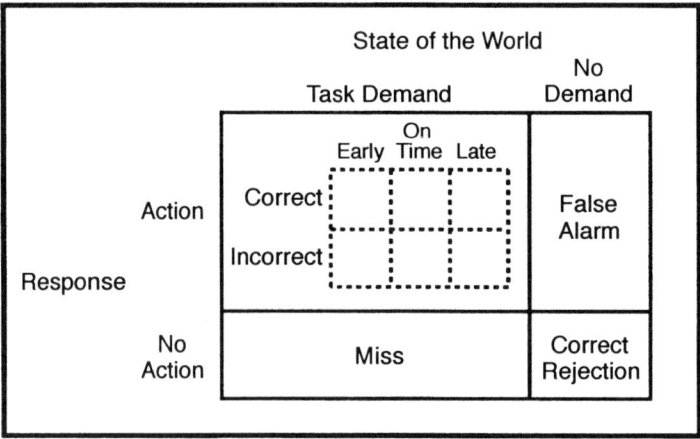

Time Window Outcomes

Figure 3. A comparison of signal detection theory outcomes and outcomes based on the time window construct.

figure, the use of time windows to measure performance expands the number of possible outcomes, or performance measures. Each action prescribed by the ROE has a specified initial time at which it becomes appropriate and a specified end time before which it must be performed. This interval is the time window associated with the action. Outcomes can be classified into the nine categories shown in the figure, contrasted with the four categories used in signal detection theory. These categories formed the basis by which we measured the performance of experimental participants. In addition, we measured other aspects of behavior such as the priority level of an action taken when there were multiple competing actions possible and the number and timing of interface manipulation sequences.

Results. Of primary interest is the performance of full-task and part-task groups in the final three GT-ASP sessions, in which there was no interaction between the participants and the trainers. For these three sessions, we found no substantive differences between the performances of the two groups. This result was striking in that it indicated that the part-task group was able to achieve the same level of performance as the full-task group with approximately half the amount of full-task training and experience. Analysis of the learning curve for the full-task group did not indicate that these participants had reached ceiling halfway through training; instead, they continued to learn the task throughout all 15 sessions of full-task training. The learning curve of the part-task group was steeper. Hodge (1997) provides a more detailed presentation of these experimental results.

It is important to stress the practical implications of this "null result." The type of part-task training system used in this experiment was extremely simple and relatively inexpensive, especially compared with the team training system we observed in our field study. Part-task trainers of this type could be readily constructed and deployed to allow naval trainees to become expert at their console manipulation and symbol interpretation tasks prior to entering tactical decision-making and crew-coordination training at the sophisticated and expensive naval training center. In addition, one can envision training systems of this type being deployed aboard ship, so that CIC personnel who are temporarily shifted to other duties can retain their skills or personnel can be rotated among a variety of CIC consoles without having to learn how to operate these consoles under operational conditions. In summary, we believe that a part-task training approach based on automaticity theory holds great promise in improving the efficiency, and thus decreasing the cost, of training decision skills in complex environments.

A Feedback Augmentation Approach

Dimensions of Feedback Augmentation

On the basis of our field study observations, we concluded that the nature and presentation of feedback provided to trainees could be improved in a variety of ways. Specifically, we observed that training efficiency and effectiveness could be enhanced by improving feedback in four areas: (a) timeliness, (b) standardization, (c) diagnostic precision; and (d) presentation mode. In the following sections, we consider each of these issues in more detail.

Timeliness. Training scenarios can last a considerable period of time, and feedback associated with particular details of a trainee's performance given in postscenario debriefing appeared to come too late for the trainee to benefit fully from this information. Although some on-line feedback is provided by human coaches while they observe an exercise, this feedback is naturally oriented toward outcome rather than process. Coaches typically wait until a critical error in tactical decision making or crew coordination has occurred before pointing out the problem to the trainee during the course of an exercise. By this point, this feedback may be too late to be of use in modifying any of the many preceding behaviors that eventually gave rise to the error.

Standardization. Although the insightful feedback provided by human experts observing trainee performance on-line is extremely valuable, we noted that this type of feedback tended to be highly idiosyncratic. Trainers appeared to have their own approaches to the task and their own priorities for what lessons are most important to learn and what error types are most severe. In particular, we observed some tendency for expert trainers to focus heavily on the strategic and perhaps political challenges posed by a scenario, even when a trainee was still facing severe challenges in accomplishing lower level, tactical, and even procedural, tasks. This situation is perhaps to be expected: The higher level strategic and political concerns pose the greatest challenges to the experts, and thus they tend to be the aspects of trainee behavior that attract the greatest attention. Because these most challenging aspects of the task must be evaluated on the basis of experiential knowledge, feedback associated with these issues tended to be highly specific to the particular operational experiences of the expert trainer.

Diagnostic precision. It is more difficult to provide useful feedback information than to provide information evaluating performance. Errors exhibited by trainees and pointed out to them as "feedback" often appeared to have resulted from a combination of preceding factors. These factors were typically a mixture of current performance limitations (e.g., in attention allocation to the highest priority events) and factors associated with the particular scenario events that brought the performance limitation to light (e.g., a particular scenario situation dealing with multiple events). Although measuring the frequency of errors may be useful in assessing the trainee's performance, communicating these errors to the performer is not likely to result in useful feedback unless the trainee can associate error information regarding specific scenario events (e.g., "you were too late in noticing Item 294 on the radar screen") with the underlying, more general performance limitation that the error reflects. The trainee not only must be informed of the error, but also must be able to infer the correct lesson to be learned from the error. For example, the trainee might erroneously attribute the error to the particular events in the scenario (e.g., "I have to do better with radar screen items coming from the northwest"), when the beneficial lesson to be learned from this error lies in more general attention allocation processes.

Presentation mode. During our observations, it became apparent to us that trainees sometimes seemed to view the need to listen to and communicate with the human on-line trainer as something akin to a "secondary task." Although the verbal feedback intermittently provided by the human trainers contained valuable information, it was only natural for trainees to view this feedback as somewhat of an interruption to their already challenging duties. This situation seemed to compromise the potential benefits that could result from expert feedback.

Feedback Augmentation Experiment

Method. We evaluated our feedback augmentation training approach using a laboratory task requiring a representative set of AAWC activities and a training system designed to provide the trainee with automated, real-time feedback embedded within the task context. The performance of participants using the automated feedback system was compared to the performance of participants without such training, who instead were trained in a manner similar to that observed in our field study of current AAWC team training.

DEVELOPMENT OF AN AUTOMATED FEEDBACK MODULE. On the basis of these observations, we designed and constructed a feedback module for GT-ASP that provides trainees with real-time, on-line feedback embedded within the GT-ASP display. This module presents a list of prioritized objects in the environment that should be acted on, as prescribed by the rules of engagement (ROE) as they apply to the details of a particular scenario situation. This prioritized list of actions to be taken is influenced by the actions of the GT-ASP trainee, in that actions are dynamically removed from the list when they are successfully taken. In this manner, the module provides evaluative feedback on the degree to which the participant is successfully taking appropriate actions and also provides information that the trainee can use to direct his or her attention to high-priority concerns as a scenario unfolds over time.

This feedback information is presented in a textual manner in a window on the GT-ASP display, as shown in Figure 4. One rationale underlying this training approach is that effective attention allocation is a crucial component of skill in GT-ASP and, we believe, in the CIC environment generally. Early in training, performers may be aware of the stated ROE, but they are limited in their ability to understand how the ROE combine with the details of a particular situation to define a priority structure over available actions or, more simply, "what should be done next." We hypothesized that a real-time, embedded feedback system would improve training effectiveness by teaching the trainee how to combine ROE information with scenario information to promote situation-assessment skills. In addition, this module has the benefit of providing standardized, timely, and potentially diagnostic feedback to the trainee in addition to whatever ver-

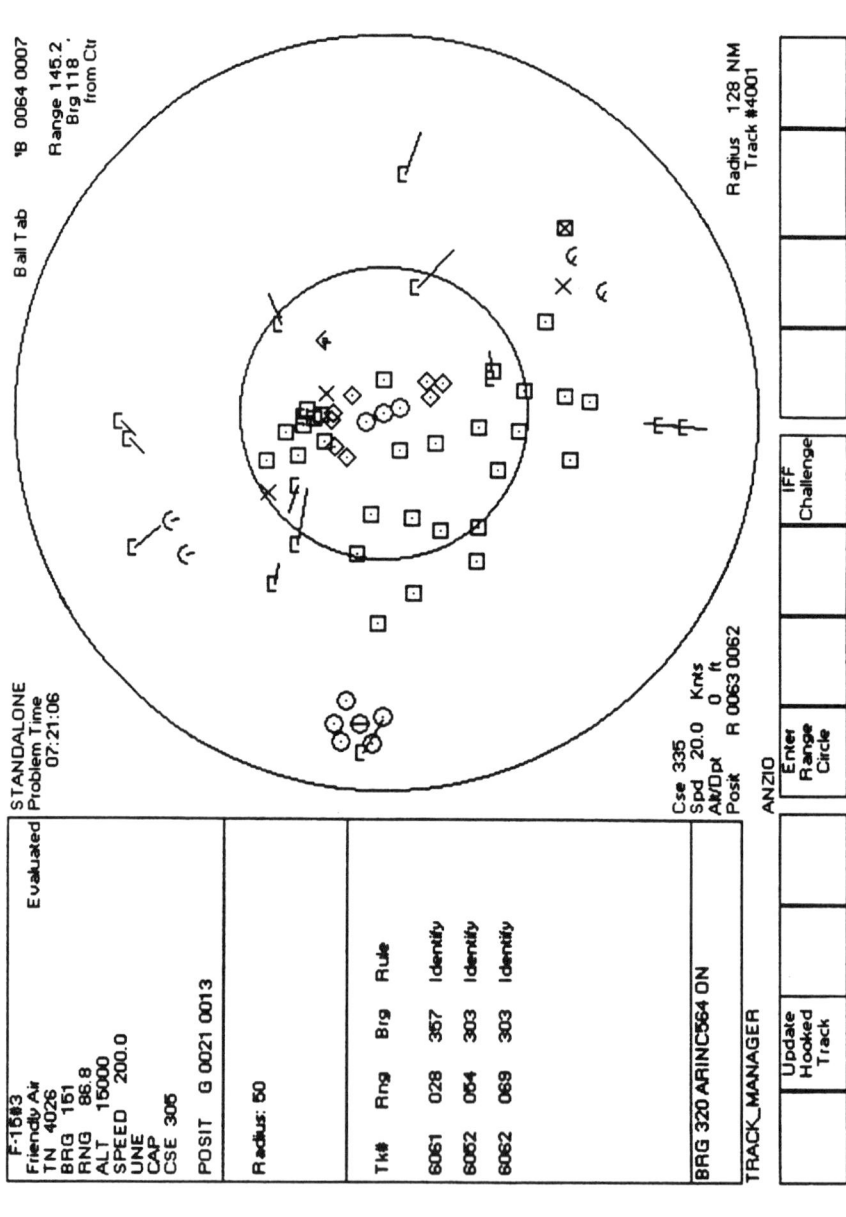

Figure 4. The GT-ASP interface with feedback augmentation (shown as the list of three recommended "Identify" actions in the left text window).

bal feedback information is provided by human, "over-the-shoulder" coaching. The following sections describe an experiment performed to evaluate this feedback augmentation training approach.

APPARATUS. A second version of the GT-ASP display was created in which a textual window provided an ordered list of the three most important actions needing to be taken from the standpoint of the ROE and the current scenario. As shown in the left text window in the situation depicted in Figure 4, this list consists of three "Identify" actions recommended for tracks (Tk#) each at a specified range (Rng) and bearing (Brg) from the ownship. The logic of the algorithm underlying this feedback augmentation was nearly identical to the logic used to measure participant performance, as described in the previous experiment. Time window information was computed in real time on the basis of the scenario, participants' actions, and the ROE, and the relative priority of an action was determined by the priority of the particular rule in the ROE that prescribed it (e.g., issuing a warning to a threatening aircraft has higher priority than identifying a newly appeared object that is 200 miles away). The baseline version of GT-ASP used in this experiment (without feedback augmentation) was identical to the version described in the previous experiment.

EXPERIMENTAL DESIGN. The experiment consisted of two groups, each containing 12 participants who were male undergraduate students at Georgia Tech. All participants attended an introductory session on the evening prior to the first day of data collection in which they were briefed on the purpose of the experiment and provided with information on the contextual foundations of the task. The baseline group received training on the full GT-ASP task in a manner modeled after the "over-the-shoulder" coaching and postscenario debriefing used in the system observed in our fieldwork. This group received 15 sessions, 30 minutes each, of full-task training and then received 3 final sessions in which there was no interaction between the participants and the human trainers. The augmentation group received the same coaching given to the baseline group over the first 15 sessions and was also provided with the display window presenting real-time feedback augmentation consisting of a dynamic, prioritized list of actions to be taken from the perspective of the ROE. This group also received 3 final sessions in which there was no interaction between participants and human trainers and in which the window providing real-time feedback augmentation was not made available.

PERFORMANCE MEASURES. The measures used to evaluate participant performance in this experiment were identical to those used in the previous experiment on part-task training.

Results. Again, of primary interest are the three final experimental sessions in which neither the baseline nor the augmentation group received any form of feedback. In these sessions, the augmentation group performed significantly better than did the baseline group on the numbers of on-time and correct actions, missed actions, and false alarms, as well as in the average priority of actions taken. On no measure of performance did the baseline group outperform the augmentation group. It is important to emphasize that the benefits accrued owing to feedback augmentation were truly benefits in learning rather than benefits owing to decision support, because the augmentation group participants outperformed the baseline group even after the on-line augmentation was taken away in the three final experimental sessions. We believe that some form of on-line feedback of the type considered here holds great promise as a way of improving the standardization, timeliness, and diagnosticity of feedback, and as a way of presenting this information in a nonintrusive fashion by embedding the feedback within the task context itself. We understand that there are currently ongoing efforts among a design group within the TADMUS effort to include feedback of this general type in future AEGIS training and display systems (Morrison, Kelly, & Hutchins, 1996).

A Diagnostic Approach to Feedback Augmentation

The third training study we performed concerned targeting the feedback provided to a trainee toward aspects of that trainee's decision-making strategy. We provide only a brief introduction and overview of this work here; see the recent dissertation by Rothrock (1995) for a complete presentation. In addition to studying the CIC environment, we have studied a variety of issues related to decision making in other complex, dynamic tasks over the past years (e.g., Kirlik, Walker, Fisk, & Nagel, 1996; Kirlik, Miller, & Jagacinski, 1993). A recurring theme arising from these studies is that in complex, dynamic tasks, decision makers develop problem-simplification strategies for coping with dynamism and complexity (see Kirlik, Rothrock, Walker, & Fisk, 1996, for an overview). Although this may not be surprising, we have found that the simplicity as well as the general effectiveness of these strategies is often remarkable, especially when the apparent complexity of these tasks is considered.

One may note, for example, the apparent complexity of our GT-ASP task in terms of the types of information that are presumably relevant to any particular object-identification decision. In principle, to reach acceptable levels of identification performance, the participant had to consider the following types of information, available from GT-ASP displays: (a) identification friend or foe (IFF) status, (b) electronic sensor emissions, (c) visual identification, (d) altitude, (e) speed, (f) course, (g) bearing, (h) range, (i) knowledge of civilian air corridors, and (j) knowledge of hostile and friendly countries. This list of potentially relevant information may be contrasted with a strategy for identifying objects as verbally reported by one of our participants in a postexperiment interview (from Rothrock, 1995):

> 1. The [sensor] is always right. ARINC564 [a sensor emission] always means assume friendly/NMIL [nonmilitary].

2. If IFF corresponds to an actual flight # and it is in a reasonable place for that flight to be, assume friendly/NMIL [nonmilitary].
3. If it has no IFF and its [sensor] is off, it is still assume[d] friendly/ NMIL if it is flying high and its speed is <400 [kts]. I usually try to VID [visually identify] these or keep checking EWS [sensor] if they're going to come close.
4. If all else fails, VID. (p. 212)

Looking beyond the cryptic terminology, one may note that the logical form of this strategy is quite simple. In addition, of the types of information described previously as being potentially relevant to this decision, this participant mentions only about half in his verbalization. Yet this was the highest performing participant in his experimental condition, and he performed a number of sessions with perfect object-identification accuracy.

There is, of course, no guarantee that a simplified heuristic strategy like the one given here will yield high performance; we have observed a number of experimental participants using heuristic strategies leading to frequent error (also see Reason, 1988). On the basis of observations such as these and our prior research in similar environments, we attempted to infer the simplification heuristics that a trainee was using from his history of decision-making behavior. Our goal in this endeavor was to provide a technique whereby a trainee's performance can be analyzed in real time to identify the heuristic decision-making strategy he or she is relying on. When the diagnosed strategy indicates that the trainee has a misconception about the task or has oversimplified the task, training exercises could be designed to target the particular misconceptions or oversimplifications, thus providing an even more extensive and targeted form of feedback augmentation.

Modeling Approach

Rothrock (1995) used a modeling approach based on genetic algorithms and optimization techniques to analyze identification decision-making data from GT-ASP participants. A genetic algorithm is an automated search procedure motivated by the phenomena of evolution and natural selection (Holland, 1975). Genetic algorithms search for rules, encoded in bit strings known as genes, by an iterative process of generating candidate rules through mutation and crossover of past genes and testing the quality of these rules for their fitness, or match, to empirical data. The essence of our use of genetic algorithms is to search for the simplest, yet sufficient, noncompensatory (if/and/nor/or/then) rule or rules that are consistent with the decisions made by a participant over the course of an experimental session. The identification decisions made by a participant were thus considered to be a simple noncompensatory function of the information available to make those decisions. The verbalized strategy offered by the participant discussed previously is one example of such a noncompensatory function. It is well known that inference problems of this type do not yield unique solutions. Rothrock described how the method he used mathematically balanced concerns of specificity, parsimony, completeness, and global effectiveness; the aim was to identify a decision-making heuristic from a participant's data, yielding a plausible account of the identification rule or rules maximally consistent with those data.

Evaluation

To evaluate this modeling approach, Rothrock considered data from one high-performing and one low-performing participant in a session from recent GT-ASP experimentation. The high-performing participant (Participant A) identified 20 of the 24 objects to be identified in the session and made no identification errors. The low-performing participant (Participant B) identified only 14 of the 24 objects and made four identification errors. In particular, three of the errors made by Participant B concerned hostile helicopters that were falsely identified as friendly. Rothrock used his model to infer the heuristic rules that were maximally consistent with these participants' identification data. For Participant B, an identification heuristic with the following object-related information was inferred: "If the object's speed is less than 200 knots, and the object's altitude is less than 18,000 feet, then identify that object as friendly" (p. 166). This rule correctly accounts for the three erroneous helicopter identifications made by Participant B, in that these three helicopters met the preconditions of this rule, were identified as friendly, yet were hostile.

Even more interesting is that the object conditions described by this simplified rule resemble the signature of a civilian airliner in the GT-ASP task environment (objects flying at less than 200 knots below 18,000 feet in GT-ASP were typically civilian airliners taking off from airports, except for the three enemy helicopters missed by Participant B). This heuristic seems to indicate that Participant B simplified the task by assuming all objects with this particular radar signature were civilian airliners, an overgeneralization that led to systematic identification errors. On the other hand, Rothrock's model of Participant A, who correctly identified the enemy helicopters, did not contain this simplified rule. The model of Participant A did not rely solely on radar information (e.g., speed, altitude) as did the model of Participant B; instead, it contained rules that also considered the sensor and IFF information needed to discriminate accurately friendly from enemy aircraft.

Although we cannot be sure that the task-simplification heuristic inferred from Participant B's decision-making data was the source of his systematic identification errors, one can certainly imagine using this diagnosis of a trainee's behavior to motivate the selection of training exercises to test the accuracy of this diagnosis and also to inform the trainee of the flaws in his strategy. We see promise in computationally based inferential modeling techniques in enabling the design of feedback augmentation focused on the specific needs of individual trainees.

Lessons Learned

The following lessons learned from our efforts may improve the effectiveness and efficiency of decision-making training in complex, dynamic environments such as the naval CIC.

1. When training high-level cognitive skills such as decision making and team coordination, one must ensure that individuals enter training with mastery of the perceptual and motor skills they will use to interact with the task environment. The maximum effectiveness of decision-making training will not be achieved if trainees must devote attention to learning and performing perceptual and motor tasks.

2. Part-task practice is an attractive and cost-efficient approach for training the perceptual and motor activities discussed in Lesson 1.

3. When perceptual and motor activities that will eventually be used in the context of higher cognitive skills must be trained, a combination of task segmentation and simplification is recommended for task decomposition. Segmentation ensures that the sequences of activity occurring in the operational environment are preserved in part-task training. Simplification ensures that full attention can be given to learning perceptual and motor activities, without distractions from demands for high-level decision making and coordination.

4. Part-task systems should be designed around consistent components identified in the analysis of task performance in the operational context.

5. Performance improvements accrued from part-task training will occur only in situations for which the mapping between perceptual information and action is the same from trial to trial.

6. To alleviate high-workload situations, consistent task components must be identified and trained to levels of automaticity.

7. To make performance reliable despite environmental stressors (e.g., time pressure, fatigue, heat, noise), training should be conducted to develop automatic task components.

8. Part-task training must be supplemented with additional full-task training to provide the trainee with an opportunity to integrate part-task skills with the cognitive activities required by the full-task context.

9. On-line, embedded feedback can improve the standardization, timeliness, and diagnostic precision of feedback provided to trainees.

10. On-line, embedded feedback can reduce the distractions caused by trainee–trainer interactions, especially in time-critical dynamic tasks.

11. The nature of on-line, embedded feedback should be consistent

with the manner in which trainee performance is ultimately evaluated.

12. Decision makers, particularly in complex and dynamic tasks, seek to develop and use task-simplification strategies to reduce the amount of information that must be obtained.

13. Decision heuristics, particularly in complex and dynamic tasks, are likely to be in the form of noncompensatory (if/then/and/or/not) rules that reduce information-integration demands.

14. Methods for automated inference of decision heuristics from empirical data hold promise for identifying individual trainees' misconceptions and oversimplifications of the task environment; they should be considered as techniques for feedback augmentation.

Conclusion

We have described three approaches to training decision making in the CIC context. The part-task approach, based on automaticity theory, provides a method for ensuring that costly and complex tactical decision-making and team-coordination training will have its maximum benefit by ensuring that personnel come to this training fully skilled in console manipulation activities. Empirical results indicate that the design of part-task training systems could result in substantial cost savings. The real-time, embedded feedback augmentation approach provides a method for improving the standardization, timeliness, and diagnosticity of feedback by augmenting the feedback provided by expert coaches with automated feedback embedded within the task context itself. Empirical results indicate that feedback augmentation can significantly improve training effectiveness in complex, dynamic environments. Finally, the diagnostic approach to feedback augmentation, although still in its concept-exploration stage, holds promise for targeting feedback to a particular trainee's misconceptions or oversimplifications of the task environment, as evidenced in his or her history of decision-making behavior.

In closing, we emphasize a methodological point that has guided our research efforts under the TADMUS program, as evidenced by the studies we describe in this chapter. Our empirical investigations were based on a laboratory simulation of the operational environment and on experimentation in that simulated environment, in an attempt to evaluate a variety of training interventions. However, the identification of important challenges and prospects for training, and the design of the simulated environment itself, were motivated by both psychological theory and initial field studies of the target operational context to which our training results were intended to apply. We encourage others involved in skills training in complex operational environments to consider these methods, which promise at least some assurance of relevance to the operational context (through field study), and of empirical rigor (through controlled experimentation using laboratory simulation). Both of these goals must be achieved to ensure that a training intervention resulting from scientific research will have both practical relevance and a sound empirical foundation.

References

Fisk, A. D., Ackerman, P. L., & Schneider, W. (1987). Automatic and controlled processing theory and its application to human factors problems. In P. A. Hancock (Ed.), *Human factors psychology* (pp. 159–197). New York: North-Holland.

Fisk, A. D., Kirlik, A., Walker, N., & Hodge, K. (1992). *Training for decision making under stress: Review of relevant literature and research plan* (HAPL-9201). Atlanta, GA: Georgia Institute of Technology, School of Psychology, Human Attention and Performance Laboratory.

Fisk, A. D., & Schneider, W. (1983). Category and word search: Generalizing search principles to complex processing. *Journal of Experimental Psychology: Learning, Memory, and Cognition, 9,* 177–195.

Hodge, K. A. (1997). *Training for decision making in a complex, dynamic task environment: An empirical investigation focusing on part-task training and feedback.* Unpublished doctoral dissertation, School of Psychology, Georgia Institute of Technology, Atlanta.

Holland, J. H. (1975). *Adaptation in natural and artificial systems.* Ann Arbor: University of Michigan Press.

James, W. (1890). *Principles of psychology.* New York: Holt.

Kirlik, A., Miller, R. A., & Jagacinski, R. J. (1993). Supervisory control in a dynamic and uncertain environment: A process model of skilled human–environment interaction. *IEEE Transactions on Systems, Man, and Cybernetics, 24,* 929–952.

Kirlik, A., Rothrock, L. R., Walker, N., & Fisk, A. D. (1996). Simple strategies or simple tasks? Dynamic decision making in "complex" worlds. In *Proceedings of the Human Factors and Ergonomics Society 40th annual meeting* (pp. 184–188). Santa Monica, CA: Human Factors and Ergonomics Society.

Kirlik, A., Walker, N., Fisk, A. D., & Nagel, K. (1996). Supporting perception in the service of dynamic decision making. *Human Factors, 38,* 288–299.

Morrison, R., Kelly, P., & Hutchins, S. (1996). Impact of naturalistic decision support on tactical situation awareness. In *Proceedings of the Human Factors and Ergonomics Society 40th annual meeting* (pp. 199–203). Santa Monica, CA: Human Factors and Ergonomics Society.

Norman, D. A. (1976). *Memory and attention: An introduction to human information processing* (2nd ed.). New York: Wiley.

Norman, D. A. (1981). Categorization of action slips. *Psychological Review, 88,* 1–15.

Posner, M. I., & Snyder, C. R. R. (1975). Attention and cognitive control. In R. L. Solso (Ed.), *Information processing in cognition: The Loyola Symposium* (pp. 55–83). Hillsdale, NJ: Erlbaum.

Reason, J. (1988). Modelling the basic error tendencies of human operators. *Reliability Engineering and System Safety, 22,* 137–153.

Rothrock, L. R. (1995). *Performance measures and outcome analyses of dynamic decision making in real-time supervisory control.* Unpublished doctoral dissertation, School of Industrial and Systems Engineering, Georgia Institute of Technology, Atlanta.

Schneider, W. (1985). Training high-performance skills: Fallacies and guidelines. *Human Factors, 27,* 285–300.

Schneider, W., & Shiffrin, R. M. (1977). Controlled and automatic human information processing: I. Detection, search, and attention. *Psychological Review, 84,* 1–66.

Shiffrin, R. M., & Schneider, W. (1977). Controlled and automatic human information processing: II. Perceptual learning, automatic attending, and a general theory. *Psychological Review, 84,* 127–190.

Zachary, W. W., Zaklad, A. L., Hicinbothom, J. H., Ryder, J. M., Purcell, J. A., & Wherry, R. J. (1991). *COGNET representation of tactical decision making in ship-based anti-air warfare* (CHI Systems Tech. Rep. No. 911015.9009). Spring House, PA: CHI Systems.

6

Training and Developing Adaptive Teams: Theory, Principles, and Research

Steve W. J. Kozlowski

Several evident trends make the improvement of team decision making (TDM) of critical importance to virtually all organizations. First, people today increasingly rely on technologically mediated systems to perform a wide array of tasks. From generating nuclear power, to traveling by air, to manufacturing automobiles, to placing a phone call, people rely on computer-mediated systems to accomplish critical and basic functions. Complex systems are not merely the stuff of science fiction; they touch everyone everywhere (Turnage, 1990).

Second, we see increasing demands for the specialization of individual skills and expertise, but also increasing reliance on teams of experts. Teams composed of distributed experts, matrixed managers, and cross-functional specialists are ubiquitous in organizations. After a decade of downsizing, restructuring, and redesign (Kozlowski, Chao, Smith, & Hedlund, 1993), teams have been emphasized as the foundation for work design in organizations. Teams composed of experts with diverse skills are intended to enhance flexibility, responsiveness, and self-management. Coordinating this diverse expertise is a central concern.

Third, we see increasing recognition that real-world problems place extraordinary demands on decision makers. Classic models of decision making generally assume static problem domains, rational analysis, and suboptimal human decision making. Thus, efforts to improve decision effectiveness often focus training on improving expertise for well-defined problems and on aiding the decision maker to be more rational. Although this approach is useful, many real-world tasks fail to conform to the as-

I wish to acknowledge the many colleagues and students who stimulated and contributed to this work. It has been, in all respects, a team effort. My thanks to Kenneth Brown, Janis Cannon-Bowers, Stanley Gully, Kevin Ford, Morell Mullins, Earl Nason, Eduardo Salas, Eleanor Smith, Rebecca Toney, and Daniel Weissbein.

115

sumptions of the rational model. Naturalistic problem domains are dynamic, ambiguous, and emergent; they cannot be completely defined in advance; and they can shift dramatically and unexpectedly. Models of naturalistic decision making (NDM) place a high premium on situation assessment, prioritization, and strategic action (Klein, 1989). The process is often rapid and emergent, necessitating adaptation of the strategy as the situation evolves and implicating coordination across all members of the team. Thus, an NDM perspective necessitates a focus on the knowledge and skills that promote adaptive expertise (Kozlowski, 1995; Smith, Ford, & Kozlowski, 1997).

Finally, there is increasing recognition that the cognitive and behavioral capabilities needed for adaptive expertise and team coordination are contextually based. Whereas basic skills can be developed in conventional training environments (i.e., the classroom), adaptive skills are fully developed and refined in the performance environment. This means shifting more training to the performance context and developing new training strategies and techniques that can be integrated into the work environment. This is not merely a call for more on-the-job training (OJT). It is a call for reconceptualizing the way professionals think about training systems; the goals of training; and when, where, how, and by whom training is conducted.

In this sense, the Tactical Decision Making Under Stress (TADMUS) project represents a major influence among several other trends that are driving a fundamental reorientation of training systems. Training designs are needed that can enhance the development of adaptive expertise for individuals and teams operating in complex systems. For optimal development of these high-level capabilities, training systems must shift in orientation from off-site, single-episode, individual-level skills delivery to multiple-episode, on-line, multilevel systems. Training must be increasingly shifted to the work environment, focused on the development of adaptive individual and team skills, and delivered as needed to meet skill demands when they arise.

The purposes of this chapter are to provide an overview of theory that lays a foundation for this reconceptualization; to highlight key principles that derive from these theoretical frameworks; and to discuss research findings, practical implications, and lessons learned that emerge from this perspective. I describe research that my colleagues, students, and I have been conducting that has been stimulated by the TADMUS project. Although my associates and I have been informed by the contributions of many other researchers, the breadth of our theoretical perspective— ranging from cognitive psychology to instructional design to team leadership and development to organizational theory—provides a unique framework for reconceptualizing training.

I first address the notion of adaptive expertise and its implications for training design and transfer. I then consider four related theoretical efforts that reconceptualize training to meet these emerging demands. Representative principles and lessons learned for team training design are highlighted. I then describe a training research paradigm and initial experiments designed to examine foundation principles that have the potential to enhance adaptive expertise for individuals and teams. Specific lessons learned from these initial efforts are derived. The chapter concludes with some considerations for further development of theory and research within this perspective.

Decision Making and Adaptive Expertise

Everyone makes mistakes; it is a universal aspect of the human condition. Often the errors are trivial, and many other more serious ones are detected and corrected. On occasion, one big mistake or many small slips combine to create a serious TDM failure. The consequences of such mistakes are dramatic and often costly in both human and material terms. The classic approach to reducing errors recognizes the limitations and biases inherent in human decision making. It assumes that true problem states can be determined. Rational analysis then yields optimal solutions. The classic approach is outcome oriented; training is predicated on teaching people to make the *right decision*. The NDM perspective challenges the notion that true states can be determined for real-world problems. Rational analysis cannot yield optimal solutions when the problem is ill-defined, information is ambiguous, and the situation is dynamic. The NDM approach is process oriented; training is predicated on teaching people to make the decision *in the right way*. The exegesis of individual and team decision making from an NDM perspective has implications for training that are at the heart of the development of adaptive expertise.

Classic Decision Making and Routine Expertise

Characteristics of the classic model of decision making influence assumptions about the nature of expertise and thus the use of training to enhance it. The classic model assumes single decision events. Decision-making scenarios are self-contained, complete, and abstracted from broader environmental and organizational contexts. The problem domain is clearly bounded. It is well structured such that specific information, knowledge, and expertise can resolve the problem. The problem event is static, with parameters fixed for the duration of the decision-making period. Often there is no relation between a decision event and subsequent decisions. In other instances there may be a linear series of decision events, but they unfold in a fairly predictable sequence (e.g., a chess game). These characteristics drive a model of decision making that is explicitly rational. Decision makers should search for information to define the problem and establish the appropriate problem parameters. Courses of action or problem solutions are weighted against desired outcomes or goal states. Then a course of action is selected that maximizes the goal. Errors occur when decision makers fail to search for all relevant information, inappropriately weigh the information cues, or employ nonrational models to combine their information.

Traditional models of expertise and training are consistent with the classic approach. In early models of expertise, it was assumed that training on general analytic skills and heuristics for problem solving would enhance decision effectiveness (e.g., Newell & Simon, 1972). However, research soon indicated that decision skills were bounded by problem content, with general skills evidencing little transfer across domains. This work recognized that experts acquire extensive and well-organized domain knowledge, and compile procedural rules for addressing conditional problem states (e.g., Anderson, 1993). Well-practiced performance can acquire the characteristics of automaticity. The development of domain knowledge, its organization in memory, and its compilation to automaticity is predicated on extensive practice and experience. This characterizes what Holyoak (1991) describes as *routine expertise*.

Training for routine expertise is predicated on developing declarative domain knowledge and providing practice on well-established normative problem situations. For complex tasks, high-fidelity simulations of realistic problems may be used to enhance the proceduralization of knowledge for given problems. Learning is usually assessed with measures of basic declarative knowledge and error-free performance in the simulated problem situations (Kraiger, Ford, & Salas, 1993). Expression of the trained skills in the work environment is taken as evidence of transfer effectiveness; that is, transfer is conceptualized as the reproduction of skills across environments, from training to the performance context. Ultimately, high levels of routine expertise accrue through cumulative experience in the performance context that is largely unguided by the training system.

Routine experts can rapidly apply domain-specific strategies and heuristics to problems for which they have had extensive exposure. They are effective at identifying the key characteristics of well-structured problems and implementing well-learned solutions. Although this is an important foundation for domain expertise, it has the inherent limitations of the classic model. In particular, routine experts have difficulty dealing with ill-structured and novel problems. Routine experts are often no better than novices at solving ill-structured problems, to which their well-learned knowledge is more difficult to apply (Devine & Kozlowski, 1995). In fact, their highly organized, static knowledge can be a liability. Sternberg and Frensch (1992) indicated that routine experts are unable to adapt when deep structural principles of their problem domain change. Indeed, habitual routines of teams can lead to disastrous consequences when the team mistakes a novel situation for a routine one and persists in applying an inappropriate strategy (Gersick & Hackman, 1990).

Naturalistic Decision Making and Adaptive Expertise

NDM models assert that most real-world problem situations do not conform to the characteristics assumed by rational models (Beach & Lipshitz, 1993). For example, the decision situation is usually embedded within a broader environmental and organizational context. The problem is ill structured, with incompatible or shifting goals. Diagnostic information is difficult to obtain and is often ambiguous or conflicting when it is available. The situation is dynamic and emergent, responsive to decision-maker actions, but also subject to unpredictable shifts. Often there are significant time pressures and high costs for mistakes. These problems are compounded when the task necessitates coordinated team actions. These characteristics are clearly nonoptimal and difficult to resolve with rational decision analysis (Orasanu & Connolly, 1993).

NDM situations call for more than the routine application of well-learned knowledge. They necessitate what Holyoak (1991) described as *adaptive expertise*. Adaptive expertise entails a deep comprehension of the conceptual structure of the problem domain. Knowledge must be organized, but the structure must be flexible. The process goes beyond procedural knowledge of an automatic sort. Adaptive experts understand when and why particular procedures are appropriate as well as when they are not. Comprehension entails active processing, allowing recognition of shifts in the situation that necessitate adaptability. Adaptive experts are able to recognize changes in task priorities and the need to modify strategies and actions.

A key factor for the development of adaptive expertise is the encouragement of active learning strategies during training (Smith et al., 1997). Active learning enhances the development of metacognitive and self-regulatory skills. Metacognition refers to executive-level processes entailing knowledge, awareness, and control of cognitive activity involved in goal attainment (Flavell, 1979). Self-regulation occurs at a micro level and entails planning, monitoring, and adjustment of cognitive and behavioral strategies necessary to accomplish subgoals. Metacognition and self-regulation are often regarded as distinct processes operating at different levels of goal specificity but operating in concert. They are relevant to both learning and performance (Flavell, 1979; Nelson, 1996), and they provide a foundation for individual adaptability (Kozlowski, 1995; Kozlowski, Gully, Smith, Nason, & Brown, 1995; Kozlowski, Gully, Smith, et al., 1996; Smith, Ford, Weissbein, & Gully, 1995; Smith et al., 1997).

In addition to cognitive and task-relevant strategies, metacognitive and self-regulatory skills entail the capability to manage motivation and affect. Learning complex tasks requires focused attention and cognitive effort. Irrelevant off-task thoughts that signify disengagement or withdrawal must be suppressed, because they draw attentional resources and reduce cognitive effort needed for learning (Kanfer & Ackerman, 1989). Complex tasks are difficult to learn. They entail the making of many errors, and frustration is common for trainees early in the learning process. The negative affect that accompanies failure to meet expectations draws attention away from learning and must be managed. Effective management of the learning process enhances self-efficacy—the self-perceived task competency that allows the individual to tackle difficult tasks and persist in the face of novel challenges. These capabilities also play an important role in maintaining motivation under challenging performance conditions (Bandura, 1991; Gist & Mitchell, 1992).

At the team level, metacognitive and regulatory processes must extend beyond the self. That is, these individual-level cognitive and behavioral skills must operate in a coherent fashion across the team. Individuals must maintain an awareness of self within the network of roles that compose the team. They must monitor the rhythm, timing, and pacing of activity to enable seamless coordination. Team members must monitor the performance of critical interdependent roles and be prepared to step in and share the workload when teammates become overloaded. They must build and maintain a sense of team efficacy to deal with challenges, and they must be capable of revising roles, tasks, and the team network when the situation demands adaptation on-the-fly (Kozlowski, Gully, Nason, & Smith, in press).

The understanding of how to train, develop, and enhance individual and team adaptability is in its infancy. It is known that active learning is an important element in the development of adaptability, and that conventional training is not explicitly designed to promote the development of metacognitive and self-regulatory skills. Behaviorally oriented training that focuses on maximizing achievement performance goals and minimizing trainee errors may enhance training performance but hinder the development of deeper skills necessary for generalization and adaptability. Effective transfer requires more than the reproduction of declarative knowledge and salient performance skills in the performance environment. It requires a foundation of knowledge and learning outcomes provided by training that can aid generalization, adaptability, and continued learning for a wide range of situations that can occur in the performance setting (Kozlowski et al., 1995, Kozlowski, Gully, Smith, et al., 1996). This necessitates a rethinking of how training for complex systems is designed.

Implications for Training

Developing adaptive capabilities entails a long-term process that provides trainees with extensive guided experience. Because adaptive skills build on a foundation of routine expertise, a considerable portion of that experience must address normative situations that will be frequently encountered. Skills that resolve consistent aspects of the task domain must be proceduralized. However, adaptive capabilities are more likely to be enhanced by variability and novelty that challenge normative skills. The key issue is how to sequence this exposure developmentally across different training experiences and environments. Training must ensure that critical learning outcomes (e.g., learning strategies, metacognitive and self-regulatory skills, knowledge structure, efficacy, and motivational skills) are acquired at both the individual and team levels.

There is evidence that adaptive skills, especially those involving team coordination, develop in the context of intact teams operating in performance settings (Cannon-Bowers, Tannenbaum, Salas, & Volpe, 1995). Skills relevant to mutual performance monitoring, pacing, load balancing, and resource sharing are contextually based (Fleishman & Zaccaro, 1992).

This raises the issue of how best to create systematic training and learning experiences in the work setting. As attention shifts to the work setting for the development of these capabilities, it implicates the key role of leaders as trainers and developers of adaptive teams. Moreover, team adaptation, much like individual expertise, can be facilitated only to a degree by explicit instructional techniques. It is impossible to anticipate and design training for every eventuality. Metacognitive and regulatory capabilities must be developed at the team level, so that the team can continuously learn. This means conceptualizing the team as an adaptive network of roles, within which team members can revise, refine, and control their interdependencies as needed to meet performance demands. The team itself becomes a self-learning system. Leaders are central to this developmental process.

Theory and Principles for Training

Our efforts to rethink the fundamentals of training systems to enhance the development of adaptive expertise for individuals and teams have generated four distinct but related theoretical frameworks driven by the implications noted previously. The four theoretical frameworks each contribute a different set of general principles relevant to a reconceptualization of training. The thematic focus of each framework is introduced here. Illustrative principles are then briefly discussed in subsequent sections.

First, *team training can be conceptualized as a sequence of developmental experiences that occur across a series of different environments.* Training can be directed at different skill levels (basic vs. advanced), different content (taskwork vs. teamwork skills), and different target or focal levels (individual vs. team). Because advanced skills accrue developmentally, training systems must be sensitive to the fact that skills acquired in one environment must be retained and transferred to the next environment, where they serve as a foundation for the next phase of skill development. To accomplish this, *different training environments must build knowledge and skills in an appropriate sequence across skill levels, content, and target levels.* This calls for a higher degree of cross-environment integration than is typically considered in training system design (Kozlowski, Ford, & Smith, 1993).

Second, *levels issues become important as the target level and content of training shift in an integrated system.* A basic assumption is that most training systems initially focus on individual-level taskwork or technical skills but shift over time to focus on developing team-level teamwork or enabling skills (e.g., communication, cooperation, and coordination). *Levels issues force a consideration of the composition of team performance (i.e., how individual outputs aggregate or compose to the team level) and the organizational context in which a team operates in determining the target level of training delivery.* Many conventional training systems deliver teamwork skills to large groups of similar specialists who belong to different teams. This individual-level target for training delivery is acceptable for generic team skills (Cannon-Bowers et al., 1995) or when team members are not interdependent. When team members are highly interdependent, necessitating adaptive coordination, training for teamwork skills is best delivered in the actual work context and to intact teams (Kozlowski & Salas, 1991, 1997). In addition, factors in the organizational and work context may set constraints on training content and transfer effectiveness.

Third, *as the focal location for individual and team skill acquisition shifts to performance environments, emphasis is placed on the role of leaders as instructors and team developers.* Leaders are responsible for melding individual technical experts together into a coherent team. Coherent teams possess shared affect (e.g., cohesion, team climate) and knowledge (e.g., role expectations, goals, and strategies) that enable coordination and adaptation in response to dynamic NDM task demands. *This model specifies training strategies to guide leaders in the development of team coherence.* The training strategies are applicable to the natural cycles of work in organizations. The model that was developed from the theory integrates a learning cycle directed by the leader into naturally occurring cycles of task intensity and the sequence of team development. Also developed is a set of leader role attributes (Kozlowski, Gully, McHugh, Salas, & Cannon-Bowers, 1996) and training guidelines (Kozlowski, Gully, Salas, & Cannon-Bowers, 1996) to help the leader in the application of the learning cycle across the developmental sequence.

Fourth, *to establish continuous learning and refinement, teams ultimately must develop their own capability to negotiate, refine roles, and make use of adaptive mechanisms.* Team development is a compilation process that proceeds from individuals to dyads to flexible networks of linked roles within the team. By specifying the learning content, process, and outcomes relevant at each phase of team development, one can guide the design of interventions with the potential to enhance long-term team adaptability (Kozlowski et al., in press). I next provide a description of the frameworks in more detail.

Team Training Systems

Team training can be conceptualized as a sequence of developmental experiences that occur across a series of different environments. A primary goal of training systems for TDM is the necessity to meld individual skills into a team capability that enables members to coordinate and adapt their expertise. This requires that the training system explicitly link the development of individual-level and team-level capabilities. From an operational perspective, this is accomplished through a series of training experiences that occur across multiple environments. Knowledge and skills acquired in one environment must be retained and transferred to the next environment to provide a foundation for the next phase of skill acquisition. Thus, training systems must be designed to link skill acquisition, transfer, and enhancement across this sequence of developmental experiences. Although many team training systems use some form of training sequence, the linkage between one training experience and another is more of an implicit than an explicit aspect of training design.

The typical training system used for TDM usually begins with classroom instruction to build individual specialist knowledge. Many trainees for the same specialty are trained together in large groups. This highly efficient training develops specialists who, if not yet fully expert, are quite knowledgeable. In the next stage, individual specialists must learn to coordinate their expertise with specialists in other areas. Often this team-coordination training is accomplished in simulated work environments that allow the team to explore and practice team skills without being subject to the costly errors that may occur in real work settings. Simulation training is often less efficient and more costly, making practice time precious and limited. Teams have an opportunity to begin developing coordination skills but little opportunity to go beyond initial acquisition. Finally, the team is assigned to the workplace, where the development of team skills continues. Indeed, this is the primary location of team skill development. However, this continuing development is based more on unstructured experience and implicit learning, and less on explicit, systematic training. Thus, the development of team skills is inefficient, and the level of skill development can vary considerably in quality across teams. Several reviewers of the team training literature have commented on the fragmentation of approaches and the absence of systematic efforts to design team training systems (Dyer, 1984; Modrick, 1986; Salas, Dickinson, Converse, & Tannenbaum, 1992).

One contributing factor is that models of training design are optimized for skill development within training environments but are not oriented toward linking developmental experiences across environments. They regard training as isolated experiences designed to deliver specific skills. Even models that are system oriented (e.g., instructional system design) treat *training as meeting focused objectives for multiple waves of different trainees over time*. These models link needs assessment, training design, and program evaluation as a closed system. The strength of this perspective is its optimization of training for specific skills within a single environment. A major limitation is its view of training experiences as independent events rather than an integrated progression of developmental episodes. Time is a key neglected issue. This view needs to be reoriented so that *training is considered to be a series of developmental experiences for the same set of trainees across multiple training episodes*. This reorientation raises issues about how best to sequence the experiences to build compatibility between acquisition and transfer, and how to target training content at the right point in the sequence to build skills as they are needed. A model illustrating this view of the training system is shown in Figure 1.

The figure presents an adapted version of the model developed by Kozlowski, Ford, and Smith (1993) that focuses on the most significant factors for purposes of this descriptive summary. The primary dimensions of the model include training concepts and principles, training content,

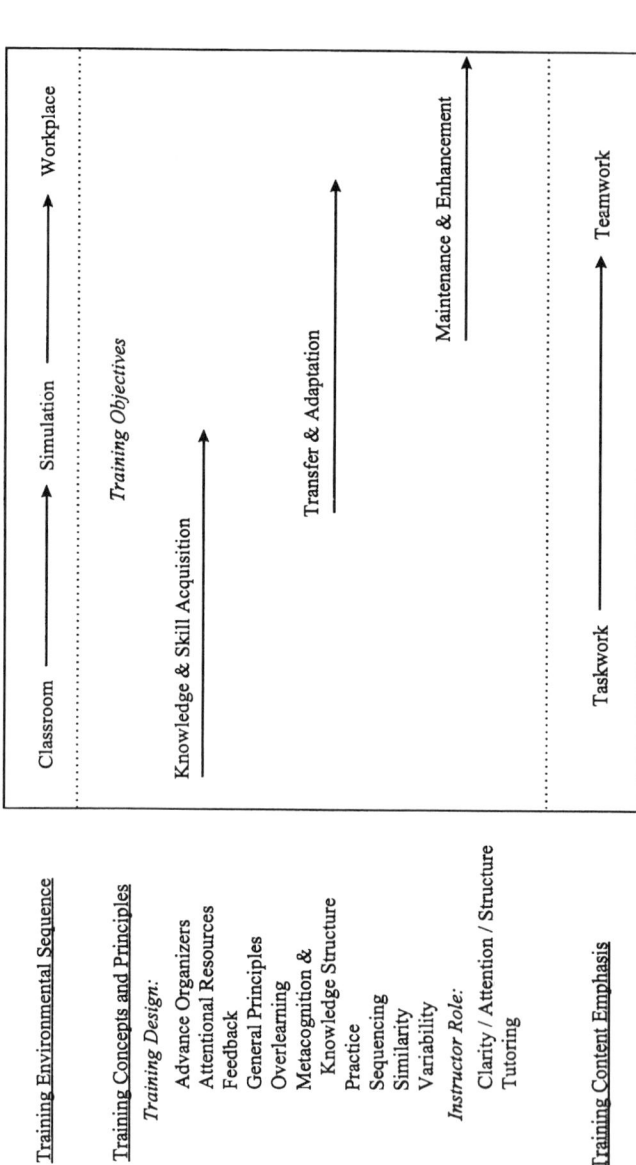

Figure 1. The team training system. From *Training Concepts, Principles, and Guidelines for the Acquisition, Transfer, and Enhancement of Team Tactical Decision Making Skills: I. A Conceptual Framework and Literature Review* (p. 13; Contract No. N61339-91-C-0117), by S. W. J. Kozlowski, J. K. Ford, and E. M. Smith, 1993, Orlando, FL: Naval Training Systems Center. In the public domain.

and the sequence of trainee development across multiple training environments. *Training concepts and principles* are categorized as those relevant to training design and those relevant to the role of the instructor during training. *Training content* refers to a focus on specific task-relevant skills, generally targeted at the individual, or a focus on team enabling skills, generally targeted at the team as a whole. The training environmental sequence considers the shift in training venue across classroom, simulation, and workplace environments. Specific *training guidelines* relevant to skill acquisition, transfer, and enhancement are linked to each stage of this sequence. The purpose of the model is to provide a foundation for reorienting training systems to improve integration of learning experiences across environments as individuals develop specialist skills, learn initial team-coordination skills, and continue to enhance their skills as they evolve as a team (Salas et al., 1992).

The derivation of concepts, principles, and guidelines was informed by the efforts of Salas and his colleagues to document research-based guidelines systematically for team training (e.g., Salas & Cannon-Bowers, in press; Salas, Cannon-Bowers, & Blickensderfer, 1997; Swezey & Salas, 1992). My colleagues and I generalized the approach to allow inferences to be drawn from individual-level theory and research in the cognitive and instructional literatures, and we extrapolated them to generate principles of team training design.

Training concepts are at a high level of abstraction and represent general training or learning strategies drawn from the literature. Concepts relevant to the model are conceptualized as categories of training and learning strategies derived from the cognitive, instructional design, and classic learning theory literatures. Design concepts derived from the model include attentional resources, feedback, general principles, introduction/advance organizers, overlearning, cognition (metacognition and knowledge structure), practice, sequencing, similarity, and variability. Instructor role concepts include structuring–clarity–directing attention and tutoring. Training design and instructor role concepts relevant to the model and our review of the literature are listed and defined in Exhibit 1.

Principles are nested under concepts and represent an effort to derive training implications from the concept-relevant literature. The derived principles are often inferences in that the research literature frequently is not specifically oriented toward training applications. Guidelines represent more specific derivations that are contextually grounded. They apply the principles to the acquisition, transfer, and enhancement linkages across the training sequence. My group's literature review included 143 sources, of which 104 were retained for the derivation of concepts and principles. Exhibit 2 provides an example of training principles and guidelines for the cognition concept. This concept area was selected for illustration because of the relevance of metacognition, self-regulation, and knowledge structure to my group's research, which is described later in the chapter, that has resulted from this theoretical effort.

Influencing the nature, amount, and quality of trainee cognition, in terms of its structure and process, is central to effective training design

Exhibit 1. Definitions of Training Design and Instructor Role Concepts

Training Design Concepts

Attentional resources: Trainees have a limited capacity of attentional resources to devote to learning a task. In addition, at early stages of learning (acquisition), trainees may have to devote more resources to learning, whereas during later stages (compilation and automaticity) the attention required may be reduced. Therefore, instructional design must take into account the short-term memory capacity of individuals at these various stages of learning.

Feedback: Feedback that is *intrinsic to the task* can be included in this category of design principles. This category should include only response-produced feedback (feedback resulting from performance of the task) or augmented feedback such as additional displays or lights that are designed into the training task to enhance acquisition.

General principles: Instruction on general rules and theoretical principles that underlie the training content.

Introduction and advance organizers: Introductory information, objectives, overviews, or advance organizers that prepare the learner for what will be learned in the lesson. Advance organizers are introductory materials at a high level of generality and inclusiveness presented in advance of the learning material.

Cognition: Cognition includes two major outcomes: metacognition and knowledge organization. First, trainees must acquire knowledge in well-organized knowledge structures (i.e., mental models). Second, trainees must develop metacognition. Metacognition is knowledge about one's own cognitions; it involves monitoring one's own thought processes (self-regulation) and using strategies for learning and remembering information (learning strategies). Also included in this category are principles concerned with teaching explicit problem-solving and performance strategies.

Overlearning: The extent to which trainees receive continued practice beyond the point at which the task has been performed successfully.

Practice: The amount of repetition allowed in the performance of skills in training; this category also includes any principles concerning massed versus spaced practice and part versus whole practice.

Sequencing: The order of presentation of knowledge or skills to be learned in training.

Similarity: The extent to which knowledge, skills, tasks, and situations share some common components from one instance to the next (fidelity issues are included here).

Variability: The extent to which components of knowledge, skills, tasks, and situations are varied from one instance to the next.

Instructor Role Concepts

Instructor cognition: The instructor's mental models, team and task knowledge, and other cognitions.

Classroom management: Administrative procedures, rules, organizational structure, and support systems that allow for the smooth functioning of the class.

Feedback: Although feedback can be used in both the structuring role and the monitoring role (see subsequent categories), it also transcends these uses. Principles concerning how to provide feedback fall under this heading. This category includes only extrinsic verbal feedback from the instructor to the trainees (feedback intrinsic to the task falls under the design principle category).

Exhibit 1 continues

Exhibit 1. *(Continued)*

Creating learning environment, motivation, and student engagment: Instructor behaviors that motivate students, keep them involved in learning, make the lesson interesting, and so on. For example, principles concerned with varying the modes of instruction that focus on variety as a motivator fit here.

Monitoring, assessment, and adjustment: The instructor moderates between training design features and trainees by observing trainee progress and modifying instruction when appropriate; this includes instructor use of questioning to monitor trainees' progress. Practice and feedback principles may also fit here if they are applied to monitor trainee's learning and adjust training.

Structuring, clarity, and directing attention: The instructor mediates between training design features and trainees by explaining, elaborating, and interpreting information. Feedback, questioning, and practice principles for the purpose of structuring or clarifying information may fit into this category.

Tutoring: The instructor or the trainee's peer teaches the trainee through one-on-one instruction. Principles focus on how to tutor, as well as how the instructor must manage the peer tutoring process.

Note. From *Training Concepts, Principles, and Guidelines for the Acquisition, Transfer, and Enhancement of Team Tactical Decision Making Skills: I. A Conceptual Framework and Literature Review* (Contract No. N61339-91-C-0117), by S. W. J. Kozlowski, J. K. Ford, and E. M. Smith, 1993, pp. 34–36. Orlando, FL: Naval Training Systems Center. In the public domain.

and instruction. Cognition refers to the quantity and type of knowledge acquired and the relationship among knowledge elements (Kraiger et al., 1993). Two major cognitive outcomes are knowledge organization and metacognition. As learning progresses, individuals develop meaningful structures for organizing knowledge. In addition, they develop metacognition, or the awareness and monitoring of one's cognition. Metacognition includes the development of learning strategies and self-regulation.

Training principles aimed at developing trainee cognition serve three main purposes: (a) to facilitate the organization of specific, detailed knowledge; (b) to provide learning strategies for the acquisition of knowledge and skill; and (c) to provide strategies for self-regulation. Training guidelines represent an effort to translate principles into application-specific recommendations. The model in Figure 1 illustrates a developmental sequence of knowledge and skill acquisition associated with shifts across training environments. Thus, the goals or purpose of training guidelines for each context must shift to reflect trainee development. Initial training in classroom or other off-site environments is targeted on acquiring basic declarative knowledge and developing knowledge of learning strategies and self-regulatory skills. Training in simulation or practice-based environments is targeted on reinforcing prior learning, proceduralizing self-regulatory skills, and developing an awareness of teamwork requirements. Simulation training is designed to enhance transfer and adaptive capabilities. Training that is embedded in the work context is targeted on skill maintenance and enhancement. For interdependent teams, this is the environment where team coordination skills are devel-

Exhibit 2.   Training Principles and Guidelines for the Cognition Training
Design Concept

Principles for Knowledge Organization

Instruction should facilitate the development of well-organized knowledge
structures that provide trainees with easy accessibility to stored information and
move them toward expert thinking.

To facilitate knowledge organization, instruction should include explicit descriptions of the functional and linking relations among the concepts, goals, and
skills to be learned and of the multilevel complexity of their organization.

Instruction on mental models should sequence these models in terms of increasing complexity, with easy transitions from one to the next; this can be accomplished by presenting exercises that are just beyond a trainee's current level of
competence.

To facilitate acquisition of the appropriate knowledge structure, instructors
should prevent, as well as identify, trainee misconceptions early in training and
guide trainees in correcting their own misconceptions later in training.

Instructors should present material to be learned in more than one modality
so that trainees will cognitively process the material in a variety of ways that
facilitate storage of and access to the knowledge structure.

Principles for Learning Strategies

Trainees should be taught explicit strategies to facilitate learning of declarative knowledge, procedural knowledge, and skills.

Instruction should help trainees to understand when and why to use various
learning strategies to facilitate transfer of these strategies to new situations:
(a) Rehearsal and mnemonic strategies are most appropriate for learning declarative knowledge; (b) visual imagery should facilitate the learning of patterns and
relationships between concepts; (c) metaphors should be used to link old or previously learned concepts to new concepts; and (d) dynamic models, analogies, and
mental simulations should be used to learn procedures and causal relations.

Principles for Self-Regulation

Self-regulation should be taught after trainees have the attentional resources
available to engage in monitoring.

Trainees should be taught how to monitor their own learning and performance by (a) establishing goals for an instructional unit or activity; (b) assessing
the degree to which the goals are being met; and (c) if necessary, modifying the
strategies being used to meet the goals.

Guidelines for Application of the Training Principles to Different Stages of
Development and Training Environments

*Knowledge and Skill Acquisition: Classroom or Other Off-Site
Training Environment*

To facilitate knowledge organization, instruction should include explicit descriptions of the functional and linking relations among the concepts, goals, and
skills to be learned, and of the multilevel complexity of their organization.

Instruction should provide problems or questions that are just beyond the
trainee's competence and trigger changes in his or her mental model.

Exhibit 2 continues

Exhibit 2. (*Continued*)

When trainees have misconceptions about the technical and other declarative knowledge being learned, instructors should guide trainees through their thought processes and employ methods such as entrapment strategies to assist trainees in identifying their own misconceptions and understanding why they are incorrect.

Mental models should be sequenced by increasing complexity and depth of information.

To facilitate knowledge organization, instructors should identify and explain the key concepts that provide the organization to the material to be learned, present information with a degree of redundancy, signal transitions between lesson parts, and review the main ideas at the end of the lesson.

Instructors should use examples, illustrations, visuals, and demonstrations to assist trainees in cognitively processing declarative knowledge in a variety of ways.

Instructors should teach learning strategies such as imagery, pattern discrimination, summarization, and analysis of key concepts to aid trainees in recalling and retaining technical knowledge.

Explicit rehearsal and mnemonic strategies should be taught to improve learning of verbal or declarative knowledge.

Trainees should be taught to use question generation as a strategy to improve comprehension of new declarative knowledge.

Instructors should teach trainees when different learning strategies are effective for learning declarative knowledge and why they should be used.

Instructors should teach self-regulation skills (e.g., goal setting) to allow trainees to monitor their own comprehension of declarative knowledge.

To teach comprehension monitoring, instructors should first model the activities; the next steps are overt and then covert rehearsal by trainees.

As self-regulation skills improve, goal specificity and difficulty should be increased.

Transfer and Adaptation: Simulation or Other Practice-Based Training Environments

Instructors or leaders should identify and define critical enabling skills (e.g., communication, coordination, performance monitoring, resource allocation) and explain how they are interrelated and relevant to team performance.

Leaders and instructors should collaborate to provide consistent knowledge structures of enabling skills to trainees.

During scenario exercises, instructors should monitor and identify trainee misconceptions about their individual-level skills and correct them.

Scenarios should be sequenced to facilitate the development of progressively more elaborate knowledge structures that specify the functional relationships between individual-level knowledge and skills and team-level knowledge and skills (i.e., sequenced to require increasing degrees of coordination).

Instructors should explain how the individual trainee's skills and task performance are related to the overall goals and functioning of the team.

Trainees should be taught team-level skills in more than one modality; verbal information during prebriefs and scenario exercises should be integrated to allow trainees to process the same information in more than one way.

Trainees should be taught mental simulation as a way to improve skills and to evaluate and develop options for decisions that must be made in time-pressured environments.

Exhibit 2 continues

Exhibit 2.   *(Continued)*

Self-regulation of skilled performance (e.g., goal setting) should be taught after trainees have acquired a sufficient declarative knowledge base and have the attentional resources to engage in self-regulation.

Trainees should be taught to evaluate their task strategies and hypotheses continually to determine when these strategies are no longer appropriate.

Once individual-level skills are mastered, trainees should be taught to monitor the performance of other team members with whom they directly interact.

Maintenance and Enhancement: Workplace or Embedded Training Environment

Leaders should review the critical team enabling skills, describing how they are interrelated and lead to team-level goals. Leaders must link these goals to different external situations.

Team members should be taught to evaluate their task strategies and hypotheses continually to determine when these strategies are no longer appropriate.

Once team members have routinized their interactions with others, they should begin monitoring the performance of the entire team.

Note. From *Training Concepts, Principles, and Guidelines for the Acquisition, Transfer, and Enhancement of Team Tactical Decision Making Skills: I. A Conceptual Framework and Literature Review* (Contract No. N61339-91-C-0117), by S. W. J. Kozlowski, J. K. Ford, and E. M. Smith, 1993, pp. 46–50. Orlando, FL: Naval Training Systems Center. In the public domain.

oped and refined. Once it is recognized that training systems need to be broader in scope and that team skills need to be developed in a more integrated fashion, the interplay of the individual and team levels of analysis have to be considered.

Levels Issues and Team Training

Levels issues become important as the target level and content of training shift in an integrated system. Training systems are much better developed for training individuals than for training teams. Training and learning theories are oriented toward individual change. Models of training design are predicated on the acquisition of skills by individuals. Although it is axiomatic that learning is an individual-level phenomenon, in most instances the effects of individual learning are intended to have impact at higher levels in the organization. How does this shift from individual-level to team-level effects occur, and what are the organizational factors that influence it? The answers to these questions have implications for the design of training and its delivery, as well as for the construction of work contexts that facilitate transfer.

The theoretical framework developed to address these issues (Kozlowski & Salas, 1991, 1997) incorporates three primary dimensions: composition, content, and congruence. Composition concerns the focal level—individual, team, or organization—at which training is intended to have effects. Whenever the focal level of training moves beyond the individual level, composition processes must be specified. Content identifies the knowledge and skill domains—technostructural (taskwork) versus enabling process (teamwork) factors—of interest for training. It enables the specification of relevant constructs and measures. Congruence concerns the alignment of other factors in the system that can either facilitate or inhibit the effects of training. Creating a consistent work context is central to facilitating transfer.

The model provides a *compositional* perspective of transfer. That is, unlike the prior model, in which transfer is viewed conventionally as a horizontal linkage across training and performance settings, in this model transfer is considered a building-up process that occurs vertically across focal levels—from individuals, to teams, to the organization. This process necessitates a clear understanding of the way in which diverse knowledge and skills are compiled across individuals to yield outcomes at higher levels in the organizational system. The theoretical specification of this combination process is termed a *composition model*. It defines how lower level units and processes compose higher level phenomena (Rousseau, 1985).

The work context concerns the issue of *congruence*, or contextual alignment. As a building-up process, the transfer of trained skills is extremely sensitive to (a) how well skills align across content domains, that is, the fit of taskwork and teamwork skill requirements, and (b) how well the newly trained skills align with situational or contextual demands, that is, the fit of training content to the context in which behavior and performance is to be embedded. Congruence is essential, and it must be considered in combination with training content and its vertical transfer to higher levels.

Figure 2 illustrates these levels issues for the organizational system. Transfer is represented as a building-up process that proceeds from the individual to higher levels. This vertical form of transfer is facilitated by congruent alignments among other relevant factors. Congruent alignments necessitate fit (a) across training content within levels and (b) across content domains across levels. The former form of congruence is relevant to the alignment of content domains (horizontal processes), and the latter form of congruence sets constraints that determine the embedding context (downward processes), which in turn influences transfer (vertical processes).

The effects of these two forms of congruence on the composition of transfer can be illustrated by example. Consider the first form of congruence. Training content at the individual level (e.g., technical and human process skills) has to be aligned such that the human process skills needed to enable specific technical expertise are compatible and coexist within trained individuals. A flexible technology that requires individual initiative in judgment cannot be effectively enabled by an operator who lacks the skills, experience, or training to be autonomous. Similarly, training operators to be autonomous when the technology is rigid and precludes discretion is futile. Consider the second form of congruence. Technical and enabling process content must be compatible with higher level constraints in order to compose vertically. Group technologies that necessitate empow-

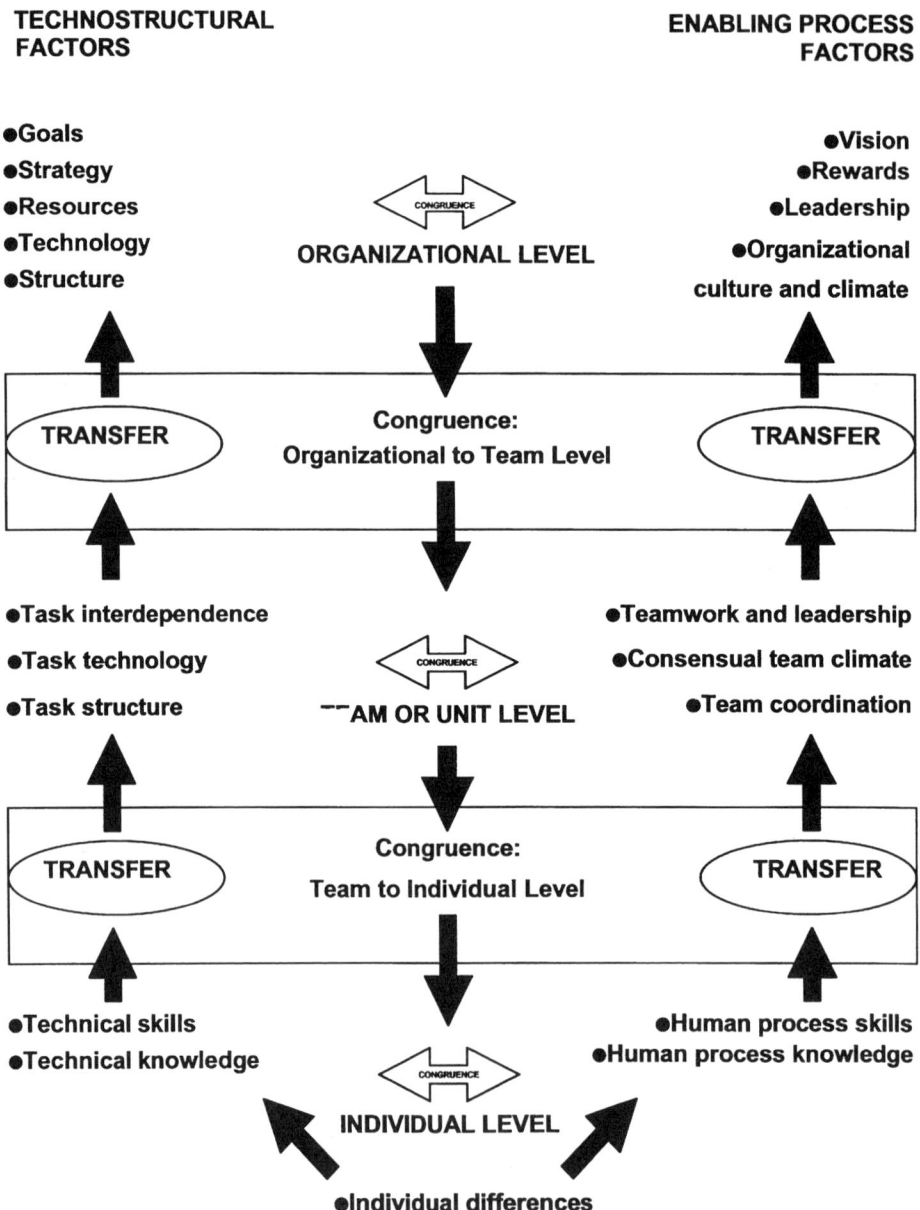

Figure 2. Team training within the organizational system. From "An Organizational Systems Approach for the Implementation and Transfer of Training," by S. W. J. Kozlowski and E. Salas, 1997, in J. K. Ford, S. W. J. Kozlowski, K. Kraiger, E. Salas, & M. Teachout (Eds.), *Improving Training Effectiveness in Work Organizations* (p. 265), Mahwah, NJ: Erlbaum. Copyright 1997 by Lawrence Erlbaum Associates. Reprinted with permission of the publisher and author.

ered, self-managing teams will be incongruent in organizations that are highly centralized and bureaucratic. Training on empowerment will not compose vertically, and transfer will not occur. Consistency with contextual factors at higher levels determines the effectiveness of vertical transfer. Although both of these examples may appear obvious, these incongruities are all too common.

Because levels issues are quite abstract, they require grounding in specific application examples. I illustrate some key implications of the model for training design by contrasting teams that vary on their interdependence requirements from simple to complex. The composition of team performance is determined by the structure of individual task-driven interdependencies (Kozlowski et al., in press). Team tasks that do not specify specialized roles and distributed task linkages carry the assumption that individual contributions to group performance are simple additive aggregates. This is a form of pooled coordination (Thompson, 1967) that allows individual contributions to be compensatory. Secretarial pools, councils, and panels represent examples of this form of structure that does not specify distinct roles or tasks for group members. Individual contributions are simply pooled together. Training delivery for group tasks of this sort can be directed at the individual level, because performance does not rest on coordinated interaction. Transfer can compose vertically even when there is considerable variation in individual effectiveness following training because individual contributions to group performance are compensatory.

When the team task structure specifies distinct roles, specialized skills, and coordinated interaction, the composition of team performance is more complex (Thompson, 1967; Van de Ven, Delbecq, & Koenig, 1976) and noncompensatory. Team performance for tasks of this type is the result of specific individual contributions that must be coordinated in time and space. The neglect or mistiming of a single individual action may jeopardize team effectiveness. Team members must monitor the performance of others, mutually adjust, and balance the load across members. They must orchestrate their actions to the timing, pacing, and rhythm of the team (Fleishman & Zaccaro, 1992). Training must be delivered to intact teams in the performance context to develop the foundation skills for this form of coordinated interaction (Cannon-Bowers et al., 1995). Moreover, transfer necessitates more uniformity in skill development across the team given the noncompensatory nature of team performance.

This theoretical rationale constructed around the composition of team performance leads to the derivation of several propositions relevant to enhancing transfer for team training. The propositions concern the implications of composition on training design, moderation by individual differences and contextual factors in the achievement of congruence, and the necessity of aligning training content with the organizational context. The main point to recognize is that complex, interdependent team tasks shift the training emphasis toward intact teams in their performance environments.

Leaders as Team Instructors and Developers

As the focal location for individual and team skill acquisition shifts to performance environments, emphasis is placed on the role of leaders as instructors and team developers. Leaders are not so much responsible for directing specific team actions as they are responsible for developing the underlying individual and team capabilities that enable teams to self-manage their action. From a training systems perspective, this reorientation generates two challenges. Given the need for contextually relevant skills, the challenge is for the leader to create learning episodes that are grounded in the natural ongoing flow of work. Given that skills are compiled from the individual to the dyadic to the team levels, the challenge for the leader is to shift his or her role and instructional activities in line with the developmental progress of the team.

Both training challenges are driven by temporal dynamics, an issue generally neglected in training design. NDM task cycles are the origin of one set of dynamics. Team tasks, particularly those of "action" teams such as aircrews, air traffic control operators, and command–control teams (Sundstrom, De Meuse, & Futrell, 1990), vary in their intensity and their corresponding load on team attentional resources. There are periods when the situation is routine and task intensity is relatively low, freeing attention for other activities, and there are periods when task intensity is high, necessitating full attention to the situation at hand. This variation in the intensity of NDM tasks creates a natural opportunity to integrate learning episodes into the natural task cycle. The team developmental sequence is the origin of the other set of temporal dynamics. New teams move through a sequence of phases as they progress through formation to development to refinement. This creates an opportunity to shift emphasis on instructional objectives and the level of focus to meet the needs of this natural developmental process.

Two related theoretical frameworks have been developed to elaborate this perspective. One framework establishes the natural compilation process for team development, in which *compilation* is conceptualized as the synthesis of knowledge and skills across levels and over time. The team compilation model specifies the focal level of development—from individuals to role-linked dyads to flexible team networks—and the learning content, process, and outcomes relevant at each phase of development (Kozlowski et al., in press). Major elements of the team compilation model are implicit in the second framework, addressing team leadership, which is the focus of discussion here. The leadership model specifies the changing role of the leader as an enhancer of this compilation process. The framework assumes that the primary function of the leader is to develop coherence around the affective (e.g., climate, cohesion) and knowledge–skill (e.g., goals and strategies, role expectations) domains that underlie team effectiveness. *Coherence* refers to the sharing, compatibility, and alignment of affect and knowledge–skill across individual team members. From this perspective, team coherence enables the communication, coordination, and adaptation needed when the team must self-manage and perform under high load and stressful task conditions (Kozlowski, Gully, McHugh, et al., 1996; Kozlowski, Gully, Salas, & Cannon-Bowers, 1996).

The training strategies by which leaders build team coherence are linked to the task cycles and developmental sequence noted previously. A model representing the strategies is shown in Figure 3. Task dynamics are linked to a learning cycle directed by the leader that has four steps: (a) setting learning objectives, (b) monitoring behavior and performance, (c) diagnosing errors, and (d) guiding process feedback. Under low-intensity task conditions, the leader discusses and establishes learning objectives to guide the team's attention in the upcoming engagement. As the task shifts to high intensity, the team engages under the watchful eye of the leader. He or she considers the source of errors and diagnoses failures to accomplish learning objectives. As the task cycles back to low intensity, the leader guides the team through process feedback. Although the leader guides this process, it is the team and its members who must diagnose the reasons for their lack of mastery. Two critical aspects of this learning cycle should be noted: (a) Goals are intended to emphasize learning and mastery, not performance, and (b) feedback is intended to focus on the process (how to), not outcomes (how well). These fundamental issues are a major focus of my group's research, discussed later.

In addition to task dynamics, Figure 3 shows shifts in the leader role and the corresponding changes in emphasis for learning objectives across the developmental sequence. At formation, when the team is new, it is no more than a collection of individuals. The situation is unstructured and social space is undefined. Here the leader serves as a *mentor*, fostering the development of shared affect and understanding of the team's mission and goals. This helps to bond individuals to the team as an entity. During development, when the team has achieved novice status, the leader takes on a more explicit *instructor* role. Members tend to be focused on their own competency. Here the emphasis is on the development of individual task knowledge, proficiency, cognitive structure, and self-efficacy. As the team members acquire and compile basic skills, becoming more expert, development shifts to a refinement phase that focuses on the team. Here the leader serves as both a coach and facilitator. As a *coach*, the leader emphasizes learning objectives that promote team efficacy, shared mental models (i.e., cognitive structure), and compatible behavior. This is a continuing process that is designed to refine team expertise and adaptive capabilities. As a *facilitator*, the leader enables team self-management. By making use of its coherence, the team is able to coordinate and adapt its actions without explicit direction from the leader. The leader facilitates this process by assessing and updating the unfolding situation to aid the maintenance of coherence, thereby providing a foundation for coordination and adaptability (Kozlowski, Gully, McHugh, et al., 1996; Kozlowski, Gully, Salas, & Cannon-Bowers, 1996).

A critical issue in the application of this training strategy is how to guide the leader across the changing role emphases and developmental shifts. This is accomplished by deriving a set of theoretical *attributes* and associated propositions (i.e., principles) to guide the shifts (Kozlowski,

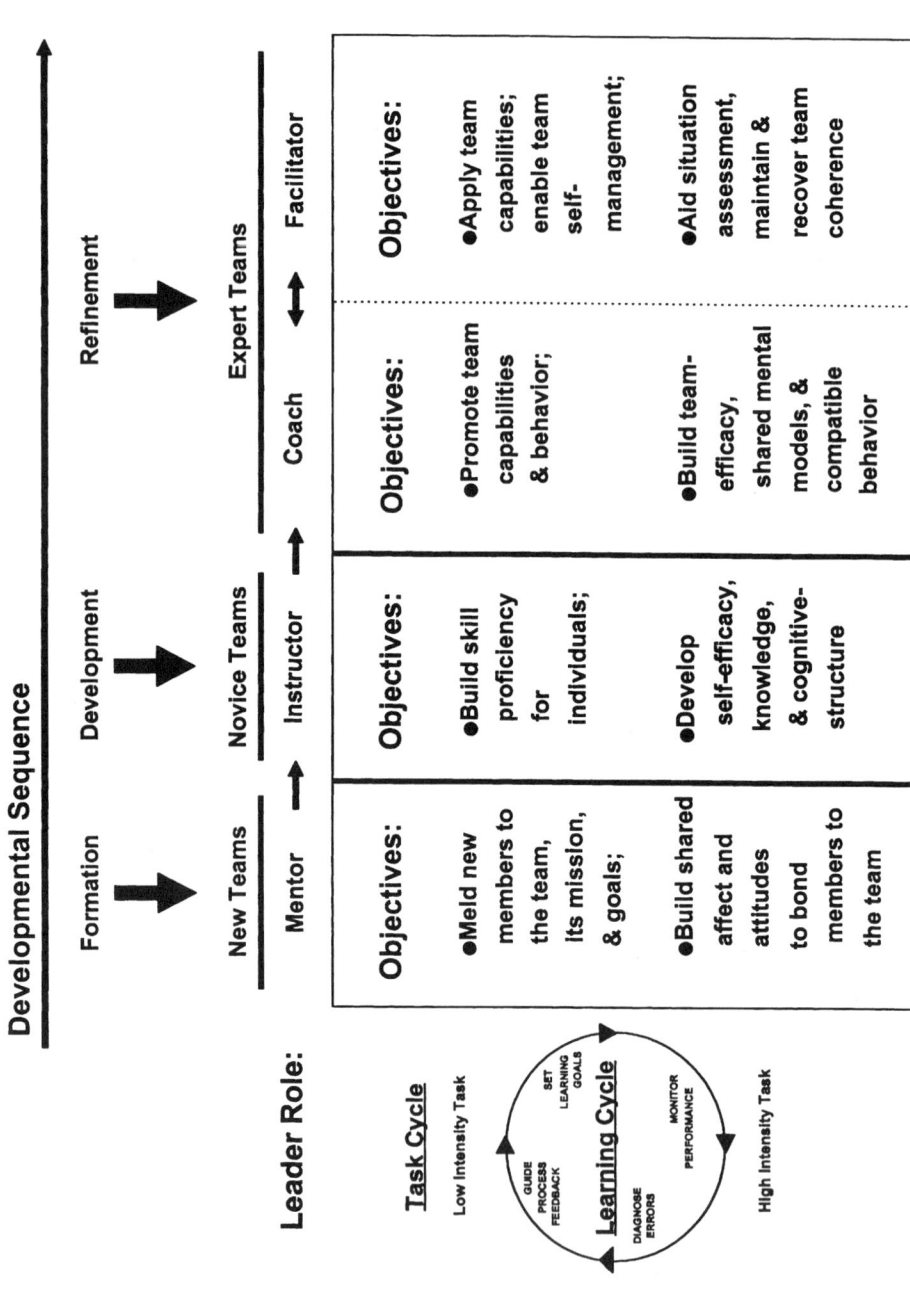

Figure 3. The role of leaders as team developers. From "Team Leadership and Development: Theory, Principles, and Guidelines for Training Leaders and Teams," by S. W. J. Kozlowski, S. M. Gully, E. Salas, and J. A. Cannon-Bowers, 1996, in M. Beyerlein, D. Johnson, & S. Beyerlein (Eds.), *Advances in Interdisciplinary Studies of Work Teams: Team Leadership* (Vol. 3, p. 262), Greenwich, CT: JAI Press. Copyright 1996 by JAI Press. Reprinted with permission.

Gully, McHugh, et al., 1996) and specifying application-oriented training guidelines to aid implementation (Kozlowski, Gully, Salas, & Cannon-Bowers, 1996). The attributes include orientation, level, depth, and elaboration. At a general level, the model specifies developmental shifts in instructional emphasis from mastery to performance, individuals to teams, proximal to distal goals, and simple to complex role linkages. The attributes and their link to team development and leader roles are illustrated in Figure 4. Selected propositions that illustrate how the attributes guide the leader are listed in Exhibit 3.

Orientation refers to the nature of goals: to master learning or to perform well. Initial emphasis is placed on mastery to enhance skill acquisition. As skills develop, the leader shifts to a performance orientation to enhance the refinement, maintenance, and transfer of well-learned skills. *Level* refers to the target of training objectives: individual or team. Leaders should emphasize individual-level instruction to enhance competence and self-efficacy early in the process. Later, they should

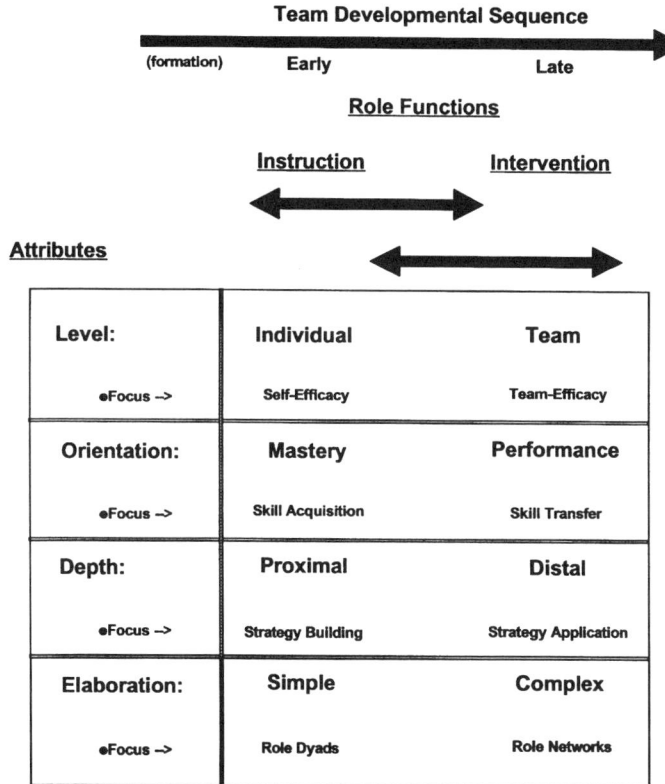

Figure 4. Leader role attributes and learning objectives. From "A Dynamic Theory of Leadership and Team Effectiveness: Developmental and Task Contingent Leader Roles," by S. W. J. Kozlowski, S. M. Gully, P. P. McHugh, E. Salas, and J. A. Cannon-Bowers, 1996, in G. R. Ferris (Ed.), *Research in Personnel and Human Resource Management* (Vol. 14, p. 277), Greenwich, CT: JAI Press. Copyright 1996 by JAI Press. Reprinted with permission.

Exhibit 3.   Leadership Propositions

Orientation

Skill acquisition will be enhanced by a mastery orientation early during team development, whereas refinement, maintenance, and transfer of well-learned skills will be enhanced by a performance orientation later in team development.

Level

Early during team development, leaders should emphasize individual-level instruction to enhance competence and self-efficacy, whereas later in team development, leaders should emphasize team-level instruction to enhance role and behavioral links, coordination, and team efficacy.

Teams with leaders who are consistent in their instructional behaviors (i.e., the cycle of goal setting, monitoring, diagnosis, and feedback) at the individual or team levels will develop coherence more rapidly.

Depth

Early in team development, the leader must specify the individual distal and proximal goals that define individual strategies for a given situation to clarify individual roles. As the team matures, the leader must shift attention to team distal and proximal goals that define the team strategy for a given situation to clarify links between each individual's role on the team and to the overall team strategy.

Leaders should provide feedback on the sequence of proximal goals when the purpose of the instructional cycle is to build strategy linkages. Leaders should provide feedback on the distal goals, which represent compiled sequences of proximal goals, when the purpose of the instructional cycle is to refine or reinforce strategies for a given situation.

Elaboration

Early in team development, leaders should provide instructional experiences that are low in elaboration to establish role and strategy links for normative situations. Late in team development, leaders should provide instructional experiences that are high in elaboration and address a wider variety of situational contingencies to generalize role–strategy links and improve team adaptability.

Note. From "A Dynamic Theory of Leadership and Team Effectiveness: Developmental and Task Contingent Leader Roles," by S. W. J. Kozlowski, S. M. Gully, P. P. McHugh, E. Salas, and J. A. Cannon-Bowers, 1996, in G. R. Ferris (Ed.), *Research in Personnel and Human Resource Management* (Vol. 14, pp. 279–289). Greenwich, CT: JAI Press. Copyright 1996 by JAI Press. Adapted with permission.

emphasize team-level instruction to enhance role and behavioral links, coordination, and team efficacy. *Depth* refers to the extent to which the goal structure is specified: proximal subgoals or distal macro goals. Generally, emphasis is placed on proximal subgoals early in the process, shifting to more distal goals as skills develop. *Elaboration* refers to how extensive an awareness of the team network the leader promotes: simple dyadic links or complex role networks. As teams develop basic knowledge of the role and strategy links for normative situations, the leader is able to increase elaboration and to introduce novel contingencies.

Lessons Learned

Salas, Cannon-Bowers, and Blickensderfer (1997) made a strong case for improving the linkage between theory and training practice. The means to achieve reciprocity is to clarify implications of theory for practical application by deriving theoretical principles, specifying application-oriented guidelines, and extracting lessons learned. Each of the theoretical frameworks described previously was designed to be consistent with this push toward reciprocity. They each carry the basic assumption that training is increasingly being shifted to the work setting and is increasingly team oriented. This assumption requires that one push training beyond the assumptions of its traditional paradigm and consider it from a broader and more integrated perspective.

Several authors have noted recent progress in training theory and research (e.g., Ford & Kraiger, 1995; Tannenbaum & Yukl, 1992), especially in the area of team training (Salas, Cannon-Bowers, & Kozlowski, 1997; Salas et al., 1992). These comments often conclude with the notation that much more needs to be done. Although I do not disagree with this conclusion, I note that much more can be done to generalize theory and research findings from other domains and apply these findings to team training. This was the aim of the theoretical frameworks described previously. The extent to which such integrative efforts inform training practice will be determined by the usefulness of the principles and guidelines generated. They are explicated in detail in the primary sources.

At a more general level, some broad prescriptive lessons can be drawn from the theoretical models that serve as basic assumptions for guiding research and application:

- *Effective team training is founded on solid individual technical training.* I do not present this as a unique observation but highlight it because it is often neglected when our attention is focused on team training. Teams are not entities unto themselves; they are composed of individuals. Individuals enact teams. When individuals lack knowledge or competence in their own areas, they cannot focus their attention effectively on team processes or performance. Team training interventions must be preceded by effective individual training. In the push to train teams, one must not forget their essential foundation.
- *The design of effective team training systems must take a developmental perspective that cuts across training episodes, programs, and settings.* An integrated team training system must explicate the development of individual specialists, training of novice teams, and refinement of expert teams. Each of these steps requires different training interventions, training objectives, and key actors that *operate as an integrated system.* Again, I do not represent this as a unique observation but note it because it is generally neglected in training design. Team training is often designed as a series of discrete and isolated experiences with specific objectives.

The missing element is the integration across the discrete experiences. Part of this lack of integration is due to the models of instructional design, which tend to be discrete. However, another contributor is the way training is administratively structured in organizations. When different operational groups are responsible for the discrete training programs, lack of integration is a predicable outcome.

- *When and how team training is delivered is dependent on team task structure.* Teams that are low on task interdependence can benefit from individually oriented training and generic team training. The absence of coordination demands allows traditional individual-level models to be useful guides for training design. However, teams with high interdependence demands require differential individual- and team-specific training (Cannon-Bowers et al., 1995; Kozlowski & Salas, 1991, 1997).

- *Team-specific training pushes the primary leverage point of team training into the workplace.* Classroom and simulation-based training are essential to the development of team-specific skills. However, it is only in the work context that individual skills combine to yield the seamless coordinated behavior indicative of effective teams. Improvement is needed in the ability to embed training experiences into the natural ongoing rhythm of work. This shift in training venue emphasizes the roles of team leaders as mentor, instructor, coach, and facilitator in building effective teams.

- *Leaders need specific guidance to enact these changing roles and to detect shifts in the team's readiness to advance.* Role enactment is aided by specific description of desired role behavior and its objectives. Interventions exist that have demonstrated effectiveness for role-based training (e.g., behavioral role modeling). To provide guidance on role shifts, leaders must be able to assess the development of individual and team skills that underlie their role shift changes. This means training researchers have to create innovative assessments of developmental change in knowledge, skills, and capabilities that can be embedded in technology systems or otherwise made available to leaders for use in team development.

- *Adaptive teams need to acquire individual- and team-level self-regulatory skills.* Adaptability requires the generalization of knowledge and skills, not merely their rote reproduction. Yet the focus in training is often driven by performance criteria that may not provide the necessary foundation for mastery, cognitive structure, and self-efficacy that is central to self-regulation and adaptability. Training should not be predicated on creating experts per se. Rather, it should be predicated on providing structured experiences that build the underlying characteristics of expertise. Expertise will develop as a natural consequence of this strategy.

The theoretical models are comprehensive in that they cover different key issues in the training system for complex team tasks, but they are united by their focus on developing adaptive capabilities for teams. They consider the developmental links across multiple training environments, and their relevance to individual skill building and transfer; they consider the compositional links across levels and their relevance to team skill building and transfer; and they consider the leader's instructional roles for the developmental acquisition of skills (from basic to strategic) and their compilation upward across levels (from individual to team) in the performance context.

Translating Theory Into Research and Application

Overview

The concluding section of this chapter describes my group's initial efforts in empirically grounding the training principles drawn from our frameworks that have high potential for translation into effective training strategies. Because of our interest in adaptability, we focused our initial efforts on active learning strategies with the potential to promote deep comprehension and the development of metacognitive and self-regulatory capabilities. These skills are essential for learning and performance for individuals and teams (see Exhibit 2; Kozlowski, Ford, & Smith, 1993) and provide the early foundation for the development of expertise. A skeletal version of the model guiding the paradigm is shown in Figure 5. Initial work has contrasted mastery- and performance-training goals, examining their differential effects on learning and adaptability. Additional active learning strategies illustrated in the figure are relevant to ongoing research efforts.

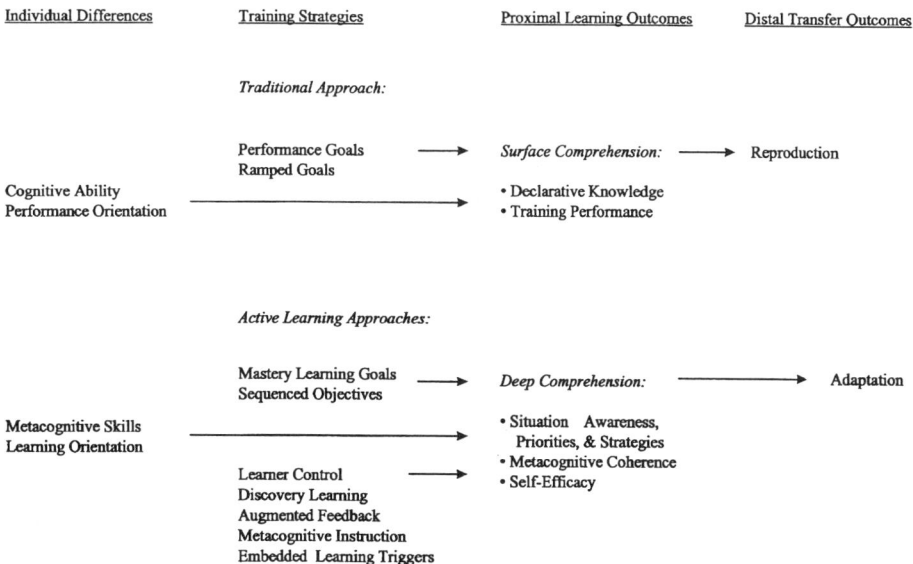

Figure 5. Adaptive training strategies.

Training Strategies for Adaptability

Traditional training design for complex tasks is typically predicated on extensive practice using a behavioral approach that reinforces performance and constrains errors. The logic of this training strategy is to guide correct responses and prevent any strengthening of incorrect behaviors. However, this strategy tends to orient trainees to salient performance skills represented in surface characteristics of the task domain. This emphasis is often enhanced by trainers or team leaders who praise successful performance and criticize failure to meet performance objectives.

Recent research has prompted a growing realization that a performance orientation during early skill acquisition may inhibit active learning and deeper comprehension of task concepts. Of greater concern is the fact that a performance-oriented strategy yields trainees who appear to perform well under training conditions but who may have difficulty adapting under realistic or more challenging task situations (Schmidt & Bjork, 1992). Performance goals orient metacognitive and self-regulatory processes toward behaviors that translate directly into goal attainment. However, this draws cognitive resources away from the self-regulation needed to promote learning. Trainees monitor goal discrepancies, but they are attempting to self-regulate around performance goals before they have developed a broad foundation of the knowledge and skills needed to do so effectively. They learn, but they learn a narrower portion of the task domain, acquire fewer skills, and develop only a limited metacognitive comprehension. In addition, a performance-oriented training strategy hinders the development of self-efficacy, a key attribute for persistence during adaptation to difficult and changing task conditions.

A mastery-oriented training strategy represents an effort to engage metacognitive and self-regulatory processes on the primary goal at hand during early skill acquisition: learning. Mastery goals orient trainees toward key learning objectives. The intent is to go beyond the development of basic declarative knowledge and surface performance skills that are the primary focus of performance-oriented training. In addition to basic knowledge and performance, learning objectives should build metacognitive coherence, a deep comprehension of task-concept relations, and self-efficacy that represents the development of self-regulation directed at learning. Mastery goals create an active, exploratory orientation that frames errors as learning opportunities. Mistakes are opportunities to self-diagnose deficits in knowledge or the application of incorrect learning strategies and to make appropriate adjustments. The focus on learning objectives prompts trainees to devote more cognitive effort to monitoring comprehension and adjusting learning strategies to acquire key skills, building a better conceptual comprehension of the task domain.

Moreover, by sequencing mastery-learning objectives from basic to strategic, knowledge and skills are compiled in an integrated fashion. Because trainees are oriented to monitor and regulate their own learning, they develop enhanced task self-efficacy. Sequenced mastery goals promote "a guided, yet self-managed, learning experience...that enhances understanding and provides a foundation for adaptive expertise" (Kozlowski, 1995, p. 8).

Research Paradigm

Naturalistic decision-making simulation. My group's interest in training for NDM task environments necessitated the development of a research paradigm suitable for an examination of learning processes in complex skill acquisition. The experimental platform we use is called *t*eam *e*vent-based *a*daptive *m*ultilevel simulation/*t*actical *n*aval *de*cision *m*aking (TEAMS/TANDEM; Gully & Kozlowski, 1996). There were three primary design goals in the development of TEAMS/TANDEM. The first was to construct a simulation with high psychological fidelity for NDM tasks by using dynamic event-based scenarios to provide a complex, shifting, and emergent task environment. Simulations that present a series of single decision situations lack the dynamics, complexity, and emergent properties of NDM. They are more consistent with a classic approach to decision making and are not well suited to examining complex skill acquisition and adaptability. The second goal was to ensure comprehensive assessment of different aspects of learning during complex skill acquisition. Training research often takes a narrow focus on simple declarative (factual) knowledge or task performance. The paradigm uses multiple assessments of cognitive, behavioral, and affective indicators (Kraiger et al., 1993). In particular, it focuses on adaptability, not just rote reproduction. The third goal was to devise a theoretically based system to allow team constructs to be examined, not just as simple *additive* or *average* individual contributions, but as a series of complex individual contributions that *compose* to represent team processes and outcomes. This feature is unique to TEAMS/TANDEM.

The simulation is based on the architecture of TANDEM, developed by the Naval Air Warfare Center Training Systems Division (NAWCTSD) and the University of Central Florida (UCF) as a low-fidelity team simulation to examine tactical decision making under stress (see chapter 3, this volume; Weaver, Bowers, Salas, & Cannon-Bowers, 1995). Design modifications made in cooperation with NAWCTSD and UCF have yielded enhanced experimental power, design flexibility, and data reporting. TEAMS/TANDEM has the capacity to (a) script event-based scenarios, (b) embed task strategies in the scenario design, (c) detect strategy application, and (d) mathematically model individual and team-level decision processes. TEAMS/TANDEM provides participants with a dynamic, self-contained, and novel task environment that is appropriate for examination of complex skill acquisition and adaptability at the individual and team levels of analysis.

Research issues. Although we are not the first researchers to examine the effects of mastery and performance orientations on training effectiveness, there are several unique contributions of our conceptual approach and research paradigm that make it more theoretically comprehensive and methodologically rigorous than other approaches. First, prior research has not focused on complex, dynamic NDM tasks. Instead, it has focused on self-contained static tasks, such as word processing (Martocchio, 1994) and classroom test performance (Ames & Archer, 1988), that are loaded on declarative knowledge. Second, prior research has typically conceptualized mastery and performance orientations as individual *traits* but has attempted to manipulate them as situational *states* (e.g., Elliott & Dweck, 1988; Martocchio, 1994). This approach confuses conceptualization and operationalization, and the source of training effects becomes equivocal. My colleagues and I distinguish the theoretical mechanisms for the traits and states, and we model them distinctly. Third, training must demonstrate learning effects above and beyond more fundamental, unmeasured causes. As a case in point, the effects of cognitive ability on training success are well established, but cognitive ability is rarely accounted for in training research in this area (e.g., Ames & Archer, 1988; Elliott & Dweck, 1988; Martocchio, 1994). Fourth, research is needed to distinguish the effects of training interventions on different types of learning (e.g., declarative knowledge vs. strategic knowledge vs. metacognitive structure) and to distinguish learning from performance. Many researchers neglect learning altogether, treat it as simple declarative knowledge, or regard learning and performance as isomorphic. Finally, the criterion for training effectiveness should be broadened to incorporate adaptability.

Research approach. Our primary research focus has been to model the development of several learning indicators—declarative knowledge, task performance, metacognitive coherence, priorities and strategies, and self-efficacy—that reflect the initial knowledge and skills that contribute to adaptive capabilities. For research purposes, the term *adaptive capabilities* refers to the ability of trainees to apply the knowledge and skills acquired during training to a task situation that is more difficult and complex. The adaptive task situation is created by increasing the workload, making the task more dynamic, and changing rules and principles. Trainees are required to draw on their knowledge and to adapt their behavior to maintain performance. Adaptation should be facilitated by a deeper comprehension of the task and by the sense of self-competence that provides resilience to stress. The paradigm controls for relevant individual differences, including cognitive ability and dispositions (learning and performance goal orientations), and examines the effects of training strategies on developmental learning indicators as well as the independent effects of the learning indicators on adaptability. It is a systematic, comprehensive, and rigorous paradigm.

Training procedures. Mastery trainees are taken through a sequence of four episodes that set learning objectives, orienting the trainees to explore different kinds of knowledge and skills. Each episode is a combination of self-instruction using written materials and hands-on practice. The training sequence is predicated on developing a foundation in basic knowledge and expanding awareness to deeper strategic skills. The first episode begins with the basic declarative knowledge needed to make effective decisions in TEAMS/TANDEM. This requires that trainees learn 15 decision cues, their value ranges, and their links to the subdecision structure. The second episode emphasizes ambiguous and conflicting cue information and how to deal with it efficiently. The third episode begins to focus on strategic skills, prompting attention to situation awareness and task dynamics that can alter priorities. The fourth episode orients trainees to consider the overall situation and to be prepared to make strategic trade-offs that satisfy best their overall objectives.

Performance goal trainees are also taken through a sequence of four training episodes. However, rather than focusing on learning goals, these trainees are oriented to achieving specific levels of performance during practice. Because difficult terminal performance goals are known to have detrimental effects on learning (Kanfer & Ackerman, 1989) and self-efficacy (Gist & Mitchell, 1992), the goals are ramped across the episodes—that is, maintained at the 85th percentile to reduce the discrepancy between the goal and initial performance (e.g., Earley, Connolly, & Ekegren, 1989). This procedure weakens the potential difference between mastery and performance conditions, but it provides a rigorous evaluation of mastery goal effects. To ensure that the only difference between training conditions is due to trainee objectives, performance goal trainees are given the exact same instructional information regarding the task (e.g., logic, strategies, procedures).

Training proceeds for 5 hours across 2 consecutive days. Trainees are not expected to be experts at the end of the training; the research focus is on early development of adaptive capabilities. A variety of individual difference, learning process indicators, and task behavior/performance measures are obtained. Cognitive ability and mastery and performance traits are measured to ensure that training effects can be distinguished from relevant individual differences. Conventional indicators of training effectiveness include declarative knowledge and task performance. Measures of declarative knowledge assess learning of basic facts and task features, whereas measures of task performance assess information-processing effectiveness, such as querying the target, interpreting cues, and making decisions. The theoretical framework for sequenced mastery goals indicates three additional learning outcomes of interest: strategic behavior, metacognitive coherence, and self-efficacy. Strategic behavior involves specific actions that reflect situation awareness, target prioritization, and task strategies. Metacognitive coherence reflects a consistent comprehension of key task concepts, task strategies linked to particular clusters of concepts, and specific task actions that implement the strategies—a kind of dynamic mental model. Self-efficacy is a task-specific self-perception of competence and capability to manage challenges.

In summary, performance goals are expected to inhibit learning, even though trainees exhibit satisfactory training performance. Because trainees are unlikely to meet their specific performance objectives, their self-efficacy will be lower. When asked to perform under more complex and dynamic situations, they may have difficulty and lack the resilience needed to adapt and adjust. Mastery learning goals should create a different learning experience. Sequenced learning goals prompt self-monitoring and self-regulation as trainees attempt to learn successively more complex task features. This process should yield improved metacognitive coherence. The iterative process of self-regulated learning should also promote the development of task self-efficacy, providing mastery trainees with resilience under more demanding conditions. Although mastery trainees may achieve lower levels of performance during training, they should be better able to maintain their performance and to adapt to changes in the task.

Findings and Implications

Initial research results have been consistent with the theoretical model and principles. Sequenced mastery-learning objectives appear to be an effective training strategy for enhancing the early development of adaptive capabilities. My colleagues and I conducted two studies directly examining the mastery-training strategy (Kozlowski et al., 1995; Kozlowski, Gully, Smith, Brown, Mullins, & Williams, 1996). The first investigation established the experimental protocol and operationalized the simulation paradigm. The second experiment improved measures, refined the simulation training–transfer scenarios, and replicated primary findings.

Results across both studies indicated that mastery goals focus trainee attention on the exploration of key task strategies. Mastery trainees evidenced greater attention to developing situation awareness, prioritizing targets, and making strategic trade-offs. Thus, mastery goal trainees developed a deeper comprehension of the task that went beyond superficial declarative knowledge, reflected in their better metacognitive coherence. Coherence reflects a consistent comprehension of key task concepts, task strategies linked to particular clusters of concepts, and specific actions that implement the strategies. Mastery trainees exhibited more rapid development of and greater levels of metacognitive coherence. Metacognitive coherence is regarded as an indicator of metacognitive processes of monitoring and self-regulation during learning. This iterative regulatory process is also expected to yield improved self-efficacy. Mastery goal trainees evidenced more rapid and sustained development of efficacy during training.

Performance goal trainees exhibited higher scores on target engagements at the end of training, as expected. Target engagement represents information-processing effectiveness (i.e., cue to decision links) and is a salient aspect of performance. However, it is a superficial indicator that captures target processing but neglects the situation awareness, prioritization, and strategic aspects of NDM effectiveness. This focus on superficial processing is reflected in the results for transfer and adaptability. Mastery and performance goal trainees were equally effective at target engagements on the transfer task. However, mastery goal trainees better retained and adapted their more complex knowledge and skills. Even under large increases in workload, greater stress, and changes in underlying task rules and principles, mastery trainees exhibited greater attention to situational awareness, prioritization, and strategies.

In addition to the direct effects of mastery goals on transfer and adaptability, other analyses modeled indirect effects on adaptability through the learning indicators. Overall, the research model accounted for nearly 75% of the variance in adaptive performance. Moreover, metacognitive coherence and self-efficacy accounted for unique portions of variance in adaptability after controlling for individual differences in ability and learning traits, the training strategies, and declarative knowledge and final training performance (Kozlowski et al., 1995). Thus, sequenced mastery goals evidence considerable promise as a training strategy for enhancing the early development of adaptive capabilities.

Lessons Learned

The initial results of the research indicate the following lessons for the design of training for complex NDM tasks:

1. Sequenced performance goals appear to improve training performance and rote reproduction but are less effective for developing a deep comprehension of the task and adaptive capabilities.
2. Sequenced mastery-training goals enhance the development of metacognitive structure.
3. Sequenced mastery-training goals enhance the development of situation awareness, prioritization, and strategic skills.
4. Sequenced mastery-training goals enhance the development of self-efficacy and provide resilience for dealing with novel and challenging task situations.
5. Sequenced mastery-training goals enhance adaptability in more difficult and complex task situations.

It seems reasonable, therefore, to conclude that sequenced mastery goals are an effective training strategy for the acquisition of complex NDM skills. From a broader perspective, the theory and research described also suggest the following lessons:

6. *Understanding the etiology of adaptive expertise necessitates research using complex, dynamic NDM tasks with high psychological fidelity.*

    Prior research on mastery versus performance goals has not focused on complex, dynamic NDM tasks. Although simpler tasks such as word processing and classroom test performance are embedded in more "realistic" settings than are synthetic tasks, they provide a limited view of learning and adaptability. Such tasks are heavily loaded on declarative knowledge, memorization, and rote reproduction. There is little requirement for the proceduralization of knowledge, metacognitive strategies, or self-regulatory skills.

7. *Research to examine the development of complex, adaptive skills must assess a broad range of cognitive, behavioral, and affective indicators of learning.*

    Training researchers often treat training performance as synonymous with learning. In other instances, declarative knowledge may be distinguished from performance, but broader indicators of trainee development are ignored. Our research has indicated that training strategies have differential impacts on learning. For example, mastery goals had their primary impacts on metacognitive coherence (cognition), priorities and strategies (behavior), and self-efficacy (motivation and affect). Had we not carefully specified the theoretical effects of our intervention and used appropriate learning indicators, the effects of the intervention would not have been detectable. Research is needed to specify the differential *theoretical mechanisms* and the *core learning indicators* by which training is expected to have its effects.

8. *Training for complex NDM tasks should focus on adaptive capabilities, not the rote reproduction of static knowledge and skills.*

    The dynamic, emergent, and shifting nature of NDM tasks distinguishes them from classic decision-making tasks. Classic models are founded on the rote reproduction of knowledge and skill: routine expertise. This is perfectly sensible in a static and closed task environment; however, NDM task domains go beyond this fixed knowledge, necessitating situation awareness, the establishment of priorities, and the implementation of appropriate task strategies as the situation unfolds. Such tasks demand adaptive expertise. Training research is needed to incorporate explicitly this emphasis on adaptive knowledge and skills.

Future Research and Practical Implications

Theory suggests that mastery goals influence learning by enhancing the metacognitive processes of self-monitoring and self-regulation. Extensions of our research are designed to examine self-monitoring, self-regulation, and affective reactions more directly. In particular, my colleagues and I are interested in how metacognition can be influenced by different forms of interpretive feedback that are designed to augment the effects of mastery goals. In a related effort, investigators are attempting to develop theoretical principles for designing scenarios for complex simulations. The focus of this effort is to create scenario designs with embedded error events, high in diagnostic potential, that prompt metacognitive processing. In addition, researchers are examining training strategies designed to stimulate metacognitive processing directly through metacognitive training, discovery learning, and learner control (Brown & Kozlowski, 1997; Brown, Mullins, Weissbein, Toney, & Kozlowski, 1997; Smith, 1997; Weissbein & Ford, 1997).

Finally, my associates and I have initiated work to extend the research paradigm to the team level of analysis, with a particular focus on team self-regulation during team training (Gully, 1997a, 1997b; Gully & Kozlowski, 1996). In addition, Nason (1995) showed that team role development has a significant impact on communication and coordination patterns and affects the development of compatible expectations among teammates. This finding suggests the early development of team mental models and the foundation for adaptability. These combined research efforts have the potential for high payoff, particularly as demands increase for training systems that are embedded in the workplace.

Mastery training shows potential as a novel training design intervention with wide applicability across a variety of training settings. No changes in existing training hardware or systems are required to take advantage of enhancements. The essential elements of mastery goals can be provided by instructors and team leaders, who can be easily trained to exhibit the necessary behaviors to emphasize this focus on key learning objectives. Because mastery goals build self-efficacy and metacognitive structure—learning outcomes that are neglected in traditional approaches to training—they provide a new path for improving training effectiveness. They have the potential for considerable flexibility as a general training design feature that can be implemented across a wide variety of educational, industrial, and military training settings.

Conclusion

Organizations are placing increasing emphasis on performance in complex systems, creating the need to improve teamwork, communication, and coordination. NDM task environments demand adaptive capabilities on the part of individuals and teams to deal with evolving and shifting demands. Demands for continuous learning and skill updating outside of training and in the performance environment are increasing. When one asks how these desired capabilities can be enhanced, the attention naturally turns to training. Yet traditional training systems are not well equipped to address these important concerns. Traditional training approaches tend to be oriented toward individuals, not teams and complex systems; toward simple reproduction of skills acquired in training, not adaptation to new situations; and toward training as a special activity that occurs off-site, not as a natural ongoing process that occurs in the workplace. There is a need to reorient the perspective of training systems—what they should do, how they do it, who does it, and where it is done—to meet these emerging needs.

The TADMUS project has been a major force stimulating this reorientation and reconceptualization of team training systems. It has directly contributed to the theoretical frameworks described in this chapter, which provide a conceptual foundation for this effort. In our perspective, training is viewed as a sequence of integrated episodes that occur across different environments, as a process designed to enhance the upward composition of skills across levels, and as a series of developmental experiences that are directed by team leaders in the performance context. The training implications of these theoretical frameworks are represented in a series of propositions and principles, many of which are illustrated in the chapter. Principles that specify new training strategies consistent with the reorientation have been subjected to research scrutiny, and initial results are promising. There is still much to do, but the translation, elaboration, and refinement of theoretical principles into new and more effective training strategies has already begun to yield tangible improvements.

References

Ames, C., & Archer, J. (1988). Achievement goals in the classroom: Students' learning strategies and motivational processes. *Journal of Educational Psychology, 80*, 260–267.

Anderson, J. R. (1993). Problem solving and learning. *American Psychologist, 48*, 35–44.

Bandura, A. (1991). Social cognitive theory of self-regulation. *Organizational Behavior and Human Decision Processes, 50*, 248–287.

Beach, L. R., & Lipshitz, R. (1993). Why classical decision theory is an inappropriate standard for evaluating and aiding most human decision making. In G. Klein, J. Orasanu, R. Calderwood, & C. Zsambok (Eds.), *Decision making in action: Models and methods* (pp. 21–35). Norwood, NJ: Ablex.

Brown, K. G., & Kozlowski, S. W. J. (1997, April). *Self-evaluation and training outcomes: Training strategy and goal orientation effects.* Paper presented at the 12th annual conference of the Society for Industrial and Organizational Psychology, St. Louis, MO.

Brown, K. G., Mullins, M. E., Weissbein, D. A., Toney, R. J., & Kozlowski, S. W. J. (1997, April). Mastery goals and strategic reflection: Preliminary evidence for learning interference. In S. W. J. Kozlowski (Chair), *Metacognition in training: Lessons learned from stimulating cognitive reflection.* Symposium conducted at the 12th annual conference of the Society for Industrial and Organizational Psychology, St. Louis, MO.

Cannon-Bowers, J. A., Tannenbaum, S. I., Salas, E., & Volpe, C. E. (1995). Defining competencies and establishing team training requirements. In R. A. Guzzo & E. Salas (Eds.), *Team effectiveness and decision making in organizations* (pp. 333–380). San Francisco: Jossey-Bass.

Devine, D. J., & Kozlowski, S. W. J. (1995). Expertise and task characteristics in decision making. *Organizational Behavior and Human Decision Processes, 64*, 294–306.

Dyer, J. C. (1984). Team research and team training: State-of-the-art review. In F. A. Muckler (Ed.), *Human factors review* (pp. 285–323). Santa Monica, CA: Human Factors Society.

Earley, C. P., Connolly, T., & Ekegren, G. (1989). Goals, strategy development, and task performance: Some limits on the efficacy of goal setting. *Journal of Applied Psychology, 74*, 24–33.

Elliott, E. S., & Dweck, C. S. (1988). Goals: An approach to motivation and achievement. *Journal of Personality and Social Psychology, 54*, 5–12.

Flavell, J. H. (1979). Metacognition and cognitive monitoring: A new area of cognitive-developmental inquiry. *American Psychologist, 34*, 906–911.

Fleishman, E. A., & Zaccaro, S. J. (1992). Toward a taxonomy of team performance functions. In R. W. Swezey & E. Salas (Eds.), *Teams: Their training and performance* (pp. 31–56). Norwood, NJ: Ablex.

Ford, J. K., & Kraiger, K. (1995). The application of cognitive constructs to the instructional systems model of training: Implications for needs assessment, design and transfer. In C. L. Cooper & I. T. Robertson (Eds.), *International Review of Industrial and Organizational Psychology* (pp. 1–48). New York: Wiley.

Gersick, C. J. G., & Hackman, J. R. (1990). Habitual routines in task-performing groups. *Organizational Behavior and Human Decision Processes, 47*, 65–97.

Gist, M. E., & Mitchell, T. R. (1992). Self-efficacy: A theoretical analysis of its determinants and malleability. *Academy of Management Review, 17,* 183–211.

Gully, S. M. (1997a, April). A cross-level analysis of the influences of cognitive ability and goal orientation on individual outcomes in a team training context. In J. E. Mathieu (Chair), *Cross-level investigations of group influences on individual outcomes.* Symposium conducted at the 12th annual conference of the Society for Industrial and Organizational Psychology, St. Louis, MO.

Gully, S. M. (1997b). *The influence of individual self-regulatory processes on learning and performance in a team training context.* Unpublished doctoral dissertation, Department of Psychology, Michigan State University, East Lansing.

Gully, S. M., & Kozlowski, S. W. J. (1996, August). The influence of self-efficacy and team-efficacy on training outcomes in a team training context. In J. George-Flavey (Chair), *Defining, measuring, and influencing group level efficacy beliefs.* Symposium conducted at the annual convention of the Academy of Management Association, Cincinnati, OH.

Holyoak, K. J. (1991). Symbolic connectionism: Toward third-generation theories of expertise. In K. A. Ericsson & J. Smith (Eds.), *Toward a general theory of expertise* (pp. 301–336). Cambridge, England: Cambridge University Press.

Kanfer, R., & Ackerman, P. L. (1989). Motivation and cognitive abilities: An integrative aptitude–treatment interaction approach to skill acquisition. *Journal of Applied Psychology, 74,* 657–689.

Klein, G. A. (1989). Recognition-primed decisions. In W. B. Rouse (Ed.), *Advances in man-machine systems research* (Vol. 5, pp. 47–92). Greenwich, CT: JAI Press.

Kozlowski, S. W. J. (1995). Enhancing the training and development of adaptive expertise. *Psychological Science Agenda, 8*(5), 7–9.

Kozlowski, S. W. J., Chao, G. T., Smith, E. M., & Hedlund, J. A. (1993). Organizational downsizing: Strategies, interventions, and research implications. In C. L. Cooper & I. T. Robertson (Eds.), *International review of I/O psychology* (Vol. 8, pp. 263–332). New York: Wiley.

Kozlowski, S. W. J., Ford, J. K., & Smith, E. M. (1993). *Training concepts, principles, and guidelines for the acquisition, transfer, and enhancement of team tactical decision making skills: I. A conceptual framework and literature review* (Contract No. N61339-91-C-0117). Orlando, FL: Naval Training Systems Center.

Kozlowski, S. W. J., Gully, S. M., McHugh, P. P., Salas, E., & Cannon-Bowers, J. A. (1996). A dynamic theory of leadership and team effectiveness: Developmental and task contingent leader roles. In G. R. Ferris (Ed.), *Research in personnel and human resource management* (Vol. 14, pp. 253–305). Greenwich, CT: JAI Press.

Kozlowski, S. W. J., Gully, S. M., Nason, E. R., & Smith, E. M. (in press). Developing adaptive teams: A theory of compilation and performance across levels and time. In D. R. Ilgen & E. D. Pulakos (Eds.), *The changing nature of work performance: Implications for staffing, personnel actions, and development.* San Francisco: Jossey-Bass.

Kozlowski, S. W. J., Gully, S. M., Salas, E., & Cannon-Bowers, J. A. (1996). Team leadership and development: Theory, principles, and guidelines for training leaders and teams. In M. Beyerlein, D. Johnson, & S. Beyerlein (Eds.), *Advances in interdisciplinary studies of work teams: Team leadership* (Vol. 3, pp. 251–289). Greenwich, CT: JAI Press.

Kozlowski, S. W. J., Gully, S. M., Smith, E. M., Brown, K. G., Mullins, M. E., & Williams, A. E. (1996, April). Sequenced mastery goals and advance organizers: Enhancing the effects of practice. In K. Smith-Jentsch (Chair), *When, how, and why does practice make perfect?* Symposium conducted at the 11th annual conference of the Society for Industrial and Organizational Psychology, San Diego, CA.

Kozlowski, S. W. J., Gully, S. M., Smith, E. A., Nason, E. R., & Brown, K. G. (1995, May). Sequenced mastery training and advance organizers: Effects on learning, self-efficacy, performance, and generalization. In R. J. Klimoski (Chair), *Thinking and feeling while doing: Understanding the learner in the learning process.* Symposium conducted at the 10th annual conference of the Society for Industrial and Organizational Psychology, Orlando, FL.

Kozlowski, S. W. J., & Salas, E. (1991, April). Application of a multilevel contextual model to training implementation and transfer. In J. K. Ford (Chair), *Training as an integrated activity: An organization system perspective.* Symposium conducted at the 6th annual conference of the Society for Industrial and Organizational Psychology, St. Louis, MO.

Kozlowski, S. W. J., & Salas, E. (1997). An organizational systems approach for the implementation and transfer of training. In J. K. Ford, S. W. J. Kozlowski, K. Kraiger, E. Salas, & M. Teachout (Eds.), *Improving training effectiveness in work organizations* (pp. 247–287). Mahwah, NJ: Erlbaum.

Kraiger, K., Ford, J. K., & Salas, E. (1993). Application of cognitive, skill-based, and affective theories of learning outcomes to new methods of training evaluation. *Journal of Applied Psychology, 78,* 311–328.

Martocchio, J. J. (1994). Effects of conceptions of ability on anxiety, self-efficacy, and learning in training. *Journal of Applied Psychology, 79,* 819–825.

Modrick, J. (1986). Team performance and training. In J. Zeidner (Ed.), *Human productivity enhancement* (Vol. 1, pp. 130–166). New York: Praeger.

Nason, E. R. (1995). *Horizontal team member exchange (HMX): The effects of relationship quality on team processes and outcomes.* Unpublished doctoral dissertation, Department of Psychology, Michigan State University, East Lansing.

Nelson, T. O. (1996). Consciousness and metacognition. *American Psychologist, 51,* 102–116.

Newell, A., & Simon, H. A. (1972). *Human problem solving.* Englewood Cliffs, NJ: Prentice-Hall.

Orasanu, J., & Connolly, T. (1993). The reinvention of decision making. In G. Klein, J. Orasanu, R. Calderwood, & C. Zsambok (Eds.), *Decision making in action: Models and methods* (pp. 3–20). Norwood, NJ: Ablex.

Rousseau, D. M. (1985). Issues of level in organizational research: Multi-level and cross-level perspectives. *Research in Organizational Behavior, 7,* 1–37.

Salas, E., & Cannon-Bowers, J. A. (in press). The anatomy of team training. In L. Tobias & D. Fletcher (Eds.), *Handbook of training.* Hillsdale, NJ: Erlbaum.

Salas, E., Cannon-Bowers, J. A., & Blickensderfer, E. L. (1997). Enhancing reciprocity between training theory and practice: Principles, guidelines, and specifications. In J. K. Ford, S. W. J. Kozlowski, K. Kraiger, E. Salas, & M. Teachout (Eds.), *Improving training effectiveness in work organizations* (pp. 291–322). Mahwah, NJ: Erlbaum.

Salas, E., Cannon-Bowers, J. A., & Kozlowski, S. W. J. (1997). The science and practice of training: Current trends and emerging themes. In J. K. Ford, S. W. J. Kozlowski, K. Kraiger, E. Salas, & M. Teachout (Eds.), *Improving training effectiveness in work organizations* (pp. 357–368). Mahwah, NJ: Erlbaum.

Salas, E., Dickinson, T. L., Converse, S. A., & Tannenbaum, S. I. (1992). Toward an understanding of team performance and training. In R. W. Swezey & E. Salas (Eds.), *Teams: Their training and performance* (pp. 3–29). Norwood, NJ: Ablex.

Schmidt, R. A., & Bjork, R. A. (1992). New conceptualizations of practice: Common principles in three paradigms suggest new concepts for training. *Psychological Science, 3,* 207–217.

Smith, E. M. (1997, April). The effects of individual differences, discovery learning, and metacognition on learning and adaptive transfer. In S. W. J. Kozlowski (Chair), *Metacognition in training: Lessons learned from stimulating cognitive reflection.* Symposium conducted at the 12th annual conference of the Society for Industrial and Organizational Psychology, St. Louis, MO.

Smith, E. M., Ford, J. K., & Kozlowski, S. W. J. (1997). Building adaptive expertise: Implications for training design. In M. A. Quinones & A. Dudda (Eds.), *Training for a rapidly changing workplace: Applications of psychological research* (pp. 89–118). Washington, DC: American Psychological Association.

Smith, E. M., Ford, J. K., Weissbein, D. A., & Gully, S. M. (1995, May). The effects of goal orientation, metacognition, and practice strategies on learning and transfer. In R. J. Klimoski (Chair), *Thinking and feeling while doing: Understanding the learner in the learning process.* Symposium conducted at the 10th annual conference of the Society for Industrial and Organizational Psychology, Orlando, FL.

Sternberg, R. J., & Frensch, P. A. (1992). On being an expert: A cost-benefit analysis. In R. R. Hoffman (Ed.), *The psychology of expertise* (pp. 191–203). New York: Springer-Verlag.

Sundstrom, E., De Meuse, K. P., & Futrell, D. (1990). Work teams: Applications and effectiveness. *American Psychologist, 45*, 120–133.

Swezey, R. W., & Salas, E. (1992). Guidelines for use in team-training development. In R. W. Swezey & E. Salas (Eds.), *Teams: Their training and performance* (pp. 329–353). Norwood, NJ: Ablex.

Tannenbaum, S. I., & Yukl, G. (1992). Training and development in work organizations. *Annual Review of Psychology, 43,* 399–441.

Thompson, J. (1967). *Organizations in action: Social science bases of administrative theory.* New York: McGraw-Hill.

Turnage, J. J. (1990). The challenge of new workplace technology for psychology. *American Psychologist, 45,* 171–178.

Van de Ven, A. H., Delbecq, A. L., & Koenig, R. (1976). Determinants of coordination modes within organizations. *American Sociological Review, 41,* 322–338.

Weaver, J. L., Bowers, C. A., Salas, E., & Cannon-Bowers, J. A. (1995). Networked simulations: New paradigms for team performance research. *Behavioral Research Methods, Instruments, and Computers, 27,* 12–24.

Weissbein, D. A., & Ford, J. K. (1997, April). The effects of metacognitive training on metacognitive reflection and learning. In S. W. J. Kozlowski (Chair), *Metacognition in training: Lessons learned from stimulating cognitive reflection.* Symposium conducted at the 12th annual conference of the Society for Industrial and Organizational Psychology, St. Louis, MO.

7

Critical Thinking Skills in Tactical Decision Making: A Model and a Training Strategy

Marvin S. Cohen, Jared T. Freeman, and Bryan Thompson

In this chapter we describe a model of decision-making skills under time stress, a training strategy based on that model, and experimental tests of the training strategy. A prime example of the kind of decision making addressed by the model is naval ship defense against an approaching aircraft whose intent is ambiguous. The *Vincennes* and *Stark* incidents are prototypical cases of this kind, illustrating two different ways that such decision making can go wrong. However, similar decision-making skills are required in many other domains, such as medicine, fire fighting, commercial airline emergencies, and even chess. In each case, the decision maker must gauge the available time for collecting and analyzing information and usually must act on the basis of an incomplete picture of the situation.

The Tactical Decision Making Under Stress (TADMUS) program encompasses several different approaches to this problem. One approach is through workload reduction and automaticity; this approach is based on the premise that practice in lower level skills may free cognitive resources for higher level decision making. Another approach is through pattern recognition, in which decision makers learn to recognize and respond appropriately to a large number of situations, bypassing deliberate decision making altogether. Still another approach involves training in team-coordination and communication skills to ensure that relevant information is communicated to the right people at the right time. These approaches improve decision making by addressing its inputs, the resources it draws on, or the alternative processes that make it unnecessary. The approach described here, by contrast, provides a direct focus on decision-making processes in novel or unexpected situations.

In the second section of this chapter, we briefly describe a model of

We thank Al Koster, Robin Waters, and John Poirier of Sonalysts, Inc.; Bill Kemple of the Naval Post Graduate School; and Steve Wolf and Laura Militello of Klein Associates for their help during this research.

155

real-world decision making under time stress and contrast it with alternative conceptualizations. The model is based on naturalistic observations of experienced decision makers in real-world contexts and on cognitive theory. The third section of the chapter provides an overview of a training strategy based on the model. The strategy makes use of instruction, demonstration, and practice to teach naval officers methods for identifying and handling qualitatively different kinds of uncertainty in assessments and plans. In the fourth section, we describe experimental tests of the training at two naval training facilities with active-duty naval officers. The effects of training on decision-making processes, on accuracy of situation assessment, and on appropriateness of action were examined. The final section provides conclusions and lessons learned from the research, including potential extensions of the critical thinking training to other domains.

A Naturalistic Framework for Decision Making

Models of Decision Making

Efforts to train people in decision making have been shaped by competing conceptions of what decision making is. The approach described here contrasts with and borrows from classical decision theory, pattern recognition, and problem-solving approaches to decision making.

Decision theory. Perhaps the most familiar framework for decision making, classical decision theory contains two main parts: *Bayesian probability theory* for drawing inferences about the situation, and *multiattribute utility theory* for selecting an optimal action. Bayesian probability theory requires that decision makers identify a set of mutually exclusive and exhaustive hypotheses (e.g., about the intent of an approaching aircraft). The next steps are to assess the probability that each hypothesis is true, identify all the potential observations that might bear on those hypotheses in the future, and quantify the impact each such observation would have. Then, as new observations occur, decision makers can use algorithms from the theory for updating belief in the hypotheses.

Multiattribute utility theory is an analogous method for making choice. It requires that decision makers specify a set of possible actions, an exhaustive and mutually exclusive set of uncertain states of the world, and a set of evaluative dimensions. The decision maker then assesses the probability of each uncertain state, the importance of each evaluative dimension, and the score of each action–state combination for every evaluative dimension. The theory enables decision makers to calculate a score reflecting the overall desirability of each action.

Along with others, we have argued that decision theory is not in general *cognitively compatible* with the way experienced decision makers work (Cohen & Freeman, 1996). Problems with the classic framework include the kinds of inputs it demands, the kind of processing it prescribes, and the outputs it produces.

First, by demanding a complete model up front, with fixed assessments of uncertainty and preference, decision theory tends to discourage the dynamic evolution of problem understanding through time, for example, as new hypotheses, options, observations, outcomes, and even goals are discovered.

Second, by reducing all uncertainty to a single measure (probability), decision theory obscures important qualitative differences in the way different types of uncertainty are handled, such as gaps, conflict, and unreliable assumptions. Decision theory, for example, treats conflicting evidence in the same way that it treats congruent evidence, essentially by taking an average. Experienced decision makers, on the other hand, may use conflict as an opportunity to identify the faulty assumptions in their beliefs that produced the conflict (Cohen, 1986). Similarly, decision theory handles conflicting goals in the same way that it handles congruent goals, by calculating an overall score for each option that is an average of the different goals. Experienced decision makers, by contrast, may try to learn from the conflict, by creating a better option or a deeper understanding of their true objectives (Levi, 1986).

Third, the output of a decision theoretic model is a statistical average (e.g., 70% chance hostile, 30% chance not hostile) rather than a single coherent picture of the situation. Decision makers cannot visualize, anticipate, or plan effectively for an abstract average.

Decision theoretic models, as they are typically applied, are not only descriptively inadequate but normatively inadequate as well. As argued by Cohen (1993), appropriate normative principles must capture the relevant qualitative features of the decision-making process. If a normative standard is to be used to identify decision-making errors, it must be close enough to actual performance for the discrepancies to be interesting.

Pattern recognition. A second approach to decision-making skill emphasizes holistic intuition rather than formal analysis. Decision-making skill is identified with the accumulation, through experience, of a set of virtually automatic responses to recognized patterns. This view has been popular in research on differences between experts and novices, beginning with Chase and Simon's (1973) work on chess. Unfortunately, pattern recognition views say little about the kinds of novel or ambiguous situations that are traditionally associated with the term *decision making*. How is situation assessment accomplished in new and changing circumstances? How are conflicting and unreliable data dealt with? How do decision makers change their minds? When do they stop thinking and act? Although pattern recognition appears to be at the heart of proficient performance, other processes may be crucial for success in less routine circumstances. For example, Klein (1993) discussed how options may be tested by mental simulation of their outcomes.

Problem solving. One way to address the difficulties described is to define decision making as a special case of problem solving. The decision maker may deploy a range of strategies to find the correct hypothesis or action, for example, dividing the problem into simpler subproblems or working backward from the goal to subgoals and the actions that achieve them. Such strategies do not figure prominently in either the decision theoretic or the recognitional approach.

There are two problems with the traditional problem-solving point of view, however. First, the approach fails to address the central role of uncertainty and risk in decision making (Fischhoff & Johnson, 1990). In this, it resembles recognitional approaches. A second problem is that general-purpose problem solving does not easily accommodate the role of experience-based recognition. In this, it resembles the decision theoretic approach. The problem-solving approach, as it now stands, is more of a promising direction than a full-fledged theory of decision making.

Implications for training. Different conceptions of decision-making skill are important because they are associated with different *training strategies*. According to Salas and Cannon-Bowers (1977), a training strategy orchestrates *tools* (such as feedback and simulation) and *methods* (such as instruction, demonstration, and practice) to convey a *content*.

From the decision theoretic point of view, the content of training is a set of general-purpose techniques (Baron & Brown, 1991). The principal tool for defining this content is, of course, decision theory, and the primary method of presentation is explicit classroom instruction. Examples of decision problems are used not as content but to motivate the formal techniques during instruction, to demonstrate their generality, and for paper-and-pencil practice (e.g., Adams & Feehrer, 1991).

At the opposite extreme, decision training based on the recognitional point of view conveys examples of decision problems and their solutions, rather than general-purpose techniques. The methods used in recognitional training tend to be demonstration and practice with a large set of illustrative problems, rather than explicit instruction, and to incorporate tools like high-fidelity simulation and outcome feedback (Means, Salas, Crandall, & Jacobs, 1993).

The development of an effective training strategy depends on an adequate conception of decision making. If both recognition and uncertainty are important, how are they related? If problem-solving skills are relevant, how do they apply to recognition and uncertainty? The present research has focused on these questions and their implications for training. The training strategy described here builds on recognition, but it focuses on problem-solving strategies for dealing effectively with uncertainty.

The Recognition/Metacognition Model

Proficient decision makers are *recognitionally skilled*; that is, they are able to recognize a large number of situations as familiar and to retrieve an appropriate response. Our observations of decision-making performance, in naval anti-air warfare as well as other domains, suggest that recognition is supplemented by processes that verify and improve its results (Cohen, Freeman, & Wolf, 1996). Because of their function, we call these pro-cesses *metarecognitional*. People use metarecognition skills to probe for flaws in recognized assessments and plans, to patch up any weaknesses that are found, and to evaluate the results.

Metarecognitional skills are analogous to the *metacomprehension* skills that proficient readers use when they construct a mental model on the basis of information in a text. For example, according to Baker (1985), skilled readers continually look for problems, such as inconsistencies or gaps, in the current state of their comprehension, and they adopt a variety of strategies for correcting problems, such as referring back to earlier parts of the text or relating the text to information already known.

To reflect the complementary roles of recognition and metacognition in decision making, we have called this framework the recognition/ metacognition (R/M) model (Cohen, 1993; Cohen, Freeman, & Thompson, 1997; Cohen et al., 1996). Cues in a situation activate an interpretation of their meaning. (For example, an aircraft popping up on radar at high speed and low altitude, heading toward a U.S. ship from an unfriendly country, is recognized as having the intent to attack.) This interpretation in turn may activate knowledge structures that organize actual and po-tential information. A *story* is one such structure, namely, a causal model that people construct to flesh out assessments of human intent (Penning-ton & Hastie, 1993). (For example, a story about the attacking airplane may include the motivation of the attacking country, the reason the at-tackers used this type of airplane, the reason they chose ownship as a target, and what they will do next.) Stories and other structures make up an evolving situation model and plan. According to the R/M model, the integration of observations into such structures may involve a set of meta-recognitional processes, including the following:

1. Identification of evidence–conclusion relationships (or *arguments*) within the evolving situation model and plan. This is simply an implicit or explicit awareness that Cue A was *observed* on this occasion, whereas the assessment (e.g., intent to attack), along with expectations of observing Cue B, were *inferred*. On some other occasion, Cue B might be observed and Cue A inferred.

2. Processes of *critiquing* that identify problems in the arguments that support a conclusion (e.g., hostile intent) within the situation model or plan. Critiquing can result in the discovery of three kinds of problems: *incompleteness, conflict, or unreliability* (Cohen, 1986). An argument is incomplete if it provides support neither for nor against a conclusion of interest. (For example, the kine-matics of the track suggest that it is a military aircraft but say nothing about the key issue, hostile intent.) Two arguments con-flict with one another if they provide support both for and against a conclusion, respectively. (For example, the heading of a track toward ownship suggests hostile intent, whereas its slow speed argues for routine patrol.) Finally, an argument is unreliable if the support it provides for a conclusion depends on unexamined

assumptions. (For example, a track's turning toward ownship suggests hostile intent until it is realized that the track is too far away to have detected ownship.) Unreliable support may shift or vanish when its premises are further considered.

3. Processes of *correcting* that respond to these problems. Correcting steps may instigate external action, such as collecting additional data, or internal actions that regulate the operation of the recognitional system. These internal actions include attention shifting and assumption revision. Shifting the focus of attention overcomes limitations on the automatic activation of information in memory and brings additional knowledge into view. Such knowledge may fill gaps in arguments, help resolve conflict, or confirm or deny unreliable assumptions. Adjusting assumptions involves operating the recognitional system as if an uncertain belief were known to be true or false. It permits what-if reasoning, exploratory searching for alternative causes and effects, and eventual adoption of a single coherent model or plan.

4. A higher level process called the *quick test*, which controls critiquing and correcting. Metarecognitional processing occurs when the expected benefits associated with critical thinking outweigh the costs. It is shaped, like other actions, by past experiences of success and failure. Recognition-based responding will be inhibited and critical thinking will be initiated when the costs of delay are acceptable, the situation is uncertain or novel, and the costs of an error are high.

Figure 1 summarizes the functional relationships among these processes. A more detailed description of the R/M model may be found in work by Cohen et al. (1996) and Cohen, Parasuraman, Serfaty, and Andes (1997).

A computer-based implementation of the R/M model has been developed (Thompson, Cohen, & Shastri, 1997). The model, which combines neural and symbolic features, uses an adaptive critic architecture for reinforcement learning. The recognitional aspect of the critic learns complex relations that enable it to predict events and trigger appropriate actions. In parallel, the metacognitive critic maintains a model of the recognitional critic and learns appropriate actions for regulating the behavior of the recognitional critic. Machine and human learning experiments in the domain of naval tactical anti-air warfare are currently being compared to test predictions of the R/M model.

Comparison to other models. The R/M model contrasts with *classic decision theory* in its approach to the inputs, the processes, and the outputs of decision making. (a) Rather than demanding that all inputs to a model be specified in advance, the R/M model predicts the incremental generation of new hypotheses, options, observations, outcomes, or goals in the course of working the problem. (b) Rather than assigning fixed numerical significance to cues or goals and then mathematically aggregating, R/M permits ongoing reinterpretation of cues and goals. (c) Instead of an

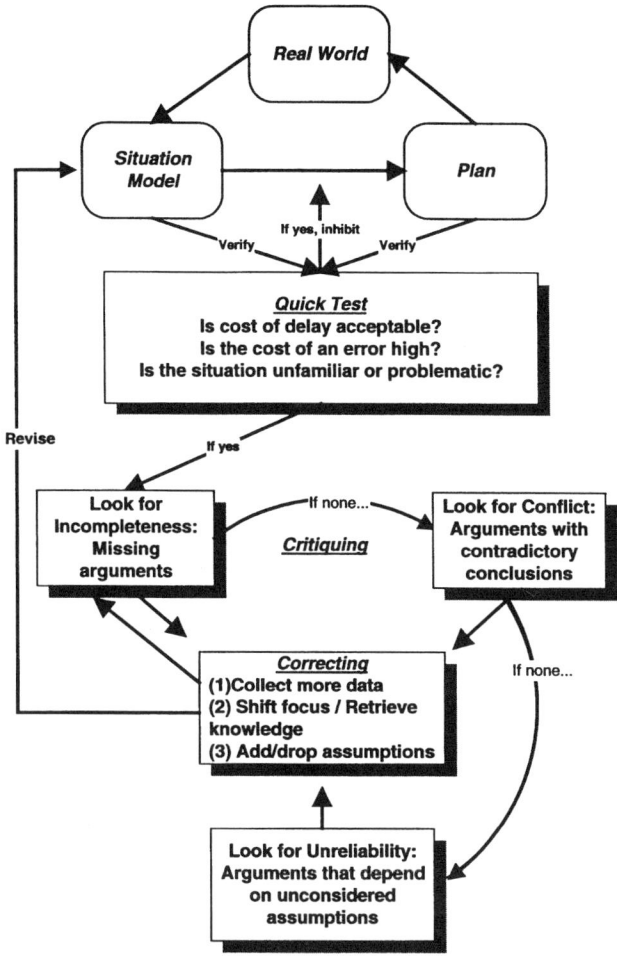

Figure 1. Components of the recognition/metacognition model. From "Meta-recognition in Time-Stressed Decision Making: Recognizing, Critiquing, and Correcting," by M. S. Cohen, J. T. Freeman, and S. Wolf, 1996, *Human Factors, 38,* p. 211. Copyright 1996 by Human Factors and Ergonomics Society. Adapted with permission.

unrealizable statistical abstraction (e.g., 70% hostile, 30% nonhostile), R/M produces a single concrete picture of a situation, together with a qualitative understanding of its specific strengths and weaknesses.

Metarecognitional processing is a highly dynamic and iterative *problem-solving* strategy. The next processing step is determined locally by the results of earlier steps, rather than by a global, fixed procedure (as in Bayesian inference or in other proposed decision heuristics). Correcting for one problem sometimes leads to identification and correction of another problem. For example, a gap in a story may be filled by collecting additional data, by remembering previously known information, or, if these fail, by making assumptions. The resulting more detailed arguments may then turn out to conflict with other arguments. Such conflict may then be addressed by examining the reliability of the conflicting arguments (e.g., shifting attention to the grounds for the arguments). This process stops when the quick test indicates that the benefits of further metarecognitional actions are outweighed by the risks of delay and that action on the basis of the current best model or plan is called for.

The R/M model reconciles *pattern recognition* with problem solving. It explains how experienced decision makers are able to exploit their experience-based intuition in a domain and at the same time handle uncertainty and novelty. They construct and manipulate concrete, visualizable models of the situation, not abstract aggregations. Uncertainty is represented explicitly at the metacognitive level by annotation of the situation model or plan to highlight points of incompleteness, conflict, and unreliability. With metacognitive strategies, one can respond to these problems, try to improve the current situation model, and plan or find better ones. To quote Dreyfus (1997, p. 28), metarecognition is "observation of one's intuitive practice-based behavior with an eye to challenging and perhaps improving intuition without replacing it."

Implications for training. The R/M model yields an approach to training that is distinct from both classical decision theory and pattern recognition models. From the point of view of the R/M model, the content of training is neither a small set of general-purpose methods (as in decision theory) nor a vast quantity of specialized patterns and responses. The focus is a moderately sized set of *domain-grounded* strategies for critical thinking. Unlike patterns, these strategies are generalizable, but unlike the methods of decision theory, they can be effectively taught only on the basis of preexisting knowledge in a particular domain.

Several aspects of metarecognitional strategies may be transferable across domains. For example, the same or similar types of uncertainty (i.e., incompleteness, conflict, and unreliability) seem to be relevant across a variety of domains. Moreover, the same or similar metarecognitional actions—collecting more data, shifting attentional focus to retrieve more information, and changing assumptions—seem appropriate for handling these types of uncertainty. Finally, the metarecognitional cycle depicted in Figure 1, with its priority of testing first for incompleteness, then for conflict, and then for unreliability, may reflect a relatively general tendency among proficient decision makers. Such decision makers try to create a complete and consistent story, and then they evaluate the plausibility of the assumptions their story demands. Nevertheless, these generalities presuppose a base of recognitional knowledge on which they operate. Metarecognitional strategies thus make little sense abstracted from a particular application area (cf. Kuhn, Amsel, & O'Loughlin, 1988).

Training Critical Thinking Skills

In this section we describe a training strategy for naval combat information center (CIC) officers that is based on the R/M model. Exhibit 1 out-

Exhibit 1.   Methods, Tools, and Content of the Critical Thinking
Training Strategy

Strategy
   Critical thinking training (for individuals)

Tools
   Cognitive task analysis (critical incident interviews)
   Simulation (DEFTT, with specifically tailored scenarios)
   Feedback (from group and instructor)
   Training guidelines
   Performance measures (process and outcome)

Methods
   Information-based: lecture and discussion
   Practice-based: guided practice, behavior modeling

Content
   Building stories in novel situations
   Detecting and handling conflicting evidence
   Generating and evaluating alternative assessments
   Adjusting to the available time

Note. DEFTT = Decision Making Evaluation Facility for Tactical Teams.

lines the crucial features of this training strategy: its tools, its methods, and its content (Salas & Cannon-Bowers, 1997).

An essential tool in the development of the training strategy is cognitive task analysis. The R/M model and the training design are based on critical incident interviews with active-duty naval officers, in which the officers described experiences in the Persian Gulf, the Gulf of Sidra, and elsewhere (Kaempf, Klein, Thordsen, & Wolf, 1996). Our analysis focused on nine incidents in which the officers decided whether to engage a contact whose intent was unknown, under conditions of undeclared hostility. We analyzed the interviews to discover the officers' thinking strategies, ways of organizing information, and decisions. Many aspects of the training are based on differences in the way that more and less experienced officers handled similar types of situations.

The training design uses both information-based and practice-based training methods. In each segment of training, officers listen to a brief verbal presentation of the concepts central to that segment, followed by questions and discussion. They then play the role of a tactical action officer in realistic scenario-based exercises. Training and test scenarios were adapted from the scenarios described by Johnston et al. (see chapter 3, this volume); modifications were made to include events from critical incident interviews that required the relevant critical thinking skills.

This scenario-based practice uses two important tools: interactive simulation and feedback. The simulation platform is the Decision Making Evaluation Facility for Tactical Teams (DEFTT), discussed by Johnston et al. (chapter 3, this volume). Feedback was provided in the form of hints from the instructor in real time as the scenario unfolded and by class discussion at the conclusion of the scenario. A third tool consists of training guidelines such as those proposed by Duncan et al. (1996) for training mental models.

A final tool is represented by a set of performance measures used to evaluate the success of the training. These measures address both critical thinking processes (e.g., the number of pieces of conflicting evidence that trainees identified and the number of factors they considered in arguments about intent) and outcomes (i.e., the agreement of trainees' situation assessments and actions with those of a subject-matter expert).

The training *content* is divided into the following four segments:

1. *Creating, testing, and evaluating stories*. This section provides an overview of the critical thinking process, called STEP. When an assessment is uncertain, decision makers can enhance their understanding by constructing a *S*tory around it. The story includes the past, present, and future events that would be expected if the assessment were true. Decision makers use the story to *T*est the assessment, by comparing expectations to what is known or observed. When evidence appears to conflict with the assessment, they try to patch up the story by explaining the evidence. They then *E*valuate the result. If the patched-up story involves too many unreliable assumptions, decision makers generate alternative assessments and stories. In the meantime, they *P*lan against the possibility that their current best story is wrong.

2. *Hostile-intent stories*. Stories contain certain typical components. Knowledge of these components can help decision makers notice and fill gaps in the stories they construct. A particularly important kind of story is built around the assessment of hostile intent. For example, a complete hostile-intent story explains the motivation for attack, the choice of a target, and the choice of an attacking platform. It also accounts for how that platform localized the target and the manner in which it will arrive at a position suitable for engaging it. The training teaches officers by practice and example how to discover story components and to let the stories guide them to relevant evidence about intent.

3. *Critiquing stories*. After a story is constructed, decision makers step back and evaluate its plausibility. This segment of the training introduces a devil's advocate technique for uncovering hidden assumptions in a story and generating alternative interpretations of the evidence. An infallible crystal ball persistently tells the decision maker that a conclusion is wrong, despite the evidence that appears to support it, and asks for an explanation of that evidence. Regardless of how confident decision makers are in their assessments, this technique can alert them to significant alternatives. It can also help them see how conflicting data could fit into a story. The technique helps decision makers expose and evaluate assumptions underlying their reading of the evidence.

4. *When to think more*. Critical thinking is not always appropriate.

Officers should not incur extreme risk to their own ship rather than engage a track. But it is necessary to evaluate the time available before taking an irreversible action. The decision maker should probably act immediately unless three conditions are satisfied: (a) The risk of delay must be acceptable, (b) the cost of an error if one acts immediately must be high, and (c) the situation must be nonroutine or problematic in some way. Training focuses on the way experienced decision makers apply these criteria. For example, they tend to use more subtle estimates of how much time is available, which are based on the specifics of the situation. They focus more on longer term objectives in estimating the costs of an error. Also, they show greater sensitivity to the mismatch between the situation and familiar patterns.

In the following four sections, we describe each of these four training segments in more detail. The discussion, like the training itself, draws on examples based on actual incidents.

Segment 1: Creating, Testing, and Evaluating Stories

Observations regarding a surface or air contact may prompt recognition of its intent. For example, in one incident a tactical action officer (TAO) in the Gulf of Sidra was notified that a track had popped up on radar at close range and was heading toward his own ship. Its speed suggested a military plane, and it did not respond to radio warning. After progressing 2 miles, it began to circle. The TAO suspected the aircraft was hostile. However, when the costs of an error (e.g., shooting at a friendly ship) are high and time is available before ownship is at significant risk, it is worth thinking critically about the assessment. This is what the TAO and his captain did.

Creating a story. Figure 2 outlines the four steps of critical thinking that make up the STEP cycle. The first step is to build a story around the current assessment. Although the term *story* sounds playful, in fact, building a story means taking an assessment seriously. The story includes what would have happened in the past, what should be happening now, and what is expected to happen in the future if the assessment is correct. The assessment can then be tested by comparing the story to the facts or (if that is not possible) by evaluating its plausibility. The training illustrates the STEP process by providing examples of increasing complexity drawn from real incidents such as this one.

The situation of the circling aircraft in the Gulf of Sidra did not fit a ready-made attack profile. As a result, the TAO tried to create a hostile-intent story, to explain what he had observed and to test the hostile-intent assessment. An attack by Libya would fit Gadhafi's objective of defending his self-proclaimed territorial waters in the Gulf of Sidra. The TAO proceeded to describe what might have happened if the contact was hostile: "I figured that the pilot took off from a Libyan base, kept his head down,

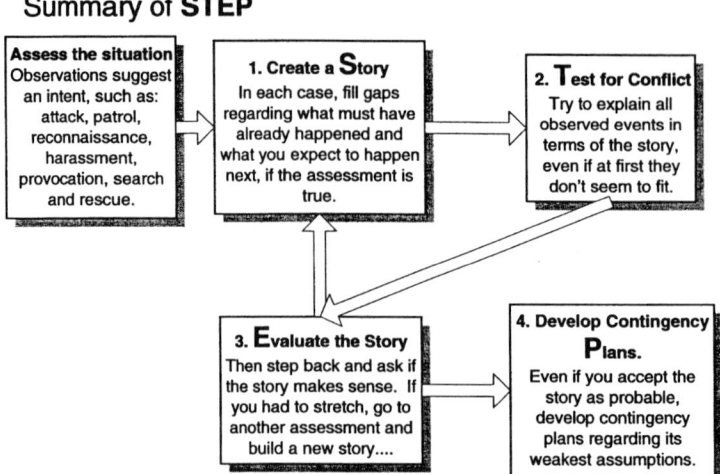

Figure 2. The STEP process for critical thinking.

and turned directly towards us. He must have wanted to seize the moment . . . attack just as we detected him and before we got our gear up."

Testing the story. The next part of the story, however, did not fit what happened: "Instead of continuing to attack, the track paused to circle. This made no sense." Another observation that conflicted with hostile intent was the absence of electronic emissions.

Did the TAO conclude, with a sigh of relief, that the aircraft could not be hostile? Certainly not. Experienced decision makers do not abandon an assessment (especially a threatening one) because it does not fit all the evidence. In many situations, no pattern fits all the evidence perfectly; the truth will therefore necessarily run counter to expectations. To give an assessment a fair chance, officers try to incorporate all the observed events into the story, even if at first they do not seem to fit.

In this incident, the TAO tried to fit the conflicting observations into the hostile-intent story. "The best interpretation I could make of this— and it wasn't too good—was that he came up to target us, but his radar had busted." This single explanation happens to account for both arguments that conflict with hostile intent. The aircraft was not emitting and was not approaching because its radar had broken.

Evaluating the story. Just because a story can be constructed, however, does not mean that the story is true. The next step in critical thinking is to evaluate: Step back and ask if the story makes sense. It is sometimes possible to gather more data to test an explanation of conflicting evidence. In other cases, it is a matter of rapidly judging plausibility. Did the officer have to stretch believability too much to make all the observations fit? Each time the decision maker explains a piece of conflicting evidence, it is like stretching a spring. Eventually, the spring resists any further efforts in that direction and snaps back; the "explained away" evidence becomes conflicting again. If time is available and stakes are high, experienced decision makers try to build a different story, on the basis of a different assessment.

In this incident, it was not very plausible that an attacking aircraft would stop and circle in plain view if its radar was not functioning. Because the TAO's story required a stretch, the captain considered the possibility that this was a friendly aircraft.

Another cycle. The captain then generated a new story on the basis of the assumption that the aircraft was friendly. "The captain . . . figured it was one of ours, his radio was off or busted, and he was trying to execute our triangular return-to-fort profile [a maneuver to signal a friendly aircraft returning to the battle group]." Unfortunately, when tested, expectations based on this story did not perfectly fit the observations either. The track did not follow the expected triangular profile closely. The captain did not abandon his assessment but tried to patch up the story to explain the discrepancy: "That pilot was doing a spectacularly lousy job of drawing that triangle."

Returning to the evaluation stage, one asks, how good is the captain's story? Like the TAO's hostile-intent story, it requires the assumption of broken equipment (radar or radio, respectively). In addition, it assumes a poorly executed maneuver. This, however, seemed more plausible than the TAO's assumption that a hostile aircraft with a broken radar would stop to circle. The captain accepted the assessment that the aircraft was probably friendly. As the TAO noted, "The captain was right."

Planning against weaknesses in the story. Although the captain believed the aircraft was friendly, he knew that he might be wrong. Therefore, he planned against this possibility by continuing to monitor the aircraft's behavior and readying relevant weapons systems.

The second and third sections of the training delve into specific aspects of the STEP cycle. In particular, the second section looks at a particular kind of story for hostile intent. The third section, on critiquing, discusses methods for helping fit discrepant observations into a story for generating alternative assessments and for making plans more robust.

Segment 2: A Hostile-Intent Story Template

Stories based on the assessment of hostile intent take on special importance when one's own ship is being approached by a contact whose purpose is unclear. In these situations, there is a consistent set of issues that experienced decision makers tend to consider. Figure 3 provides a causal structure, or template, for a hostile-intent story that incorporates those issues.

The central element in this structure is the current intent of the enemy: to attack with a particular asset (or assets) against a particular tar-

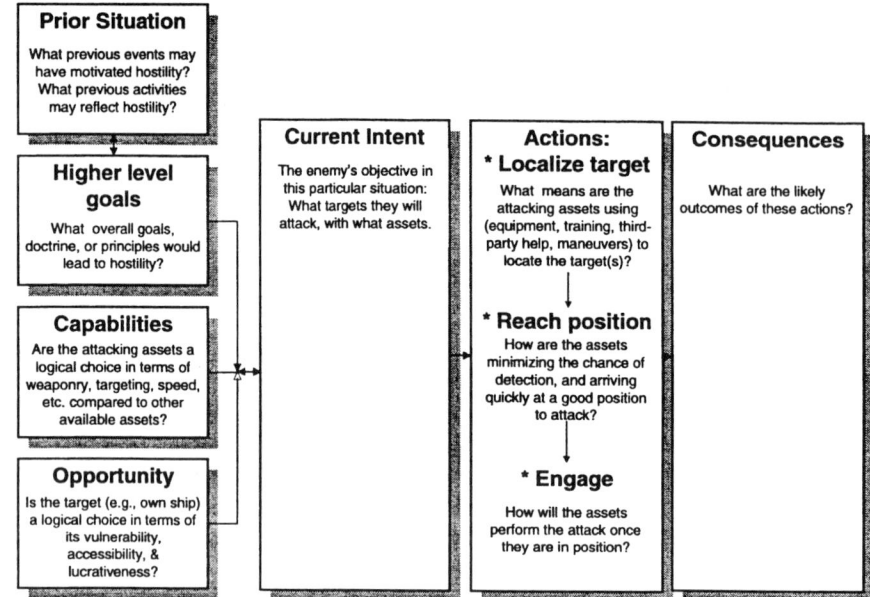

Prior Situation

What previous events may
have motivated hostility?
What previous activities
may reflect hostility?

**Higher level
goals**

What overall goals,
doctrine, or principles would
lead to hostility?

Capabilities

Are the attacking assets a
logical choice in terms of
weaponry, targeting, speed,
etc. compared to other
available assets?

Opportunity

Is the target (e.g., own ship)
a logical choice in terms of
its vulnerability,
accessibility, &
lucrativeness?

Current Intent

The enemy's objective in
this particular situation:
What targets they will
attack, with what assets.

**Actions:
* Localize target**

What means are the
attacking assets using
(equipment, training, third-
party help, maneuvers) to
locate the target(s)?

* Reach position

How are the assets
minimizing the chance of
detection, and arriving
quickly at a good position
to attack?

* Engage

How will the assets
perform the attack once
they are in position?

Consequences

What are the likely
outcomes of these actions?

Figure 3. Hostile-intent story template. From "Meta-Recognition in Time-Stressed Decision Making: Recognizing, Critiquing, and Correcting," by M. S. Cohen, J. T. Freeman, and S. Wolf, 1996, *Human Factors, 38*, p. 208. Copyright 1996 by Human Factors and Ergonomics Society. Reprinted with permission.

get (or targets). The left side of the structure represents prior causes of the intent and the right side represents the effects or consequences of the intent in the current situation. The point of telling a story is not simply to fill the slots. It is to try to *make sense of*, or *argue for*, the hostile-intent assessment from the vantage point of each of these causes and effects.

The causal factors at the left in Figure 3 (goals, opportunity, and capability) make up what might be called "the big picture," and they often shade the way kinematic cues are interpreted. Interviews suggest that more experienced officers try to create a complete story about hostile intent, incorporating all these factors. Less experienced officers tend to be more myopic: They often focus only on past and present kinematics, that is, on the speed, altitude, range, and heading of a track, rather than on the larger context and future predictions.

The training employs examples from the interviews and simulated scenarios for practice with each component of the hostile-intent story template.

Higher level goals. In what way is the country owning the platform motivated to attack a U.S. ship? Evidence that a country's goal is to attack U.S. ships can take different forms that vary in strength, ranging from prior incidents of actually engaging U.S. ships, to intelligence regarding an incipient attack, to public threats and increased tensions. In one practice scenario, for example, trainees usually decide not to engage rapidly approaching F-4s from Iran, on the basis of their understanding of the rules of engagement (ROE). Later in the same scenario, they are fired on by an Iranian boat. Still later, they are again rapidly approached by a similar group of F-4s. This time, most choose to engage the F-4s, even though the literal application of the ROE to this situation is the same as before (when they chose not to shoot). When such overt evidence is lacking, officers often rely on prior intelligence regarding a planned attack. In several incidents, officers cited the lack of such prior intelligence as a key factor in causing them to doubt the hostile intent of a contact.

Opportunity. How is ownship a logical target for attack given the country's goals and other potential targets that it could have chosen? Opportunity can be a subtle cue regarding intent. In one practice scenario, an air contact on its way toward ownship passes a U.S. command and control ship. The latter is at least as lucrative a target as ownship and is more accessible to attack. The failure to select the most favorable target does not disprove hostile intent but argues against it. The failure may have a good explanation, for example, the contact did not detect or have intelligence regarding the command and control ship, or there may be some unsuspected reaon that one's own ship is a more desirable target. A complete hostile-intent story must include some such explanation, which must then be tested or judged for plausibility. Conversely, the presence of a lucrative target such as a flagship provides support for, but does not prove, the assessment of hostile intent.

Capability. How is the track a logical choice as an attack platform given the goals and available assets of the attacking country? In a number of incidents, officers puzzled over the employment of less capable platforms, such as a gunboat, helicopter, or light aircraft, against a U.S. AEGIS cruiser. Again, this argues against hostile intent but does not disprove it: The conflict may have a plausible explanation, such as unwillingness to sacrifice expensive resources or willingness to conduct a kamikaze raid.

Localization. How did the attacking platform detect ownship, or how is it attempting to do so? Localization is a surprisingly frequent concern among experienced officers, and it heavily influences the reading of more standard kinematic cues. For example, officers tend to regard a contact that emerges from a hostile nation, turns toward ownship, and speeds up as hostile. But if this contact was too far away to have detected ownship, these cues must mean something else. The hostile-intent story can be patched up, perhaps by assuming third-party targeting support or improved equipment or training. These explanations can then be tested or evaluated. Conversely, in other incidents, when a track appeared too slow or too high to fit a hostile profile, its behavior was sometimes reconciled with a hostile-intent story with the assumption of a need to localize a target.

Generalizing the story concept. Stories contain characteristic events associated with an intent. This idea can be generalized to assessments other than hostile intent. In a form of discovery learning (Collins & Stevens, 1983), instructors ask trainees to imagine that one of the aircraft in a practice scenario is on a search-and-rescue mission. Trainees are then asked to tell a story around that assessment, which includes past, present, and future events. As the events volunteered by trainees are recorded on a whiteboard and causal arrows drawn between them, a set of typical components and relationships emerges. Stories typically specify goals, opportunities to achieve goals, methods or capabilities for achieving the goals, actions preparatory to achieving the goal, and execution of the intended action.

For example, the components of the search-and-rescue story produced by one class included the following elements. (a) Opportunity: There is something in the water to be rescued. (b) Goals: Rescue is not overriden by risks such as ongoing combat. (c) Capability: The organization and capabilities are present to mount a search-and-rescue operation in the time available (i.e., the rescue is not so fast as to seem staged). The most appropriate available platform for search and rescue is chosen. (d) Arriving in position: The platform's speed, altitude, and flight pattern are appropriate for search. The contact will indicate by radio response that its mission is search and rescue. (e) Execution: The contact is engaging in, or will engage in, actions appropriate for rescue.

A similar story can be built around any assessment of intent, for example, a lost friendly aircraft, harassment, provocation, or attack, each of which has its own set of characteristic components.

As the STEP process implies, stories are not fixed patterns or checklists associated with particular intents. Stories are constructed when no pattern fits all the observations. A story template is filled in differently each time it is used, for example, as the decision maker tries to explain the conflict. The uniqueness of stories makes the evaluation step crucial. No single element is conclusive by itself; the decision maker must look at the whole story. Hostile intent is supported when the available information fits easily within the hostile-intent template or when the assumptions required to make it fit are tested and confirmed. The information weighs against hostile intent if a large number of unusual and untested assumptions are needed to make it fit.

Segment 3: Critiquing Stories

Overcoming overconfidence. Once a story has been constructed, the decision maker evaluates its plausibility. The simplest approach to evaluation is direct; one may ask, for example, "How confident am I that this platform intends to attack my ship?" Decision theory requires that answers be given in the form of numerical probabilities. There is evidence, however, that this approach can be seriously misleading. People, including experts, tend to be overconfident when they provide direct probabilistic estimates of confidence (Lichtenstein, Fischhoff, & Phillips, 1982). Overconfidence, in turn, can cut thinking short before key issues have been explored, and it may be one reason for unfortunate surprises or overhasty decisions.

A more useful approach to the estimation of confidence does not presuppose that the decision maker is already aware of all the factors contributing to uncertainty. In the R/M model, confidence is assessed through active ferreting out of the important assumptions underlying an assessment. In the remainder of this section, we describe a powerful strategy for identifying assumptions called the *crystal ball technique*. Subsequently, we describe the role it can play in various phases of the STEP cycle.

The crystal ball technique. An argument may appear plausible at first blush but have weaknesses that are not immediately recognized. Such weaknesses may be revealed by deliberate critiquing. A powerful strategy for exposing weaknesses is for decision makers to *imagine* that the arguments for a conclusion are false—that the observations are true but the conclusion is wrong—and search for an explanation. Each explanation generated in this way reflects an assumption underlying the original argument; that argument is valid only if no such explanation is true.

This segment of the training introduces a devil's advocate method called the crystal ball technique, which consists of four steps:

1. Select an important conclusion, such as the intent of a contact, or the belief that a track cannot localize ownship at its present distance.
2. Imagine that an infallible crystal ball (or, equivalently, a perfect intelligence source) tells you that this conclusion is wrong—despite the observations (or reports or analyses) that suggested it was true.
3. Explain how this could happen, that is, how the conclusion could be wrong despite the evidence supporting it. How does this change the way the evidence is interpreted?
4. The crystal ball now tells you that this explanation is wrong and sends you back to Step 3, to devise another possible explanation of the evidence. (Continue until the set of exceptions to your original conclusion seems representative of the ways the conclusion could be wrong.)

Testing an assessment. In the test phase of STEP, the decision maker identifies arguments *against* an assessment and then uses the crystal ball technique to probe these arguments critically. As a result, decision makers identify assumptions that would produce a coherent story despite apparently conflicting observations.

The assessment may be the decision maker's favored hypothesis, or it might be an alternative hypothesis that happens to be under consideration. In both cases, the crystal ball forces decision makers to explain how the assessment can be true despite conflicting evidence. In the former case, the crystal ball can prevent decision makers from giving up a favored hypothesis prematurely. In the latter case, the crystal ball forces the decision maker to consider seriously how an alternative assessment might account for evidence that appeared to support the favored hypothesis. In both cases, the crystal ball helps the decision maker go beyond the initial, recognized interpretation of evidence to find possible alternative meanings. In both cases, the assessment can then be evaluated if one looks at the assumptions required to save it.

Consider, as an example, an incident in which an officer tried to patch up a hostile-intent story. An aircraft was approaching an AEGIS cruiser on a straight course from the direction of Iran at slow speed and low altitude. The aircraft was not emitting or responding to identification-friend-or-foe (IFF) challenges. Every one of these observations—flying toward U.S. ships, emerging from Iran, low altitude, failure to respond to IFF—suggested an attack, except for slow speed.

The anti-air warfare coordinator (AAWC) had a possible explanation for some of these observations. Friendlies might appear to originate from Iran and fail to respond to IFF challenges if they had turned off their IFF transponders as a precaution before flying near hostile territory and then forgot to turn them back on coming out. (In some cases, deliberate critiquing of an argument is not necessary. Conditions under which the argument fails may be so familiar that they are recognized virtually at the same time as the argument presents itself.)

The AAWC, however, continued to take seriously the possibility that the aircraft was hostile. To create a coherent hostile-intent story, it was necessary to find an explanation of its slow speed. The officer forced himself to imagine that the aircraft was hostile despite its slow speed. This led to the consideration and rejection of a series of possible explanations:

CANDIDATE ASSUMPTION 1. Slow speed could be consistent with hostile intent if the aircraft was trying to locate its target. This explanation could be tested. If the aircraft were trying to locate a target visually, it would be flying a search pattern rather than a straight course, but it was not ("He's not scanning visually for anybody because that's a straight line").

CANDIDATE ASSUMPTION 2. Because the preceding explanation was disconfirmed, the AAWC forced himself to generate another one. The aircraft might be flying toward ownship if it had prior intelligence on shore regarding the location of the target ("I wonder what their intelligence capability is? . . . Do they know where I am?"). However, this prediction was also at least partly disconfirmed, because the aircraft was not flying directly toward own ship.

CANDIDATE ASSUMPTION 3. The AAWC then forced himself to find still another explanation: Perhaps the aircraft was trying to locate a target electronically. In that case, however, its radar would be emitting detectable signals—which it was not.

CANDIDATE ASSUMPTION 4. The only explanation the AAWC could think of now was that the aircraft planned to shoot blind! ("I can't imagine him shooting blind. They could, though—they did it once, shot something off and hoped it hit something. . . . So you're thinking, well, he probably won't shoot, but he might.") This explanation was a last-ditch effort to save the assumption that the aircraft was hostile. It could not be directly tested, but it was judged implausible.

Because he had failed to construct a plausible hostile-intent story, despite heroic efforts, the AAWC delayed engagement of the approaching contact. Just in case it was hostile, however, he warned it with fire control radar. The aircraft immediately turned on its IFF transponder and squawked a friendly response.

Evaluating an assessment. In the evaluation phase of STEP, decision makers step back to ask if a story makes sense. In doing so, they judge the plausibility of the assumptions required by the story. In this phase, they may also generate alternative assessments and begin a new STEP cycle. In subsequent evaluation phases, they can compare the assumptions they must accept if the different assessments are true.

The crystal ball technique can be used to probe the arguments *for* a favored assessment. The crystal ball says that a hypothesis other than the favored one can account for the evidence, and it forces decision makers to generate such hypotheses. Koriat, Lichtenstein, and Fischhoff (1980) found that overconfidence was reduced by forcing participants to search for reasons that they might be wrong.

We employ an actual incident as a classroom exercise, in which students use the crystal ball technique both to flesh out a hostile-intent story and to discover possible nonhostile intents. An AEGIS cruiser, escorting a flagship through the Strait of Hormuz, detected two Iranian fighters taking off from an Iranian air base and identified them as F-4s from a brief radar transmission. Typically, these aircraft would be expected to head north or south along the Iranian coast for a routine patrol. In this case, however, the planes began circling the airport. These circles gradually widened until the aircraft were coming within their weapons range of the cruiser. While the aircraft were circling, their search radar was kept on continuously. As the circles widened, the aircraft switched to fire control radar, locking on the cruiser during the portion of the orbit when they were pointed toward it. At the point in each orbit when the lock was broken, the pilot of one of the aircraft switched back to search mode.

The observation of circling aircraft in broad daylight leads to the recognitional conclusion that the intent is not hostile. In a normal attack, according to the captain, the F-4s would "come screaming at me" fast and low. However, here they were, "in broad daylight; they know we're here, we know they're there."

Nevertheless, this behavior did not fit the pattern of a routine patrol. The captain therefore tried to put together a hostile-intent story. In fact, a disturbing number of elements for such a story were present. There was appropriate motivation—Iranian hostility toward the United States—and appropriate capability—F-4s armed with antiship missiles. There was also appropriate opportunity: The presence of the flagship "obviously heightened our interest," according to the captain. Localization could be explained by the deliberate use of search radar. (Turning on this radar was noteworthy, the captain said, because, "the Iranians did not have the maintenance capability to fly their electronics [routinely] and burn them steadily.") Gradually widening circles and locking on put the aircraft in position to engage. The captain noted that by illuminating own ship, the aircraft had already satisfied the ROE.

As a class exercise, the instructor tells the trainees that an infallible crystal ball has determined that the aircraft are hostile, despite the observation that they are circling in broad daylight, and the trainees are asked to explain how this could be. This exercise typically elicits a number of potential explanations of the circling; for example, the aircraft might be planning to fire a standoff weapon. The crystal ball now says that the aircraft have hostile intent but that they do not intend to fire a standoff weapon. Other suggestions are now forthcoming; for example, the aircraft may intend to divert the cruiser from attack by other aircraft or surface vessels. This process, repeated several times, elicits still more suggestions. For example, the aircraft may be waiting to rendezvous with other aircraft for a concerted attack. The aircraft may be waiting in order to synchronize their activity with that of other aircraft. They may be targeting for other aircraft. Their radar may work better at high altitude. They may be having rudder or communications problems—and so on.

Knowledge of these possibilities provides the captain with a basis for evaluating the plausibility of the hostile-intent assessment. If the captain believes the aircraft to be hostile, he must be prepared to assume that at least one of the explanations (or some similar one) is the case.

Next, the crystal ball declares that the aircraft are not hostile and asks for explanations of the circling pattern and the emissions. A variety of explanations are forthcoming here as well: For example, the aircraft might be circling as part of a test of the radar: their radio may be broken and they may be signaling the airport; they may be testing U.S. ROE; they may be harassing U.S. ships; and so on. If the captain believes the aircraft are not hostile, he must be willing to accept one of these explanations, or some similar one. In this incident, the captain concluded that the intent of the aircraft was to harass rather than to attack his cruiser.

Planning against weaknesses in an assessment. The crystal ball exercise, although it takes only a few minutes, can have practical consequences. Many of the explanations can be tested. In the example of the circling F-4s, explanations implying diversion or coordination with other aircraft or ships in the area may lead to heightened vigilance, and they may be either confirmed or disconfirmed by observations. Other explanations (e.g., those regarding the weapons of the aircraft or their radar characteristics) may be confirmed or disconfirmed by intelligence. Some explanations may be dismissed as implausible. Some can provide the basis for contingency planning in case they turn out to be true. In this incident, the captain developed a contingency plan for the possibility of attack by ordering internal ship defensive systems to a high state of readiness.

Generalization of critiquing skills. The most important result of the crystal ball exercise is a heightened appreciation of the fallibility of judgment. Trainees are typically surprised at the number of ways an apparently compelling argument can fail (e.g., the argument that circling aircraft do not intend to attack). After about 30 minutes of practice with different scenarios, trainees start considering exceptions to conclusions spontaneously, without explicitly invoking the device of a crystal ball.

The crystal ball should reduce overconfidence, but its purpose is not to undermine confidence. Rather, it teaches decision makers that all assessments require assumptions of some sort. The existence of alternative interpretations or of conflicting evidence is not proof that an assessment is wrong. Encouraging decision makers to compare the assumptions (or stories) implied by different assessments helps them to select the best overall account. After the crystal ball exercise, officers may believe their original conclusion even more strongly than before, or they may change their minds. However, the best way to earn confidence in an assessment is to take seriously the possibility that it is wrong.

The crystal ball method can be used to uncover hidden assumptions in plans as well as in stories. In this case, the crystal ball attacks the connection between actions and goals instead of the connection between evidence and conclusions. It says that the planned actions will be attempted but the goals of the operation will not be achieved, and it demands an explanation. This can lead to a greater understanding of the weaknesses in the current plan. The result may be a changed plan, an elaborated set of contingencies, acceptance of risk, or greater confidence in the existing plan.

Segment 4: When to Think More

Critical thinking is not always appropriate, yet decision makers in combat cannot afford to spend valuable time thinking about whether to think. What we call the *quick test* is used to decide rapidly and without excessive overhead when to critique and improve an assessment and when to go ahead and act on it. The quick test requires a balance among the costs of delay, the costs of error if one acts without further critical thinking, and the degree to which the situation is either unfamiliar or problematic (Cohen, Parasuraman, et al., 1997). Experienced officers seem to differ from less experienced officers in the way they approach each of these judgments.

Costs of delay. Less experienced officers typically base judgments of available time on the enemy's doctrinal weapons range. For these officers, the cost of delay becomes excessive as soon as ownship is within range of the contact's weapons, at which time these officers stop thinking and are ready to shoot. By contrast, more experienced officers do not settle for stereotypical or doctrinal estimates of weapons range. They buy more time for decision making by considering factors that are specific to this enemy and to this situation. For example, they consider history (e.g., the ranges at which hostiles have in fact launched in past training or combat, rather than how far the manual *says* they can shoot) or visibility conditions at the time (e.g., whether it is a moonless night). They may also consider actions the enemy must take before launching (such as changing altitude or communicating) that will tip them off that an attack is imminent. At the same time, they may develop trip wires and contingency plans to make own ship's response to an attack as rapid as possible.

In sum, experienced officers explicitly ask themselves, "How much time do I have before I must act?" In addition, they buy time for decision making by estimating available time more precisely and planning more carefully than less experienced officers.

Stakes of an error. Less experienced officers tend to focus on immediate goals, such as survival of ownship and destruction of hostiles. More experienced officers give more consideration to higher level and longer term stakes. For example, they place more weight on avoidance of an international incident or other organizational objectives, including damage to their own career.

Situation typicality. An unidentified aircraft heading toward ownship at high speed (i.e., the expected kinematics of an attack) gets the attention of a junior officer just as quickly as it gets the attention of a senior officer. Less experienced officers also learn patterns such as commercial air schedules, corridors, and speeds, or routine patrol routes, speeds, and altitudes, and they are good at detecting behavior that fits or does not fit such patterns.

More experienced officers, by contrast, are more sensitive to the possibility that a situation may not fit *any* pattern perfectly. They are less likely to conclude that a contact intends to attack simply because it fails to match a routine patrol checklist and does match a few aspects of an attack pattern (e.g., heading and speed). They are quicker to notice that it *also* fails to match the attack pattern perfectly (e.g., the platform is unable to localize ownship from that distance; this is not a logical target for that platform). In sum, more experienced decision makers realize that a situation is ambiguous in cases in which less experienced decision makers do not.

Experimental Tests

Critical thinking training has now been tested at two Navy training facilities. Results of the two studies are described together. However, some important differences between them are summarized in Table 1. Study 1 was conducted at the Surface Warfare Officers School (SWOS), in Newport, Rhode Island, and Study 2 was conducted at the Naval Post Graduate School (NPGS), in Monterey, California. The two studies represent a trade-

Table 1.   Differences Between Study 1 and Study 2

Feature	Study 1	Study 2
Location	Surface Warfare Officers School, Newport, RI	Naval Post Graduate School, Monterey, CA
Number of participants	$N = 60$	$N = 35$
Experience of participants	All active-duty Navy	Half in Navy, half in other services
Design	Trained (40) versus control (20) × pretest–posttest	Pretest–posttest (no control group)
Duration of training	90 minutes	4 hours
Practice tools	Paper and pencil (although testing involved simulation)	Simulation (DEFTT)
Scheduling	All training and testing in a single 8-hour session	Training and testing in 2-hour sessions over 5 days

Note. DEFTT = Decision Making Evaluation Facility for Tactical Teams.

off between the number and experience level of participants (Study 1 was better) and the duration and quality of training (Study 2 was better).

In both studies, training focused on decision-making skills of individual officers, although the officers received the training in classes of five or six.

Method

Design. Both Study 1 and Study 2 involved a pretest–treatment–posttest design, and both studies used two scenarios, which were counterbalanced between pretest and posttest across groups. In addition, Study 1 varied the treatment condition (training vs. control) across groups. Study 2 did not have a control group.

Participants. Sixty officers at SWOS participated in Study 1 (40 in the training condition, 20 as controls). All were in the Navy, with an average of 9.5 years of military service. These officers were being trained to serve as department heads in engineering, operations, or weapons. Ninety-two percent had performed shipboard duty in the combat information center (CIC), and 32% had served as TAO.

At NPGS, 35 officers took part in Study 2. Like those at SWOS, these officers averaged 9.5 years of military service. However, only 51% were in the Navy, whereas 14% were Marines, 29% were in the Army, and 6% were in the Air Force. Forty-six percent of the officers at NPGS had worked in the CIC or in similar tactical positions in the Marines, Air Force, or Army. Only 14% of officers at NPS had served specifically as TAO.

Materials. The critical thinking training materials used in these experiments included a training text, brief explanatory lectures, discussions, and exercises. In Study 1, all practice scenarios were presented for class discussion by the instructor. In Study 2, four practice scenarios were simulated on the Decision Making Evaluation Facility for Tactical Teams (DEFTT) training system (see chapter 3, this volume). These scenarios were modified to make critical thinking more appropriate, that is, to reduce the total number of tracks while increasing uncertainty about the intent of some of the remaining tracks. In most cases, these modifications replicated the situations that had been described in critical incident interviews. Participants performed in these scenarios individually, acting as TAOs. Feedback from the instructor was provided in real time as the scenario unfolded and, after it concluded, in group discussion.

The pretest and posttest scenarios in both studies were DEFTT simulations. These simulations began with oral and written briefings concerning the geopolitical context of the scenario and the military forces involved. Each participant then turned to a personal computer that simulated a command and display (C&D) station. The C&D presents symbology concerning the identity, speed, bearing, and range of air and surface tracks, as well as textual details concerning these characteristics and the track's response to electronic interrogation (IFF). Virtually all tracks in these scenarios except ownship and accompanying surface craft were symbolically marked as unknown (rather than as friend or foe) and were unresponsive to electronic interrogation. Internal and external audio communications were simulated. Most communications concerned the location of tracks, their presumed identity (e.g., F-4, Mirage), and electronic warfare (EW) data received from the tracks (such as search radar or fire control radar emissions). In Study 2 these communications were edited so that they could be understood by non-Navy officers. In Study 1 EW data were also provided on a large display in the center of the classroom.

Participants performed scenarios in groups of five or six, but each participant worked independently. Participants could not consult with their classmates during tests or take any actions that would alter events in the scenario for themselves and their classmates. Specifically, they could not maneuver ownship, fire weapons, interrogate tracks, or initiate communications during scenarios, although they could indicate their intent to perform such actions in response to questions during test breaks.

Procedure. For participants in Study 1, the experiment began with discussion and practice to refamiliarize them with the DEFTT. For participants in Study 2, who were generally less experienced, the experiment began with a presentation concerning the function of the combat information center (CIC), the role of the tactical action officer (TAO), and the operation of the DEFTT simulator.

The pretest and posttest scenarios were each paused at three points. During each break, participants received a test booklet consisting of the following five instructions:

1. Assess the intent of a single [experimenter-designated] track and defend that assessment.
2. Generate possible alternative intent assessments and estimate confidence in each.
3. Select an assessment of the designated track that you do not agree with and defend it.
4. Identify evidence that conflicted with an [experimenter-designated] intent assessment and then defend the assessment.
5. Describe actions you would take at this time in the scenario.

Participants did not know which track would be the focus of attention or which intent assessment would be designated for consideration until after the relevant segment of the scenario was completed and the break began.

These test questions specifically described the desired behavior. For example, participants were asked to identify and explain conflicting evidence and to generate alternative assessments, rather than simply to evaluate a hypothesis. The intent was to mitigate any effects of experimenter demand on the results. The expected behavior should be as clear on the pretest (and in the control group) as on the posttraining test. Any differences in performance should be attributable to the effects of training on the relevant metarecognitional skill.

Measures based on these questions covered every phase of the STEP process: stories—the number and variety of issues considered when evaluating an assessment in Instructions 1 and 3; test—the amount of conflicting evidence that a participant identified and the number of explanations generated to patch up stories in Instruction 4; evaluate—the number of alternative assessments generated in Instruction 2 and the accuracy of the assessment favored by the participant in Instruction 1; and plan—the frequency with which contingency plans were created in Instruction 5.

Following the pretest, participants received training, except for members of the Study 1 control group. The latter completed a psychological battery, listened to a lecture on problem solving and knowledge representation strategies, and discussed challenging problems in their jobs as weapons officers, engineers, or operations officers. They appeared to find the control condition enjoyable and interesting. After the training or control treatment, participants executed the posttest. In addition, all participants completed a biographical survey concerning military experience, and all except controls evaluated the training.

As noted previously, the experiments differed in the duration and realism of training. In Study 1, the pretest, training, and posttest occurred in a single day, with only 90 minutes for the training itself. By the time of the posttest, signs of fatigue were evident. In Study 2, on the other hand, events were broken into five 2-hour sessions over 2 weeks. (The introduction to the CIC and DEFTT occupied the first session; the pretest occurred in the next session; two training sessions followed; and the posttest was administered in the final session.) Training in Study 2 used DEFTT scenarios and was therefore both more realistic and more similar to test conditions than in Study 1.

Table 2.   Summary of the Effects of Training in Studies 1 and 2 (Averaged Across Test Scenarios)

Variable	Study 1	Study 2
Number of issues considered in assessment	+7%	+30%
Number of conflicting pieces of evidence identified	+52%	+58%
Number of explanations of conflict generated	+26%	+27%
Number of alternative assessments generated	+10%	+41%
Accuracy of assessment (agreement with SME)	+42%	+18%
Consensus on assessment	+14%	+41%
Confidence in assessment	+13%	+20%
Frequency of contingency planning	+217%	—
Subjective evaluations of training	73% positive	71% positive

Note. For Study 1, numbers represent the performance of trained participants as a percentage of performance by untrained participants. These performance measures are corrected by covariate analysis with pretest scores. For Study 2, numbers represent the percentage change in performance on the posttest compared with the pretest. For both studies, a plus sign indicates an improvement associated with training. Dash indicates the variable was not analyzed. SME = subject-matter expert.

Results

Successful critical thinking training should have an impact both on decision processes and, through such processes, on the accuracy of decisions. Table 2 summarizes the main results from both studies. In Study 1, in which training was shorter and the posttest came at the end of a long day, we found either trends or significant effects on all of the critical thinking skills. In Study 2, training had a significant effect on all the critical thinking skills. For a more complete description of the results for Study 1, see Cohen, Freeman, Wolf, and Militellino (1995).

Effects of Training on Decision Processes

FILLING OUT STORIES. One of the objectives of training with story templates is to increase the scope of the factors that officers consider when evaluating an assessment. The factors considered should go beyond present and past kinematics of the track to include goals, opportunity, capability, localization, deceptive aspects of the approach (i.e., alternative interpretations of kinematics), and predictions regarding future kinematics. This analysis pertains to Instructions 1 and 3, in which officers were asked to defend an assessment they favored and an assessment they did not favor, respectively.

In Study 1, there was a trend for training to increase the number of factors considered in an assessment of intent. The number of arguments generated by trained participants was 6.5% greater for favored hypotheses (Instruction 1) and 8.6% greater for disfavored hypotheses (Instruction 3), compared to untrained participants (for Instructions 1 and 3 combined, $F(1, 47) = 2.95$, $p = .092$).

These effects became much larger in the more favorable training and testing regime of Study 2. Training increased the number of arguments participants presented in defense of their favored assessments (Instruction 1) by 22%. The number of arguments grew from a mean of 5.12 per break on the pretest to a mean of 6.26 on the posttest, $t(33) = 3.81$, $p = .001$. The percentage effect was larger when trained participants defended assessments they did not agree with (Instruction 3): an increase of 43% from a mean of 3.01 per break on the pretest to a mean of 4.31 on the posttest, $t(33) = 3.81$, $p = .001$.

Did training simply increase the quantity of issues considered, or did it also influence the type and variety of issues that officers thought about? We analyzed the distribution of arguments across story components in Study 2. Training reduced the percentage of arguments that reflected present and past kinematics from 61% on the pretest to 51% on the posttest, and it correspondingly increased the percentage of arguments reflecting other factors, $\chi^2(1) = 10.82$, $p = .001$. Taking these other factors separately, we found an increase owing to training in every nonkinematics story component: goals, capabilities, opportunity, localization, deceptive features of the track's approach, and predicted future actions.

As the quantity of arguments increased, did quality decline? If so, training might simply lower the threshold for reporting or thinking about an issue, rather than expanding the scope of understanding. This, however, was not the case. A subject-matter expert (the retired naval officer who designed the test scenarios) did a blind rating of the relevance and impact of each argument provided by each participant in Study 2. There was no effect of training on the average quality of arguments. The mean quality of arguments was 5.3 on the pretest and 5.4 on the posttest (on a 10-point scale).

IDENTIFYING CONFLICTING EVIDENCE. Another important objective of training was to improve an officer's ability to identify evidence that conflicts with an assessment. Instruction 4 specifically asked participants to list evidence that conflicted with an assessment designated by the experimenter.

In Study 1, participants who received training identified 52% more items of conflicting evidence than did controls. Trained participants identified an average of 1.4 items per break, whereas controls identified 0.9 items, $F(1, 55) = 6.24$, $p = .015$. Training increased the amount of conflicting evidence identified whether or not these participants happened to agree with the experimenter-designated assessment (agreement was determined by comparing the designated assessment with the assessment favored by the participant in Instruction 1). In Study 2, training boosted the amount of conflicting evidence identified by 58%, from an average of 1.6 items on the pretest to 2.6 items on the posttest, $t(32) = 5.48$, $p < .001$.

EXPLAINING CONFLICTING EVIDENCE. In ambiguous and complex situations, almost any assessment conflicts with some of the evidence. Yet one assessment, however implausible it may seem, must turn out to be true. To discard an assessment simply because there is evidence that conflicts with it, then, would mean rejecting all assessments and never finding the truth. More constructively, conflict can be taken as a cue to think more deeply about assumptions underlying one's interpretation of the evidence. Conflicting evidence may point to an exceptional circumstance that explains the conflict. Instruction 4 asked officers not only to identify evidence that conflicted with an experimenter-designated assessment, but also to defend the assessment by generating explanations of the conflicting cues.

In Study 1 trained officers generated 70% more explanations (0.68 explanations per posttest break) than controls (0.40 explanations per posttest break). However, variation between test scenarios was quite large, so that this positive pattern was not statistically reliable. In Study 2, however, training boosted the number of explanations significantly, by 27%, from 2.57 on the pretest to 3.25 on the posttest, $t(32) = 4.92, p < .001$.

GENERATING ALTERNATIVE ASSESSMENTS. Generating alternative assessments is an important part of evaluating a story. By suggesting alternative interpretations of observations, one exposes hidden assumptions in the current story, which can be tested or judged for plausibility. In some cases, an alternative assessment may be found that is better than the current hypothesis.

In Study 1, there was a trend for trained participants to generate more alternative assessments than controls did. Officers who received training generated 9% more assessments per break (3.6, on average) than controls (3.3), $t(59) = 1.5, p = .140$.

The effect of the more extensive training in Study 2 was highly reliable. The number of alternative assessments generated on a given break rose 41%, from 2.69 on the pretest to 3.79 after training, $t(34) = 5.88$, $p < .001$. This increase in quantity was not accompanied by a decrease in quality. The subject-matter expert's blind rating of the plausibility of assessments fell a nonsignificant 3% between pretest and posttest, $t(34) = 0.57, p = .574$.

CONTINGENCY PLANNING. The final phase of the STEP process is to plan against the possibility that one's favored assessment is wrong. Such planning is a way of buying time for critical thinking or for collecting more data.

In Study 1 actual engagements were rare among both trained and control participants. However, trained participants were twice as likely as controls to identify explicit contingencies or trip wires for engagement. An average of 6% of the control participants developed contingency plans for engagement on each break, compared to 13% of the trained participants, $F(1, 57) = 8.36; p = .005$. (Planning was not analyzed for Study 2.)

CONFIDENCE IN ASSESSMENTS. The training successfully teaches officers to question assumptions, notice conflicting evidence, and generate alternative assessments. It is natural to worry that such training might diminish officers' confidence in their assessments of enemy intent and their decisiveness in taking action. However, this is only a surface view of what the training is designed to accomplish. First, the officers are taught that critical thinking is appropriate only under special circumstances, when time, stakes, and uncertainty warrant it. Second, they can stop at any time and act on their current best assessment. Third, and on a deeper level, the training gives officers a better understanding of the reasons for confidence in an assessment. Trainees are taught that even though no story is perfect, some story, however imperfect, will turn out to be true. Hence, training emphasizes the importance of evaluating and selecting among stories, and it shows trainees how this can be done. Exploring the assumptions underlying assessments should lead to the conclusion that the assessment ultimately chosen, although imperfect, is the best available.

As a metric of confidence, we took the difference between confidence ratings for the two assessments in which a participant expressed the most confidence on Instruction 2. This reflected the participant's ability to discriminate between the preferred assessment and the second best. In Study 1, confidence ratings rose 12.5% with training. Although it was not a statistically reliable increase, the result indicates at the least that training did not *lower* confidence. Moreover, in Study 2 confidence ratings rose 20% from pretest to posttest, a marginally significant result, $t(33) = 1.99$, $p = .055$.

Effects of Training on Decision Quality

The preceding findings demonstrate that training based on the R/M model alters the ability of officers to generate, defend, and rebut assessments. However, these findings do not speak to the ultimate outcome: Do these critical thinking processes increase the accuracy of situation understanding and enhance the success of actions guided by that understanding?

As a first step in this analysis, we examined whether training changed the types of assessments officers generated. The assessment in which each trainee was most confident was categorized as either *hostile*, *not hostile*, or *unknown*. For officers in both studies, training reliably affected the category of assessment that participants preferred, broken down by scenario and test break: Study 1, $\chi^2(8) = 24.17$, $p = .002$; Study 2, $\chi^2(6) = 24.05$, $p = .001$.

We evaluated the quality of assessments by (a) comparing them with the assessments of a subject-matter expert (SME) and (b) measuring consensus among participants. In addition, we examined whether the actions officers proposed changed with their assessments.

ACCURACY OF ASSESSMENTS. The standard of accuracy was the assessment of tracks at each break by the retired senior Navy officer who designed the scenarios. In both experiments, training produced large improvements in accuracy on one of the two test scenarios, but no change in either direction in the other. We focus on the scenario in each experiment that elicited effects. We found that 77% of the trained officers in Study 1 were in agreement with the assessments of the SME, compared with only 43% of controls, an improvement of 79%, $\chi^2(2) = 6.34$, $p = .013$. Among officers in Study 2, training boosted agreement with the SME by 35%— from 60% on the pretest to 81% on the posttest (although the increase was significant at Break 1 only, $\chi^2(2) = 6.79$, $p = .034$).

CONSENSUS. An alternative index of accuracy is the level of consensus among participants regarding their assessments. Training that improves accuracy should raise consensus among participants as they converge on a common interpretation of events. We used as a measure of consensus a metric from information theory called "average uncertainty" (Garner, 1962), which is defined in this way:

$$U = -\Sigma p(x)\log[p(x)].$$

Here, $p(x)$ is the relative frequency with which members of the group picked Hypothesis x. U (uncertainty) is zero when members of a group all agree; it grows larger with disagreement.

Training appeared to increase consensus among trained officers in both studies. Among officers in Study 1, average uncertainty was 14% lower overall with training ($U = 0.91$) than without (1.04). Training lowered average uncertainty 41% among officers in Study 2, from 0.31 to 0.22.

ACTIONS. Assessments of the intent of a track may be expected to influence actions. In the Study 1 scenario that elicited training effects on assessment quality, the intent of the approaching track could have been participation in a search and rescue, or the search-and-rescue operation may have been a cover to close on own ship and attack. Control-group participants were more likely than trained officers to assess tracks as hostile, and they took actions that reflected this, such as vectoring Combat Air Patrol and illuminating the tracks, $F(1, 28) = 2.64$, $p = .081$. Trained participants (and the SME) were more likely to offer assistance in the search and rescue, $F(1, 28) = 3.38$, $p = .077$. In sum, training improved the accuracy of situation assessments, and officers' actions changed accordingly. (Actions were not analyzed in Study 2.)

Subjective Evaluations of Training

The participants in both experiments provided quantitative and qualitative evaluations of the training. During debriefing, they rated the training on a scale from 1 (*strongly negative*) to 5 (*strongly positive*).

Seventy-three percent (29 of 40) of the officers in the trained group of Study 1 gave the instruction a positive rating (4 or 5). There were seven neutral ratings (3), four negatives (2), and no strongly negative ratings (1). The average rating among participants in Study 1 was 3.7. Officers with tactically oriented specialties (weapons and operations) gave the training a higher rating than did engineers or deck officers, $F(1, 38) = 4.06, p = .051$.

In Study 2, 71% (25 of 35) of the participants rated the training positively (4 or 5). Six participants were neutral (3), two were negative (2), and one was strongly negative (1). The average rating by officers in Study 2 was 3.7. Officers with some tactical experience in the Navy or other military services tended to rate the training more positively than officers with no such experience.

Qualitative evaluations of the training were also similar for the two studies. Most participants found the training useful in solving the test problems and anticipated that it would be useful in the field. For example, participants said the training would help "organize what I have been doing previously and take it to another level," "stop me from making assumptions," "reinforce the concept that the obvious answer may not be the correct answer," and "keep tunnel vision to a minimum." Participants mentioned favorably the processes of organizing information in stories and using the devil's advocate to generate alternative interpretations of evidence.

Lessons Learned: Guidelines for Implementing Critical Thinking Training

Lesson 1: Development of Critical Thinking Training Starts With Cognitive Task Analysis

Although the overall strategy of critical thinking training is generalizable across domains (see Lesson 7), the training must be based on substantive content that is specific to a particular domain. Analysis of critical incident interviews is a valuable tool for identifying critical thinking requirements in a domain, in particular, by comparing the performance of decision makers at different experience levels. The same critical incidents can be used for generating demonstration, practice, and test materials for training in that domain.

Lesson 2: Critical Thinking Skills Are Important in Tactical Decision Making

Critical incident interviews provide evidence for the importance of critical thinking skills in proficient tactical decision making. More experienced decision makers differed from less experienced ones along a variety of dimensions. These skills include going beyond pattern matching to create plausible stories for novel situations, noticing conflicts between observations and a conclusion, elaborating a story to explain a conflicting cue rather than simply disregarding or discounting it, sensitivity to implausible assumptions in explaining away too much conflicting data, ability to generate alternative stories, planning against the possibility that the current assessment is wrong, and paying careful attention to the time available for decision making. These critical thinking skills presumably help experienced decision makers handle uncertainty effectively without abandoning the recognitional abilities they have built up.

Lesson 3: Effective Critical Thinking Training Combines Instruction With Realistic Practice

An effective strategy for training these critical thinking skills combines information-based instruction on critical thinking concepts, demonstration of critical thinking processes, and guided practice in realistic problems. Instruction sensitizes trainees to the concepts to be learned and helps them assimilate lessons during practice. Simulation exercises provide an opportunity to demonstrate critical thinking skills and to provide feedback in real time. Such experience is crucial if trainees are to learn to use critical thinking skills in a realistic, time-stressed situation.

This strategy is effective in teaching critical thinking processes. In the second of two studies, training increased the number of factors officers considered in assessing the intent of a track by 30%, increased the amount of conflicting evidence they noticed by 58%, increased the number of assumptions they identified underlying that evidence by 27%, and increased the number of alternative assessments they generated by 41%.

There is evidence that this critical thinking training strategy can improve not only processes but outcomes. Agreement of assessments with those of a subject-matter expert increased significantly in two out of four test scenarios in the two studies, by 79% and 35%, respectively. At the same time, the training tended to increase officers' confidence in their assessments. In addition, most subjective evaluations of the training were positive.

Lesson 4: Critical Thinking Training May Be Taught in Segments

Both instruction and practice in critical thinking can be conveniently divided into segments. The first segment provides an overview of critical thinking processes and introduces the idea of building, testing, and evaluating a story, and planning against its weaknesses. The second segment focuses on particular kinds of stories (such as the hostile-intent story). The third segment provides a devil's advocate strategy for evaluating a story and finding alternative interpretations of observations. A fourth segment provides strategies for deciding when to think critically and when to act immediately.

Lesson 5: Critical Thinking Training Is Reasonably Robust Over Training Conditions

Some benefits of critical thinking training were observed even with a short period of time for training, and with testing and training compressed into a single day (Study 1). However, greater benefits were realized when there was more time for training and the training and testing were spread out (Study 2).

Lesson 6: Critical Thinking Training Can Be Extended to Teams

A team-oriented training strategy for critical thinking is now being tested at SWOS. The training content includes two specifically team-based skills: (a) Team leaders are taught to articulate periodic situation updates (Entin, Serfaty, & Deckert, 1994) that mention problems with assessments, such as missing, unreliable, or conflicting evidence. Such updates not only provide team members with a *shared mental model* of the tactical situation (Entin et al., 1994), but also foster a *shared metacognitive model* of ongoing uncertainties in the situation model. The shared metacognitive model prompts team members to volunteer relevant information or insights. (b) Team members are taught to play the part of devil's advocate (crystal ball) for one another to generate new assessments and interpretations of evidence.

Lesson 7: Critical Thinking Training Is Generalizable Across Domains

A similar training strategy has been applied in the development of critical thinking training for Army battlefield command staff (Cohen, Freeman, & Thompson, 1997). This domain differs in a variety of ways from naval anti-air warfare; in particular, significantly more time is available for decision making and far more complex plans can be generated.

Critical thinking training has also been applied to the interaction of users with decision aids in the Rotorcraft Pilot's Associate program. This training focuses on techniques for monitoring the reliability of decision aid conclusions and the use of different interaction strategies as a function of trust in the aid, stakes, and available time (Cohen, Parasuraman, et al., 1997). Recent research explored applications of the model to the design of decision aids for automated target recognition (Cohen, Thompson, & Freeman, 1997). Another domain in which critical thinking training is currently being explored is commercial airline pilot decision making.

Successful application in these domains suggests that the critical thinking training strategy is transferable across domains. Future research should explore potential applications to business and other nonmilitary applications more intensely, further test the R/M model, apply it to decision aid design, and explore potential synergies between critical thinking training and decision support.

Conclusion

It has become commonplace to warn that more information does not necessarily mean better decisions. As new technologies supply decision makers with vast quantities of information, decision makers' ability to exploit that information effectively may be overwhelmed. In this environment, critical thinking skills are of growing importance. Critical thinking includes the ability to sort out what is truly important, to address conflicts in the information that is available, to ferret out and refine the assumptions required to interpret the information, and to manage time wisely so that action is taken in a timely manner. These skills, while implicit in the experience that decision makers accumulate over years of practice in a domain, need not be left to chance. The research reported here demonstrates that they can be effectively taught, and that, once learned, they can measurably improve not only decision processes, but the outcomes of decisions.

The success of the training provides additional support for a naturalistic definition of skilled decision making, drawn from observations of how proficient decision makers actually do their jobs, rather than from formally derived algorithms. There is encouraging evidence that training concepts developed in this way generalize to a wide variety of domains where decision makers have a limited amount of time to make high stakes decisions under uncertainty.

References

Adams, M. J., & Feehrer, C. E. (1991). Thinking and decision making. In J. Baron & R. V. Brown (Eds.), *Teaching decision making to adolescents* (pp. 79–94). Hillsdale, NJ: Erlbaum.

Baker, L. (1985). How do we know when we don't understand? Standards for evaluating text comprehension. In D. L. Forrest-Pressley, G. E. MacKinnon, & T. G. Waller (Eds.), *Metacognition, cognition, and human performance* (Vol. 1, pp. 155–205). San Diego, CA: Academic Press.

Baron, J., & Brown, R. V. (Eds.). (1991). *Teaching decision making to adolescents*. Hillsdale, NJ: Erlbaum.

Chase, W. G., & Simon, H. A. (1973). The mind's eye in chess. In W. G. Chase (Ed.), *Visual information processing* (pp. 215–281). San Diego, CA: Academic Press.

Cohen, M. S. (1986). An expert system framework for non-monotonic reasoning about probabilistic assumptions. In J. F. Lemmer & L. N. Kanal (Eds.), *Uncertainty in artificial intelligence* (pp. 279–293). Amsterdam: North-Holland.

Cohen, M. S. (1993). The naturalistic basis of decision biases. In G. A. Klein, J. Orasanu, R. Calderwood, & C. E. Zsambok (Eds.), *Decision making in action: Models and methods* (pp. 51–99). Norwood, NJ: Ablex.

Cohen, M. S., & Freeman, J. T. (1996). Thinking naturally about uncertainty. In *Proceedings of the 40th annual meeting of the Human Factors and Ergonomics Society*. Santa Monica, CA: Human Factors and Ergonomics Society.

Cohen, M. S., Freeman, J. T., & Thompson, B. T. (1997). Training the naturalistic decision maker. In C. E. Zsambok & G. Klein (Eds.), *Naturalistic decision making* (pp. 257–268). Mahwah, NJ: Erlbaum.

Cohen, M. S., Freeman, J. T., & Wolf, S. (1996). Meta-recognition in time stressed decision making: Recognizing, critiquing, and correcting. *Human Factors, 38*, 206–219.

Cohen, M. S., Freeman, J., Wolf, S., & Mitellino, L. (1995). *Training metacognitive skills in naval combat decision making*. Arlington, VA: Cognitive Technologies.

Cohen, M. S., Parasuraman, R., Serfaty, D., & Andes, R. (1997). *Trust in decision aids: A model and a training strategy*. Arlington, VA: Cognitive Technologies.

Cohen, M. S., Thompson, B., & Freeman, J. T. (1997). *Cognitive aspects of automated target recognition interface design: An experimental analysis.* Arlington, VA: Cognitive Technologies.

Collins, A., & Stevens, A. L. (1983). A cognitive theory of inquiry teaching. In C. M. Reigeluth (Ed.), *Instructional-design theories and models: An overview of their current status.* Hillsdale, NJ: Erlbaum.

Dreyfus, H. L. (1997). Intuitive, deliberative, and calculative models of expert performance. In C. E. Zsambok & G. Klein (Eds.), *Naturalistic decision making* (pp. 17–28). Mahwah, NJ: Erlbaum.

Duncan, P. C., Rouse, W. B., Johnston, J. H., Cannon-Bowers, J. A., Salas, E., & Burns, J. J. (1996). Training teams working in complex systems: A mental model-based approach. *Human Technology Interaction in Complex Systems, 8,* 173–231.

Entin, E. E., Serfaty, D., & Deckert, J. C. (1994). *Team adaptation and coordination training* (Tech. Rep. No. 648-1). Burlington, MA: Alphatech.

Fischhoff, B., & Johnson, S. (1990). The possibility of distributed decision. [*Distributed decision making: Report of a workshop*]. Washington, DC: National Academy Press.

Kaempf, G. L., Klein, G., Thordsen, M. L., & Wolf, S. (1996). Decision making in complex command-and-control environments. *Human Factors, 38,* 220–231.

Klein, G. A. (1993). A recognition-primed decision (RPD) model of rapid decision making. In G. A. Klein, J. Orasanu, R. Calderwood, & C. E. Zsambok (Eds.), *Decision making in action: Models and methods* (pp. 138–147). Norwood, NJ: Ablex.

Kuhn, D., Amsel, E., & O'Loughlin, M. (1988). *The development of scientific thinking skills.* San Diego, CA: Academic Press.

Levi, I. (1986). *Hard choices.* Cambridge, England: Cambridge University Press.

Lichtenstein, S., Fischhoff, B., & Phillips, L. D. (1982). Calibration of probabilities: The state of the art to 1980. In D. Kahneman, P. Slovic, & A. Tversky (Eds.), *Judgment under uncertainty: Heuristics and biases* (pp. 306–334). Cambridge, England: Cambridge University Press.

Means, B., Salas, E., Crandall, B., & Jacobs, T. O. (1993). Training decision makers for the real world. In G. A. Klein, J. Orasanu, R. Calderwood, & C. E. Zsambok (Eds.), *Decision making in action: Models and methods* (pp. 306–326). Norwood, NJ: Ablex.

Pennington, N., & Hastie, R. (1993). A theory of explanation-based decision making. In G. A. Klein, J. Orasanu, R. Calderwood, & C. E. Zsambok (Eds.), *Decision making in action: Models and methods* (pp. 188–201). Norwood, NJ: Ablex.

Salas, E., & Cannon-Bowers, J. A. (1997). Methods, tools, and strategies for team training. In M. A. Quinones & A. Ehrenstein (Eds.), *Training for a rapidly changing workplace: Applications of psychological research* (pp. 249–279). Washington, DC: American Psychological Association.

Thompson, B. T., Cohen, M. S., & Shastri, L. (1997). *A hybrid architecture for metacognitive learning* (Hybrid Architectures for Complex Learning Program, Office of Naval Research, Contract No. N00014-95-C-0182). Arlington, VA: Cognitive Technologies.

8

Stress Exposure Training

James E. Driskell and Joan H. Johnston

Imagine that you are assigned a task to sit at your desk and mentally add a series of numbers. Now, imagine that you invite several people into your office, turn a radio and television on, and have someone flick the light switch on and off. For good measure, let's let through all those calls from telephone salespeople asking you to switch your phone service. Furthermore, let's add some consequences to the task: If your calculations are incorrect, you must pay a penalty of $1 for each number that you are off. It is difficult enough to perform even a moderately complex task in a normal or benign performance environment, as the literature on training and skill acquisition attests. However, it would be far more difficult to perform a task effectively under the high-stress or high-demand conditions of this imaginary scenario.

The military offers numerous real-world examples of complex and demanding task environments. Today's ships, airplanes, and tanks are technologically advanced systems that greatly extend the range of human capabilities. For example, military anti-air warfare systems aboard modern naval ships allow military personnel to detect aircraft at great distances. On the other hand, these systems also increase the demands on the operator: The amount and complexity of information that must be processed in a short period of time once a target has been detected is enormous. Therefore, although modern military systems have greatly extended the military's capabilities, they have both increased the stress under which personnel must perform and increased the potential for catastrophic errors. The informational complexity, task load, and time pressure inherent in this environment increases the potential for error, such as the 1988 downing of an Iranian commercial aircraft by the USS *Vincennes*.

Furthermore, there are a large number of applied settings outside of the military that share the commonality of a potentially high-stress, high-demand performance environment. These settings are found in the fields of aviation (Prince, Bowers, & Salas, 1994), emergency medicine (Mackenzie, Craig, Parr, & Horst, 1994), mining (Perrow, 1984), diving (Radloff & Helmreich, 1968), parachuting (Hammerton & Tickner, 1969), bomb disposal (Rachman, 1983), police work (Yuille, Davies, Gibling, Marxsen, & Porter, 1994), and fire fighting (Markowitz, Gutterman, Link, & Rivera, 1987). These stereotypically high-stress environments impose a particularly high demand on those who work in them, and there is a substantial potential for risk, harm, or error. People who work in these environments often perform under extreme pressures and demands. Emergency or crisis conditions may occur suddenly and unexpectedly, and the consequences of error are immediate and often catastrophic. Although major disasters such as Three Mile Island and Chernobyl are etched in our collective memory, the daily newspapers provide evidence of the almost commonplace occurrence of accidents or near accidents involving ships, trains, and airplanes in which increased environmental demand inevitably plays a role.

Whereas errors that occur in these types of settings are often broadcast on the evening news, on a more personal level everyone is faced at one time or another with having to perform under the pressure of deadlines, while juggling multiple tasks, and in the face of various distractions. In everyday settings such as working in an office or driving home, we may be subjected to stressors such as time pressure, noise, novel or threatening events, demands or requests of others, and other distractions that may disrupt task performance and increase errors.

The impact of stress on the individual has become a primary concern in industry (Spettell & Liebert, 1986), the military (Driskell & Olmstead, 1989), aviation (Prince et al., 1994), sports (Jones & Hardy, 1990), and other applied settings in which effective performance under stress is required. Therefore, the development of effective training interventions to ameliorate the negative effects of stress on performance has taken on increased importance in the training community (see Driskell & Salas, 1991; Ivancevich, Matteson, Freedman, & Phillips, 1990). The goal of stress-exposure training is to prepare personnel to perform tasks effectively under high-demand, high-stress conditions. In this chapter we present a model of stress-exposure training, describe empirical research that supports this approach, and derive guidelines for implementing stress-exposure training.

What Is Stress Training and Why Is It Needed?

We use the term *stress* to describe a process by which environmental demands (e.g., time pressure, novel or threatening events, industrial noises) evoke an appraisal process in which perceived demand exceeds resources and that results in undesirable physiological, psychological, behavioral, or social outcomes (Driskell & Salas, 1996). Evidence indicates that stress is a costly health-related issue, in terms of individual performance and well-being as well as organizational productivity (Ilgen, 1990). Accordingly, a great deal of research has been conducted to examine interventions to reduce the negative outcomes of stress on the individual.

It is important to distinguish between *training* and *stress training*. The primary goal of training is skill acquisition and retention. Therefore, most training takes place under conditions designed to maximize learning: a quiet classroom, the practice of task procedures under predictable conditions, uniformity of presentation, and so forth. In this manner, the traditional classroom or lecture format, supplemented with skills practice, typically is satisfactory for promoting initial skill acquisition.

However, some tasks must be performed in conditions quite unlike those encountered in the training classroom. For example, high-stress environments include specific task conditions (such as time pressure, ambiguity, increased task load, distractions) and require specific responses (such as the flexibility to adapt to novel and often changing environmental contingencies) that differ from those found in the normal performance environment. Research has shown that, for some tasks, normal training procedures (training conducted under normal, nonstress conditions) often do not improve task performance when the task has to be performed under stress conditions (Zakay & Wooler, 1984). These results suggest that, under certain conditions, the transfer of training from classroom conditions to operational conditions may be poor when there are no stress-inclusive simulations or training.

In brief, the primary purpose of *training* is to ensure the acquisition of required knowledge, skills, and abilities. The primary purpose of *stress training* is to prepare the individual to maintain effective performance in a high-stress environment. Therefore, *stress training* is defined as an intervention to enhance familiarity with the criterion environment and teach the skills necessary to maintain effective task performance under stress conditions.

It may be valuable to consider the general objectives to be met by stress training. The primary purpose of stress training is to prepare the individual to perform effectively in the stress environment. In broad terms, there are three overall goals of stress training: (a) gaining knowledge of and familiarity with the stress environment, (b) training those skills required to maintain effective performance under stress, and (c) building performance confidence. These objectives are outlined in the following sections.

Training Objective 1: To Convey Knowledge of the Stress Environment

In a study of World War II combat aircrews, Janis (1951) found a marked reduction in stress reactions when information on air attacks was provided in advance. More recently, a National Research Council study on enhancing military performance concluded that "stress is reduced by giving an individual as much knowledge and understanding as possible regarding future events" (Druckman & Swets, 1988, p. 21). Providing knowledge about stress effects during training has several beneficial consequences: (a) It enables the individual to form accurate expectations regarding the stress environment, thereby increasing predictability; (b) it decreases the distraction involved in attending to novel sensations and activities in the stress environment; and (c) it allows the individual to identify and avoid performance errors that are likely to occur in the stress environment.

Training Objective 2: To Emphasize Skill Development

Numerous stress effects have been documented in the research literature. Stress may result in *physiological changes* such as quickened heartbeat, labored breathing, and trembling (Rachman, 1983); *emotional reactions* such as fear, anxiety, frustration (Driskell & Salas, 1991), and motivational losses (Innes & Allnutt, 1967); *cognitive effects* such as narrowed attention (Combs & Taylor, 1952; Easterbrook, 1959), decreased search behavior (Streufert & Streufert, 1981), longer reaction time to peripheral cues and decreased vigilance (Wachtel, 1968), degraded problem solving (Yamamoto, 1984), and performance rigidity (Staw, Sandelands, & Dutton, 1981); and changes in *social behavior* such as a loss of team perspective (Driskell, Salas, & Johnston, 1997) and decrease in prosocial behaviors such as helping (Mathews & Canon, 1975). All of these stress effects can affect task performance. Therefore, one objective of stress training is to overcome these decrements. For example, to address the degradation that stems from having to juggle multiple tasks in a high-demand stress environment, stress training may provide practice in time-sharing multiple tasks and in prioritizing critical task demands. Therefore, one primary focus of stress training is to train people in the behavioral and cognitive skills that allow the trainee to maintain effective performance under stress.

Training Objective 3: To Build Confidence in the Ability to Perform

A third goal of stress training is to build confidence in the ability to perform one's task. Research indicates that stress training is effective only when the trainee experiences success or a sense of task mastery during training (Keinan, 1988). Because the stress environment is an extremely high-demand performance environment, individuals can develop either positive or negative expectations regarding their capacity to perform in that environment. Individuals who appraise the task environment in positive terms will have more confidence in their ability to perform and are likely to suffer fewer negative stress effects. They will be less aroused physiologically, less distracted by task-irrelevant concerns, and more likely to focus attention on the task. Research has shown self-efficacy to be a strong predictor of performance (Bandura, Reese, & Adams, 1982; Locke, Frederick, Lee, & Bobko, 1984). Therefore, stress training should build trainee confidence in the ability to perform in the stress environment.

Stress Exposure Training: A Model for Integrated Stress Training

There have been a number of attempts to implement different types of stress training in both civilian and military environments. For example, the military has implemented confidence courses and water-survival training to build confidence and provide familiarity with stress environments. Some training techniques, such as overlearning, have been mentioned as potentially effective candidates for stress training (Driskell, Willis, & Copper, 1992). However, most of these applications have been stand-alone attempts to improve performance under stress. For example, one study may attempt to implement overlearning as a stress training technique, and another may attempt to impose relaxation training. Some of these efforts have been successful and some have not. Although these efforts allow the cumulation of knowledge on stress training techniques, what has been lacking is an integrated approach to developing stress training. Such an approach would provide a structure for implementing stress training *programs*, not just hit-or-miss techniques. An integrated model of stress training allows the training designer to address critical questions such as these: When should stressors be introduced in training? How should training be sequenced? When should skills training be introduced? What information should be presented regarding stress effects?

The development of a model for integrated stress training, which we term stress-exposure training (SET), provides a structure for designing, developing, and implementing stress training. The SET approach is defined by a three-stage training intervention: (a) an initial stage, in which information is provided regarding stress and stress effects; (b) a skills training phase, in which specific cognitive and behavioral skills are acquired; and (c) the final stage of application and practice of these skills under conditions that increasingly approximate the criterion environment. Table 1 provides an outline of the stress-exposure training model.

During the initial phase of SET, *information provision*, information is provided regarding stress and likely stress effects. The purpose of this phase of training is to emphasize the value and goals of stress training and to provide trainees with basic information on stress, stress symptoms, and likely stress effects in the performance setting. The first phase of stress-exposure training, therefore, provides trainees with preparatory information regarding what to expect in the stressor environment; typical physiological, emotional, and cognitive reactions to stress; and how stress is likely to affect performance in the operational environment. Preparatory information regarding the stressor environment increases a sense of controllability and increases confidence in the ability to perform; enables the trainee to form accurate expectations regarding reactions to stress and

Table 1. Structure of Stress-Exposure Training

Phases	Activities
1. Information provision	Indoctrination and preparatory information
2. Skills acquisition	Behavioral and cognitive skills training
3. Application and practice	Practice of skills under conditions that gradually approximate the stress environment

events likely to occur in the stress performance environment; and decreases the distraction of attending to novel sensations and activities in real time in the stress environment, thus increasing attention devoted to task-relevant stimuli. In addition, preparatory information regarding the effects of stress on performance allows the individual to anticipate performance errors that are likely to occur in the stress environment.

During the second phase of training, *skills acquisition*, specific cognitive and behavioral skills are taught and practiced. Training at this stage may include a wide array of training techniques such as training attentional focus, overlearning, and decision-making skills. The specific training techniques implemented vary according to the specific requirements of the task setting. The goal of training at this stage is to build the high-performance skills that are required to maintain effective performance in the stress environment.

The final phase of SET involves the *application and practice* of these skills while trainees are gradually exposed to task-relevant stressors. Effective task performance requires not only that specific skills be learned, but that they be transferred to the operational setting. This requires practice of skills under operational conditions similar to those likely to be encountered in the stress setting. Allowing skills practice in a graduated manner across increasing levels of stress (from moderate stress scenarios or exercises to higher stress exercises) allows skills learned to be practiced in increasingly realistic task environments.

There are several characteristics of the SET approach that should be emphasized. First, SET is a model for stress training rather than a specific training technique. The SET model describes three stages of training, each with a specific overall objective. However, the specific content of each stage will vary according to the specific training requirements. Both the type of stressors and the skills required for effective performance depend on the specific task setting. Therefore, stress training must be context specific; a training approach applicable to one setting may not be relevant to a different setting. For example, consider a complex decision-making tasks such as Navy shipboard combat information center (CIC) operations. Stress training for this environment must address the particular stressors relevant to that environment in Phase 1 of training (e.g., auditory distraction, time pressure); it must involve skills training relevant to the task in Phase 2 (e.g., decision-making training, time sharing); and it must provide practice of these skills in an environment that simulates these conditions in Phase 3. The design of stress-exposure training for emergency medical technicians will likely involve different stressors, different types of skills training, as well as different types of realistic practice and simulations than that provided in the naval CIC stress training. In brief, the SET approach, which incorporates knowledge of the stressor environment, skills training, and graduated exposure and practice in the simulated stress setting, provides an integrated structure for stress training. Stress-exposure training does not prescribe one type of training that must be applied in all settings, but provides a model to guide the design of stress training for any task environment.

Second, the three-stage SET model is patterned on a cognitive–behavioral approach to stress management called *stress inoculation training* (Meichenbaum, 1985). However, stress inoculation training was originally developed as a clinical treatment program to teach clients to cope with physical pain, anger, and phobic responses. The stress inoculation training approach retains several clinical emphases that may limit its application in applied training environments: (a) the intensive therapeutic involvement of a skilled facilitator; (b) one-on-one individualized treatment; and (c) a primary emphasis on alleviating anxiety, depression, and anger (Johnston & Cannon-Bowers, 1996). Nevertheless, results of a meta-analysis by Saunders, Driskell, Johnston, and Salas (1996) indicated that, in the settings in which it is used, stress inoculation training is an "effective means for reducing state anxiety, reducing skill-specific anxiety, and enhancing performance under stress" (p. 170). Therefore, a modification of this three-stage approach, adapted for an applied training environment, should hold considerable promise for stress training.

Third, SET addresses the three major objectives of stress training: enhancing familiarity with the stress environment, building skills to maintain effective performance under stress, and boosting confidence in the ability to perform. Familiarity is enhanced by providing trainees with accurate information on stress and on specific stress effects that are likely to occur in the operational environment. The acquisition of skills to support effective performance in the operational setting takes place in Phase 2, and these skills are rehearsed in Phase 3 of the SET approach. Finally, trainee confidence can be enhanced by providing the opportunity to practice skills in a setting that gradually approximates the stress environment.

Implementing Stress-Exposure Training

In the following sections, we examine training events that take place within each stage of SET. We describe the empirical research that underlies the activities that constitute each stage and derive guidelines for implementing stress-exposure training.

Phase 1: Information Provision

The primary goal of Phase 1 of stress-exposure training is information provision. Phase 1 includes two primary components: (a) indoctrination, or discussion of why stress training is important, and (b) preparatory information describing what stressors are likely to be encountered in the task environment, the likely effects of stress on how the trainee may feel, and the likely effects of stress on how the trainee may perform.

Indoctrination is aimed at increasing the attention and motivation necessary to acquire the skills required by the particular stressful task setting. Indoctrination often emphasizes the rewards and costs of effective and ineffective performance in the stress environment in order to under-score the value of training (see Hoehn & Levine, 1951). Indoctrination may be provided by discussing operational incidents in which environmental stress is prevalent. These may include case histories and lessons learned from military and industrial accidents, and other incidents in which fac-tors such as extreme time pressure and task load had a significant impact on performance. This type of indoctrination is standard fare in military training, but it may be particularly relevant for SET because stress train-ing is training above and beyond basic technical training, and thus the concept of "user acceptance" becomes important. That is, the user (in this case, the trainee) must understand the purpose and value of the stress training and be motivated to undertake the training.

The second primary component of Phase 1 is the provision of prepar-atory information. Stressful, threatening, or demanding situations can lead to a number of undesirable consequences, including heightened anx-iety and decrements in performance (see Driskell & Salas, 1991, 1996; Keinan, 1987). There is some evidence that preparatory information can lessen negative reactions to stress. Although the bulk of the existing re-search has been performed in clinical or medical settings (see Taylor & Clark, 1986), some studies have suggested the efficacy of preparatory in-formation on enhancing performance in applied task environments. In ex-amining the training of soldiers for nuclear combat, Vineberg (1965), noted that the communication of accurate information was critical for clarifying misconceptions, reducing fear of the unknown, and increasing a sense of control in this type of environment. More recently, a National Research Council study on enhancing military performance concluded that "stress is reduced by giving an individual as much knowledge and understanding as possible regarding future events" (Druckman & Swets, 1988, p. 21). However, these authors also noted that this approach often runs counter to military practice, which is to give the individual the least amount of information necessary for a given situation.

It is likely that preparatory information mitigates negative reactions to stress in several different ways. First, preparatory information, by pro-viding a preview of the stress environment, renders the task less novel and unfamiliar (Ausubel, Schiff, & Goldman, 1953). This may lead to a more positive expectation of self-efficacy, which research has shown to be a strong predictor of performance (Bandura, Reese, & Adams, 1982; Locke, et al., 1984). Second, knowledge regarding an upcoming event increases predictability, which can decrease the attentional demands and distraction of having to monitor and interpret novel events in real time (Cohen, 1978). Third, preparatory information may enhance the sense of behavioral or cognitive control over an aversive event by providing the individual with an instrumental means to respond to the stress (Keinan & Friedland, 1996; Thompson, 1981).

High-stress events, such as an airplane crew responding to a mechan-ical failure in flight, a power plant operator reacting to a system accident, or a work team deliberating a task under extreme deadline pressure, share several common characteristics. First, those involved are likely to experience a number of novel and unpleasant sensations, such as a pounding heart, muscle tension, and feelings of anxiety, confusion, or frustration. Second, stress may produce a qualitative change in the task environment, in which stressors such as noise, time pressure, threat, and other demands occur suddenly, and individuals are faced with a transition from routine conditions to emergency conditions. The nature of the task environment may change dramatically, from a relatively benign environment to one that is fast paced, aversive, or threatening. Finally, in many cases, these stressors and the individual reactions to stress disrupt goal-oriented behavior, and task performance must be adapted to meet these new demands. For example, to maintain effective performance under stress, individuals may need to attend selectively to task-relevant stimuli to counter the attentional overload imposed by stress conditions (see Singer, Cauraugh, Murphey, Chen, & Lidor, 1991).

Therefore, a comprehensive preparatory information strategy should address how the person is likely to feel in the stress setting, describe the events that are likely to be experienced in the transition from normal to stress conditions, and provide information on how the person may adapt to these changes. We may define three types of preparatory information: sensory, procedural, and instrumental. *Sensory information* is information regarding how the individual is likely to feel when under stress. Under stress, the individual may perceive a number of intrusive physical and emotional sensations. Typical physiological reactions include increased heart rate, sweating, shallow breathing, and muscle tension; emotional reactions to stress may include fear, frustration, and confusion. Although the relationship of physiological and emotional state to performance is complex, these reactions are common and are, at the least, a source of interference and distraction to the task performer.

Furthermore, research has suggested that individuals under stress or novel conditions tend to overinterpret stress symptoms; that is, they assign a heightened importance to physical symptoms such as an increased heart rate. Second, they tend to misinterpret these "normal" stress reactions as catastrophic (Clark, 1988). The problem in this case is not that people experience these symptoms; the problem is that they experience "normal" stress symptoms, but because of the novelty or unfamiliarity of these symptoms, they expend a disproportional amount of attentional capacity attending to them, which distracts from task-focused activity. Worchel and Yohai (1979) found that individuals who were able to label or identify physiological reactions (i.e., individuals who were able to attribute their physiological reactions to some reasonable cause) were less distressed or aroused by those reactions. Therefore, it is likely that providing personnel with accurate information regarding normal physiological symptoms and responses to stress will reduce the distraction of having to interpret or attend to these unfamiliar reactions in the operational task environment.

Procedural information describes the events that are likely to occur in the stress environment. Procedural information may include a description of the setting, the types of stressors that may be encountered, and the effects the stressors may have. For example, procedural information provided to a novice parachutist may include a description of the activities that will take place prior to a jump, the noises and time pressure that may be present, and the distraction and lapses of attention that these stressors may cause. In an early study of combat aircrews, Janis (1951) found a reduction in negative stress reactions when descriptive information on air attacks was provided in advance.

Finally, the third type of preparatory information, *instrumental information*, describes what to do to counter the undesirable consequences of stress. For example, Egbert, Battit, Welch, and Bartlett (1964) provided individuals with information on how they would feel following a medical operation, as well as what they could do to relieve the discomfort. Preparatory information may be most effective, especially in a performance environment, if it has instrumental value, that is, if the information provides the individual with a means to resolve the problems posed by the stress environment. For example, it may be of value to know not only how noise may contribute to distractions during task performance, but also what one can do to overcome the effects of these distractions.

In a recent study of the effectiveness of preparatory information on enhancing performance under stress, Inzana, Driskell, Salas, and Johnston (1996) implemented a comprehensive preparatory information intervention that incorporated sensory, procedural, and instrumental information. This study is unique in that it examined the effect of preparatory information on task performance in a real-world decision-making environment. Participants performed a computer-based decision-making task under either high-stress or normal-stress conditions, and they were given either general task instructions or instructions that included specific preparatory information regarding the stress environment. Results of this study indicated that those who received preparatory information before performing under high-stress conditions reported less anxiety, were more confident in their ability to perform the task, and made fewer performance errors than those who received no preparatory information.

Phase 2: Skills Acquisition

The primary goal of Phase 2 of SET, *skills training*, is skill acquisition and rehearsal. Training activities in Phase 2 are focused on developing cognitive and behavioral skills that are required to maintain effective performance under stress. Whereas some skills are somewhat generic and are likely to be relevant to most tasks that may be performed under stress conditions (e.g., cognitive control strategies), the implementation of other types of skills training (e.g., decision-making training) is dependent on the specific tasks to be performed in the criterion setting. Therefore, the specific skills-training techniques to be implemented in this phase of stress-exposure training vary depending on the requirements of the task.

The types of skills to be taught can be grouped into two broad categories. Some stress-training strategies are intended to make the trainee more resistant to stress effects. For example, if an individual overlearns a task, he or she will be less likely to become distracted by novel or stressful events and more resistant to other stress consequences such as narrowing of attention that can lead to error. These approaches attempt to minimize the effects of stress on the individual. On the other hand, in other stress training approaches, these stress-related performance decrements (i.e., narrowed attention) are taken as a given, and attempts are made to train individuals to compensate for these losses in the task situation. For example, the attentional-focus training approach proposed by Singer et al. (1991) represents an attempt to train individuals to maintain attentional focus on task-relevant stimuli in the face of external distractions. In the following sections, we examine several types of stress training strategies that may be incorporated into the skills acquisition phase of SET.

Cognitive control strategies. The term *cognitive control* subsumes a number of cognitive coping strategies that have the purpose of providing the trainee with control over distracting or dysfunctional thoughts and emotions that arise under stress conditions. The primary emphasis of these interventions is to replace negative or distracting cognitions with task-focused cognitions.

Evidence suggests that experiencing novelty in the immediate environment may lead to greater self-attention (Wegner & Giuliano, 1980). Moreover, novel or stressful stimuli may instigate an active search for meaning that involves a turning inward, or self-focus. A primary emphasis of the appraisal process is the evaluation of what novel stimuli mean to *one's self* (Lazarus & Folkman, 1984). Thus, under stress, individuals begin to "time-share" cognitive resources between (a) the task and (b) worrying about the stress itself and how the stressor will affect their own well-being. Performance suffers as attention is distributed between task-relevant and task-irrelevant cognitions.

The primary technique used in cognitive restructuring is the training of individuals to recognize task-irrelevant thoughts and emotions that degrade task performance and to replace them with task-focused cognitions. The key focus of this training approach is to train the individual to regulate emotions (e.g., worry and frustration), regulate distracting thoughts (self-oriented cognitions), and maintain task orientation. As Wine (1971) noted, "performance may be improved by directing attention to task-relevant variables, and away from self-evaluative rumination" (p. 100).

Wine (1971) and Tryon (1980) argued that many stress training techniques may be effective in reducing self-reported anxiety and negative affect but generally show little impact on performance. Therefore, interventions directed at reducing negative emotional response may get participants to feel better; however, these interventions ignore the cognitive components of stress reaction (i.e., interfering self-oriented cognitions and attention reduction), which seem to be more directly related to performance.

Wine (1971, 1980) concluded that this perspective suggests the benefits of *attentional training*, which would focus on the narrowing of attention and self-focused thoughts inherent in performance anxiety, with the goal of focusing attention to task-relevant stimuli. In a recent study of training in attention-focusing skills, Singer et al. (1991) found that this training resulted in improved task performance when participants worked under noise stress. This approach included training to describe when, why, and how attention may be distracted during task performance, as well as practice in performing the task under high-demand conditions, focusing attention, and refocusing attention after distraction. Training that concentrates directly on enhancing attentional focus may help alleviate the distraction and perceptual narrowing that occur in stress environments.

Physiological control strategies. Some training strategies are intended to provide the trainee with control over negative physiological reactions to stress. Progressive muscle relaxation training techniques involve a series of muscle tensing and relaxing exercises, often supplemented with imagery (Burish, Carey, Krozely, & Greco, 1987). The goal of relaxation training is to train the individual to control muscle tension and breathing. The basic premise of relaxation training is that relaxation and stress are incompatible; that is, if someone is physiologically relaxed, he or she is less likely to experience the negative physiological responses brought on by the stress environment. The value of this training is that it attempts to train people in the responses characteristic of effective performers under stress conditions: calmness, relaxation, and control.

Several other types of behavioral training interventions can be identified. For example, biofeedback involves training an individual to control physiological responses (such as systolic blood pressure or heart rate) by using external monitoring devices to indicate when a desired change occurs. Using these devices, individuals can learn to bring their physiological processes under conscious control (e.g., Dobie, May, Fischer, & Elder, 1987; Jones, 1985). Autogenic-feedback training has been used successfully in alleviating space motion sickness and space adaptation syndrome (Cowings & Toscano, 1982). This technique involves the individual in an active effort to monitor and regulate internal cues that signal and exacerbate motion sickness. All of these techniques have in common an effort to increase the extent to which the person's physiological reactions are under conscious control.

Overlearning. Almost all basic military training is intended to reduce the disruptive effects of stress in combat through the use of repetitive drill, providing soldiers with a set of habitual responses that are less vulnerable to stress decrement. The term *overlearning* refers to deliberate overtraining of a performance beyond the level of initial proficiency (Driskell et al., 1992). For example, Schendel and Hagman (1982) examined the effects of overlearning on retention of a military procedural task (disassembly and assembly of an M60 machine gun). They found that the overtrained group made 65% fewer errors than a control group when retested after 8 weeks.

A number of researchers have argued that overlearning is a particularly potent training procedure for the stress environment (see Cascio, 1991; Deese, 1961; Fitts, 1965; Weitz, 1966). Janis (1949) noted that "drill of this type, when repeated so that the response is overlearned, tends to build up an automatic adaptive response" (p. 222). Given that one effect of the stress environment is to reduce or restrict the attentional capacity of the individual, behaviors that are more automatic should be more resistant to degradation. Geen (1989) noted that automated processing of information occurs as tasks become well rehearsed and performance becomes routinized or more automatic. Automated tasks require less active attentional capacity and are less subject to disruption by increased attentional demands.

In a meta-analysis of research on the effects of overlearning, Driskell et al. (1992) found that overlearning resulted in a significant increase in retention. However, they also noted several cautions regarding the use of overlearning in stress training. First, overlearning can lead to rigidity of response. The repetition of a single behavior or response over a large number of trials may result in a loss of flexibility and the tendency to persist with a single response even when the behavior is no longer correct. Second, it is critical that the behavior that is trained or overlearned in the training setting closely reflect the task that is required in the actual performance setting. To the extent that stress in the real-world task environment changes the nature of the task or the types of behavior required for successful performance, the overlearning of a task in a training environment that does not incorporate these factors can lead to the reinforcement of inappropriate or ineffective behavior. For example, Zakay and Wooler (1984) found that training conducted under normal conditions improved decision performance under normal conditions but did not improve performance when subjects performed under time pressure. It may be detrimental to overlearn a set of responses during training, therefore, if the behavior called for in the real-world environment requires a different type of response. In summary, whereas overlearning can lead to enhanced performance, the training designer must ensure that the task that is overlearned is the task called for in the performance setting. When the concern is preparing personnel to perform under stress conditions, the practice conditions provided during training must approximate the stress environment.

Mental practice. *Mental practice* refers to the cognitive rehearsal of a task in the absence of overt physical movement (Richardson, 1967). When a musician practices a passage by "thinking it through" or when an athlete prepares for an event by visualizing the steps required to perform the task, he or she is engaging in mental practice. Mental practice has been studied most extensively in educational and sports research, and it is a component of many cognitive stress reduction techniques (see Meichenbaum, 1985; Meichenbaum & Cameron, 1983).

In a typical implementation of this procedure, participants mentally practice or mentally rehearse performing a task. Instructions are to sit quietly and mentally practice performing the task successfully from start to finish. The overall results of research are encouraging: In a recent meta-analysis of the mental practice literature, it was concluded that mental practice was an effective means for enhancing performance, although somewhat less effective than physical practice (Driskell, Copper, & Moran, 1994). Driskell et al. noted that mental practice offers the opportunity to rehearse behaviors and to code behaviors into easily remembered words and images to aid recall. Mental practice does not offer direct knowledge of results or visual and tactile feedback, as does physical practice. However, Driskell et al. (1994) concluded that mental practice may be a particularly effective technique for training complex cognitive tasks, for rehearsing tasks that are dangerous to train physically, and for training tasks in which the opportunity for physical practice seldom occurs.

Training time-sharing skills. High-stress or high-demand performance environments often involve increases in task load and time pressure. Increased task load may result from the imposition of additional tasks (e.g., an air traffic controller whose task suddenly increases from monitoring one aircraft to monitoring multiple aircraft in a given airspace) or from having to attend to novel or unfamiliar stimuli (e.g., a worker whose task is to monitor a visual display while alarms are blaring and other people are running about). Broadly speaking, task load refers to the pressure or demand of performing multiple tasks, and studies show that performing multiple tasks often carries a penalty. Most stressful environments also involve time pressure, or the restriction in time required to perform a task. Research has suggested that time pressure may degrade performance because of the cognitive demands, or information overload, imposed by the requirement to process a given amount of information in a limited amount of time (Wright, 1974).

A number of studies have shown that highly practiced tasks can be performed jointly with little interference. For example, Spelke, Hirst, and Neisser (1976) and Hirst, Spelke, Reaves, Caharack, and Neisser (1980) found that performance dropped dramatically when participants were asked to read prose aloud while taking dictation. However, with substantial dual-task practice—more than 50 hours—participants could more readily read while taking dictation, achieving reading and comprehension rates similar to those of the single-task control groups. However, it is important to note that even extensive practice on each of two tasks performed separately does not seem to enhance performance when those tasks are later performed concurrently (Damos, Bittner, Kennedy, & Harbeson, 1981). Researchers have questioned the existence of a generalized time-sharing ability that is independent of the particular tasks to be performed. Time-sharing is considered a task-specific skill that must be practiced in context. If tasks are likely to be performed together in the operational environment, they must be practiced together in the training environment.

Finally, it may be beneficial to train prioritization skills in multiple-task environments. One vivid example of the need for this type of training is provided by the 1972 crash of an Eastern Airlines L-1011 in the Florida

Everglades just west of Miami. This aircraft was on an approach for landing, a period of high workload for the aircrew, when a landing gear light failed to illuminate. Over the next four minutes of flight, the crew was so preoccupied with this malfunction that they failed to monitor other critical flight activities and literally flew the plane into the ground (National Transportation Safety Board, 1973). In high-workload conditions, individuals often by necessity focus on some tasks to the exclusion of others, and often attention is devoted to low-priority or irrelevant tasks. Training that allows the individual to practice task prioritization under these conditions should prove beneficial.

Training decision-making skills. Some performance strategies work effectively in the training setting but poorly under operational conditions. For example, it may be easy to learn to drive a car by lining up the left fender with the center road stripe, but this strategy is inefficient for more advanced driving performance. In a similar vein, decision-making processes that may be effective in less stressful or less time-limited task situations may be inefficient in high-demand stress environments.

Many situations allow sufficient time to make a structured decision. The normative, *analytic* decision-making process is one in which the decision maker undertakes a systematic, organized information search, considers all available alternatives, generates a large option set, compares options, and successively refines alternative courses of action to select an optimal outcome. Analytic decision making makes efficient use of the resources at the decision maker's disposal and can result in well-informed decisions. Traditionally, deviations from this pattern of decision making have been viewed as indicative of a breakdown in decision making. Beach and Mitchell (1978) noted that, "in general, people in our culture regard the more formally analytic strategies as the ones most likely to yield correct decisions" (p. 445).

However, other researchers, such as Payne, Bettman, and Johnson (1988) and Klein (1996), have argued that the effectiveness of a particular decision-making strategy is dependent on many task and context variables. Under high-demand, time-pressured conditions, there is little time to gather all available information and evaluate each alternative solution. Payne et al. (1988) emphasized the contingent nature of decision making, arguing that under certain task conditions such as increased time pressure, use of a less analytic decision strategy may be adaptive. Klein and colleagues examined decision making in real-world operational environments such as in the command center of naval ships (Kaempf, Klein, Thordsen, & Wolf, 1996), among airline crews (Orasanu, 1993), and among firefighters (Klein, 1989). Klein (1996) argued that increased time pressure may prevent the use of analytic decision strategies, but that this is little cause for concern because analytic strategies are rarely used in these settings. In naturalistic task settings (in which decisions are made under time pressure and data are ambiguous or conflicting) decision makers do not have the luxury to search painstakingly for information, weigh all available alternatives, and eliminate each systematically to arrive at a solution. Within this context, what has been termed *hypervigilant decision making* (Janis & Mann, 1977, p. 11)—the consideration of limited alternatives, nonsystematic information search, accelerated evaluation of data, and rapid closure—may not represent a defect in the decision-making process but instead may represent an adaptive and effective response given the nature of the decision-making task. In brief, it may be easy to go through a stepwise procedure to generate options for decision making during training in which conditions are relatively relaxed, but this strategy may be inefficient in more complex, real-world settings. Decision-making processes that are effective in less stressful or less time-limited task situations may be inefficient in high-demand stress environments.

Johnston, Driskell, and Salas (1997) tested this hypothesis, arguing that in a naturalistic task environment, the use of a hypervigilant decision strategy would lead to more effective decision making than the use of a more analytic or vigilant strategy. The results of this study indicated that those who were trained to use a hypervigilant decision strategy did indeed exhibit the characteristics of nonanalytic decision making: (a) consideration of limited alternatives, (b) nonsystematic information search, (c) rapid evaluation of data, and (d) limited review of alternatives before making a decision. However, the results further indicated that, on a naturalistic task, this type of decision-making pattern led to better performance than did a more analytic strategy. The results of this study do not imply that a disorganized pattern of decision making is superior to an organized pattern; they demonstrate that a hypervigilant pattern of decision making, which has been described by others as disorganized and simplistic, can under some conditions be an effective decision strategy. Moreover, these conditions (e.g., time pressure, ambiguous data) characterize many applied, real-world tasks.

The results of this study have clear implications for training. Orasanu (1993) warned that the tendency to impose a normative model as a standard basis for decision-making training is seductive. Encouraging the decision maker to approximate a normative model could undermine behavior that may more adequately fit the requirements of the task situation. Johnston et al. (1997) concurred and suggested that training should not encourage the adoption of a complex analytic strategy under the conditions that characterize many naturalistic task environments. Thus, one goal of decision-making training for stressful environments is to emphasize the use of simplifying heuristics to manage effort and accuracy, and to improve the capability of the decision maker to adapt decision-making strategies to high-demand conditions.

Training team skills. Real-world incidents provide anecdotal but distinct illustrations of how team performance may falter under stress. United Airlines Flight 173 crashed near Portland, Oregon, in December, 1978, as it ran out of fuel while the crew attempted to deal with a landing gear malfunction. The National Transportation Safety Board (1979) report cited a breakdown in teamwork as a primary cause of this accident. The report indicated that the captain was preoccupied with an individual task, that "the first officer's main responsibility is to monitor the captain" and this was not done, and that "the flight engineer's responsibility is to monitor the captain's and first officer's actions" and that this was not done (p. A-5). In a review of crew performance in aviation, Foushee (1984) noted that a majority of accidents are related to breakdowns in crew or team coordination.

Why does group coordination become more problematic under stress or emergency conditions? Research at the *individual* level of analysis has suggested that individuals' focus of attention shifts from the broad to the narrow when under stress, and we believe this phenomenon may have significant implications for group interaction. One of the better established findings in the stress literature is that as stress or arousal increases, the individual's breadth of attention narrows (Combs & Taylor, 1952; Easterbrook, 1959). Perhaps the earliest statement of this phenomenon was by William James (1890) who believed that the individual's field of view varied, from a broader perspective under normal conditions to a more narrow, restricted focus under stress. Pennebaker, Czajka, Cropanzano, & Richards (1990) provided an empirical test of the hypothesis that normal thought processes and attentional focus are restricted under stress; they found that individuals confronted with uncontrollable noise tended to move from higher to lower levels of thought—from a broad to a narrow perspective—when under stress. Other researchers have shown that stress may increase individual self-focus; for example, Wegner and Giuliano (1980) found that increased arousal led to greater self-focused attention.

Driskell, Salas, and Johnston (1997) extended this research to the group level of analysis by arguing that group members under stress may become less *group* focused. They held that the narrowing of attention, or "tunnel vision," that occurs under stress may include a restriction of social stimuli as well, and that under stress, group members may adopt a narrower, more individual perspective of task activity. With this narrowing of perspective, team members' cognitions shift from a broader, team perspective to a narrower, individualistic focus.

Driskell, Salas, and Johnston (1997) conducted a study in which three-person teams performed a decision-making task under normal or high-stress conditions. Results indicated that stress causes a narrowing of team perspective. Team members in the stress condition were less likely to develop a strong team identity and to adopt a team-level task perspective. Furthermore, team perspective was found to be a significant predictor of task performance. These results suggest that, under stress, team members may lose the collective representation of group activity that characterizes interdependent team behavior.

To learn how to counter the effect of stress on the narrowing of team perspective, a follow-up study was conducted on the effects of enhancing or strengthening team perspective. In this study, Driskell, Salas, and Johnson (1997) implemented a SET-type intervention to train and reinforce teamwork skills. This intervention consisted of (a) information on the importance of teamwork skills and how they can be affected by stress,

(b) training in team skills such as providing feedback to other team members, and (c) practice of these skills in a realistic task simulation. Teams that received this training maintained a broader team perspective and performed better under stress than teams without training. For team tasks that must be performed under stress or emergency conditions, therefore, results suggest that it may be useful to implement training to reinforce teamwork skills.

Phase 3: Application and Practice

> It is immensely important that no soldier ... should wait for war to expose him to those aspects of active service that amaze and confuse him when he first comes across them. If he has met them even once before, they will begin to be familiar to him. (Clausewitz, 1976)

As part of the classic American Soldier studies conducted during World War II, military researchers asked combat veterans who served in the Italy and the North African campaign this question: "What type of training did you lack?" The most frequent response from these soldiers was that they lacked training under realistic battle conditions (Janis, 1949, p. 229).

Training generally occurs in a calm and relatively benign environment. This type of environment is designed to be conducive to learning, and it allows trainees to acquire initial skills in an efficient manner. Yet, actual task conditions are often quite unlike those found in the training setting. In fact, the extreme time restrictions, novelty, ambiguity, and confusion that occur under stress conditions often create a substantially different task environment than that experienced in a normal training setting. Thus, the novelty of performing even a well-learned task in a high-stress environment can cause severe degradation in performance.

One crucial aspect of maintaining effective performance in a stressful environment is providing practice and exercise of critical tasks under operational conditions similar to those likely to be encountered in the real-world setting. Training that allows some degree of preexposure to the stress operational environment should reduce the extent of performance decrement encountered in the operational setting. This strategy has been successful in a number of military applications, including water-survival training, flight emergency training, and fire-fighting training.

The primary goal of Phase 3 of SET, therefore, is to provide for the application and practice of task skills in a simulated stress environment. This strategy has several benefits. First, it allows trainees to perform tasks in the simulated stress environment and to experience the type of performance problems encountered in this setting. Furthermore, use of the skills taught in Phase 2 and now practiced in Phase 3 should allow trainees to adapt performance to this environment. Second, preexposure to criterion-like stressors reduces uncertainty and anxiety regarding these events and increases confidence in the ability to perform in this setting.

Third, events that have been experienced during training will be less distracting when faced in the operational environment.

In the SET model, the third phase of training requires the practice of skills in a simulated stress environment. However, one issue that has been deliberated by researchers is the timing and manner in which realistic stressors should be introduced in training. High-fidelity stressors are those that are "just like" those encountered in the operational environment. Some researchers have argued for high-fidelity simulation of stress in training (Terris & Rahhal, 1969; West, 1958; Willis, 1967). Others have argued that the high demand, ambiguity, and complexity of the stress environment is not conducive to the early stages of learning and that exposure to stressors too early in training may interfere with initial skill acquisition (Lazarus, 1966).

Reigeluth and Schwartz (1989) noted that: "the design of the instructional overlay for any simulation begins with making sure not to overload the learner. The real situation is usually quite complex ... to begin with so many variables in the underlying model will clearly impede learning and motivation" (p. 4). Elaborating this point, Regian, Shebilske, and Monk (1992) claimed that it is not necessarily true that higher fidelity always leads to better training and that many training strategies reduce fidelity early in training to reduce complexity. Friedland and Keinan (1986) found evidence to support the effectiveness of *phased training* as an approach to manage training for complex environments. On the basis of the assumption that a high degree of complexity in the training environment may interfere with initial skill acquisition, phased training was used to maximize training effectiveness by partitioning training into separate phases: During initial training, trainees learned basic skills in a relatively low-fidelity, or low-complexity, environment, and later stages of training incorporated greater degrees of complexity, or realism. Keinan and Friedland (1996) noted that allowing skills practice in a graduated manner across increasing levels of stressors (from moderate stress exercises to higher stress exercises) satisfies three important requirements: It allows the individual to become more familiar with relevant stressors without being overwhelmed; it enhances a sense of individual control and builds confidence; and it is less likely to interfere with the acquisition and practice of task skills than is exposure to intense stress. The SET model incorporates this aspect of phased training in the overall three-phase training approach.

One question that has substantial applied consequences as well as theoretical implications is the extent to which stress training is generalizable, from stressor to stressor and from task to task. First, consider the question of generalization from stressor to stressor. In a SET training intervention, trainees receive (a) specific stress information, (b) skills training, and (c) practice of the task under simulated stress conditions. Will the positive effects of training that addresses one type of stress (e.g., time pressure) generalize to a task situation involving a novel stress (e.g., noise)? Will the skills learned in the stress training (e.g., how to focus attention under time pressure) generalize to a novel stress setting? In other words, do trainees learn a specific skill in training (how to focus attention under time pressure), or do they learn a more generalizable stress adaptive skill (how to focus attention under stress conditions)?

A related question is whether stress training generalizes from task to task. Again, consider a stress training intervention in which trainees receive (a) specific stress information, (b) skills training, and (c) practice of the task under simulated stress conditions. Will the benefits gained in stress training generalize when the trainees face not Task A (the training task), but a novel task, Task B? In other words, do trainees learn specific stress skills that are applicable only to the particular task that is practiced in training, or do they learn generalizable skills that would transfer to novel tasks?

Driskell, Johnston, and Salas (1997) conducted a study to demonstrate the efficacy of a brief SET intervention in decreasing self-reported stress and enhancing performance under stress. However, the primary focus of the study was to determine whether stress generalized from stressor to stressor (Experiment 1) and whether the effects of training generalized from task to task (Experiment 2).

In Experiment 1, there were three performance trials for each participant: (a) Performance was assessed pretraining, when participants performed the task under either time pressure or noise conditions; (b) performance was assessed posttraining, when participants who received "noise stress" training performed under noise stress, and those who received "time pressure" training performed under time pressure; and (c) performance was assessed under novel stressor conditions, when participants who received noise stress training now performed under time pressure and those who received time pressure training now performed under noise stress. The design of Experiment 2, which assessed generalization of training from task to task was similar: (a) Performance was assessed pretraining, when participants performed either Task A or Task B under stress; (b) performance was assessed posttraining, when participants who received stress training and practice for Task A then performed Task A under stress and those who received stress training and practice for Task B then performed Task B under stress; and (c) performance was assessed under novel task conditions, when participants who received stress training for Task A now performed Task B under stress and those who received stress training for Task B now performed Task A under stress.

Overall results indicated that (a) the SET intervention resulted in decreased subjective stress and improved performance and (b) the beneficial effects of stress training were maintained when participants performed in the presence of a novel stressor and performed a novel task. This study has significant consequences for the design of stress training in that the exact types of stress inherent in many real-world task environments (e.g., the threat, noise, and time pressure present in a flight emergency) are often difficult to create or anticipate fully during training. These findings suggest that the skills learned from stress training are generalizable to novel task and stressor settings.

Lessons Learned: Guidelines for Implementing Stress-Exposure Training

Lesson 1: High-Demand, High-Stress Conditions Often Lead to Disrupted Performance

We state this truism because it clarifies our interest in stress and performance. Stress affects physiological, cognitive, emotional, and social processes, and these effects may have a direct impact on task performance. Stressors such as time pressure, task load, information complexity, and ambiguity occur in many applied settings, such as aviation; military operations; nuclear, chemical, and other industrial settings; and everyday work situations. When individuals face stressors that disrupt goal-oriented behavior, performance effects may include increased errors, slowed response, and greater variability in performance.

Lesson 2: Technical Skill Is a Necessary but Not a Sufficient Condition to Support Effective Performance in the Stress Environment

Some tasks must be performed under conditions quite unlike those encountered in the training classroom. For example, high-stress environments include specific task conditions (such as time pressure, ambiguity, increased task load, distractions) and require specific responses (such as the flexibility to adapt to novel and often changing environmental contingencies) that differ from those required in the normal performance environment. Preparing personnel to perform under high-stress conditions requires that the task performer be highly skilled, be familiar with the stress environment, and possess the special knowledge and skills necessary to overcome the deficits imposed by high-stress or high-demand conditions.

Lesson 3: The Three-Phase SET Training Format Is an Effective Approach for Enhancing Performance Under Stress

The SET approach is defined by a three-stage training intervention: (a) an initial stage in which information is provided regarding stress and stress effects; (b) a skills training phase, in which specific cognitive and behavioral skills are acquired; and (c) the final stage of application and practice of skills learned under conditions that increasingly approximate the criterion environment. Research has shown that this three-stage stress training intervention is an effective approach for reducing anxiety and enhancing performance under stress (Saunders et al., 1996). It is likely that each phase of training contributes to overall training effectiveness.

Lesson 4: Preparatory Information Regarding the Nature of the Stress Environment Can Lessen Negative Stress Effects and Enhance Performance in the Operational Environment

One objective of the first phase of SET is to provide information on the nature of the stress environment and individual reactions to stress. Research showed that those given preparatory information before performing in a stressful environment made fewer errors, were less likely to feel stressed, and were more confident in their ability to perform the task (Inzana et al., 1996). A comprehensive preparatory information intervention should include sensory, procedural, and instrumental information.

Lesson 5: Stress Training Should Focus on Developing the Cognitive and Behavioral Skills Required to Maintain Effective Performance Under Stress

During the second phase of SET, trainees acquire and practice stress-management skills to enhance the capability to respond effectively in the stress environment. The type of skills training implemented varies according to the specific training requirements but may include cognitive control techniques that train the individual to regulate negative emotions and distracting thoughts, as well as training to enhance physiological control (i.e., awareness and control of muscle tension and breathing). Other training strategies that have been shown to be effective in enhancing performance include overlearning (Driskell et al., 1992), mental practice (Driskell et al., 1994), and training in decision making-skills (Johnston et al., 1997).

Lesson 6: One Crucial Aspect of Maintaining Effective Performance in a Stressful Environment Is Providing Practice and Exercise of Tasks Under Operational Conditions Similar to Those Likely to Be Encountered in the Real-World Setting

The final phase of training provides the opportunity to apply and practice task skills in a setting that approximates the real-world stress environment. Providing skills practice in a graduated manner across increasing levels of stress (from moderate stress exercises to higher stress exercises) enhances a sense of control and confidence and is less likely to interfere with the acquisition and practice of task skills than does initial exposure to more intense stress.

Lesson 7: Stress-Exposure Training May Be Presented as a Component of Initial Technical Training or as a Part of Recurrent or Refresher Training

If SET is presented as a component of initial technical training, it should be introduced after basic technical skills are developed. The introduction of SET too early in the training curriculum may interfere with initial skill acquisition. If SET is presented as a component of refresher training, the trainer should ensure that trainees have the opportunity to practice basic skills before stress training exercises are presented.

Lesson 8: Absolute Fidelity in Training Is Not Possible nor Necessarily Desirable

Fidelity refers to the degree to which characteristics of the training environment are similar to those of the criterion setting. Many bemoan the fact that training will never approach or capture the "life-threatening" feel of the real-world setting (such as an aircrew emergency when lives are on the line). This is true; trainees are generally aware that they are in a "safe" training environment. However, a well-designed training simulation can be quite involving and can "feel" like the real thing without imposing extreme or dangerous levels of stress. Moreover, absolute fidelity in stress training is not often desirable. If stress is too high in training (e.g., if time pressure is too high), there may be little chance of successful task performance, and trainees may receive a negative training experience. Research has suggested that stressors introduced at a moderate level of fidelity during training can provide an effective and realistic representation of the stress environment.

Conclusion

Spettell and Liebert (1986) described the high task demands and pressures under which personnel perform in high-stress settings such as in the nuclear power and aviation industries, and they proposed that "well-established psychological training techniques [such as] stress inoculation may help to avoid or neutralize these threats" (p. 545). However, other researchers have cautioned that the stress-management literature has suffered from a lack of rigorous evaluation and that proof of the effectiveness of these programs is difficult to obtain (Newman & Beehr, 1979; Wexley & Latham, 1991). This chapter has presented a model of stress-exposure training, described the empirical research that supports this approach, and derived some guidelines for implementing stress-exposure training. A reasonable interpretation of the results is that the stress-exposure training model is an effective method for reducing anxiety and enhancing performance in stressful environments. The results of this analysis should clearly encourage further application and research.

References

Ausubel, D. P., Schiff, H. M., & Goldman, M. (1953). Qualitative characteristics in the learning process associated with anxiety. *Journal of Abnormal and Social Psychology, 48,* 537–547.

Bandura, A., Reese, L., & Adams, N. E. (1982). Microanalysis of action and fear arousal as a function of differential levels of perceived self-efficacy. *Journal of Personality and Social Psychology, 43*, 5–21.

Beach, L. R., & Mitchell, T. R. (1978). A contingency model for the selection of decision strategies. *Academy of Management Review, 3*, 439–449.

Burish, T. G., Carey, M. P., Krozely, M. G., & Greco, F. A. (1987). Conditioned nausea and vomiting induced by cancer chemotherapy: Prevention through behavioral treatment. *Journal of Consulting and Clinical Psychology, 55*, 42–48.

Cascio, W. F. (1991). *Applied psychology in personnel management.* Englewood Cliffs, NJ: Prentice-Hall.

Clark, D. M. (1988). A cognitive model of panic attacks. In S. Rachman & J. D. Maser (Eds.), *Panic: Psychological perspectives* (pp. 71–89). Hillsdale, NJ: Erlbaum.

Clausewitz, C. von. (1976). *On war* (M. Howard & P. Paret, Trans.). Princeton, NJ: Princeton University Press.

Cohen, S. (1978). Environmental load and the allocation of attention. In A. Baum, J. E. Singer, & S. Valins (Eds.), *Advances in environmental psychology* (Vol. 1, pp. 1–29). Hillsdale, NJ: Erlbaum.

Combs, A. W., & Taylor, C. (1952). The effect of the perception of mild degrees of threat on performance. *Journal of Abnormal and Social Psychology, 47*, 420–424.

Cowings, P. S., & Toscano, W. B. (1982). The relationship of motion sickness susceptibility to learned autonomic control for symptom suppression. *Aviation, Space, and Environmental Medicine, 53*, 570–575.

Damos, D. L., Bittner, A. C., Kennedy, R. S., & Harbeson, M. M. (1981). Effects of extended practice on dual-task tracking performance. *Human Factors, 23*, 627–631.

Deese, J. (1961). Skilled performance and conditions of stress. In R. Glaser (Ed.), *Training research and education.* Pittsburgh, PA: University of Pittsburgh. (AD 263 439)

Dobie, T. G., May, J. G., Fischer, W. D., & Elder, S. T. (1987). A comparison of two methods of training resistance to visually-induced motion sickness. *Aviation, Space, and Environmental Medicine, 58*, 34–41.

Driskell, J. E., Copper, C., & Moran, A. (1994). Does mental practice enhance performance? *Journal of Applied Psychology, 79*, 481–492.

Driskell, J. E., Johnston, J., & Salas, E. (1997). *Is stress training generalizable to novel settings?* Manuscript in preparation.

Driskell, J. E., & Olmstead, B. (1989). Psychology and the military: Research applications and trends. *American Psychologist, 44*, 43–54.

Driskell, J. E., & Salas, E. (1991). Group decision making under stress. *Journal of Applied Psychology, 76*, 473–478.

Driskell, J. E., & Salas, E. (Eds.). (1996). *Stress and human performance.* Hillsdale, NJ: Erlbaum.

Driskell, J. E., Salas, E., & Johnston, J. H. (1997). *Does stress lead to a loss of team perspective?* Manuscript submitted for publication.

Driskell, J. E., Willis, R., & Copper, C. (1992). Effect of overlearning on retention. *Journal of Applied Psychology, 77*, 615–622.

Druckman, D., & Swets, J. (1988). *Enhancing human performance.* Washington, DC: National Academy Press.

Easterbrook, J. A. (1959). The effect of emotion on cue utilization and the organization of behavior. *Psychological Review, 66*, 183–201.

Egbert, L. D., Battit, G. E., Welch, C. E., & Bartlett, M. K. (1964). Reduction of postoperative pain by encouragement and instruction of patients. *New England Journal of Medicine, 270*, 825–827.

Fitts, P. M. (1965). Factors in complex skill training. In R. Glaser (Ed.), *Training research and education* (pp. 177–197). New York: Wiley.

Foushee, H. C. (1984). Dyads and triads at 35,000 feet: Factors affecting group process and aircrew performance. *American Psychologist, 39*, 885–893.

Friedland, N., & Keinan, G. (1986). Stressors and tasks: How and when should stressors be introduced during training for task performance in stressful situations? *Journal of Human Stress, 12*, 71–76.

Geen, R. G. (1989). Alternative conceptions of social facilitation. In P. B. Paulus (Ed.), *Psychology of group influence* (pp. 15–52). Hillsdale, NJ: Erlbaum.

Hammerton, M., & Tickner, A. H. (1969). An investigation into the effects of stress upon skilled performance. *Ergonomics, 12*, 851–855.

Hirst, W., Spelke, E. S., Reaves, C. C., Caharack, G., & Neisser, U. (1980). Dividing attention without alternation or automaticity. *Journal of Experimental Psychology: General, 109*, 98–117.

Hoehn, A. J., & Levine, A. L. (1951). *The development and maintenance of motivation in training and career development* (HBM 200/1 Appendix 94). Washington, DC: Research and Development Board, Department of Defense. (AD 618 538)

Ilgen, D. R. (1990). Health issues at work: Opportunities for industrial/organizational psychology. *American Psychologist, 45*, 273–283.

Innes, L. G., & Allnutt, M. F. (1967). *Performance measurement in unusual environments* (IAM Technical Memorandum No. 298). Farnborough, England: RAF Institute of Aviation Medicine.

Inzana, C. M., Driskell, J. E., Salas, E., & Johnston, J. H. (1996). Effects of preparatory information on enhancing performance under stress. *Journal of Applied Psychology, 81*, 429–435.

Ivancevich, J. M., Matteson, M. T., Freedman, S. M., & Phillips, J. S. (1990). Worksite stress management interventions. *American Psychologist, 45*, 252–261.

James, W. (1890). *The principles of psychology* (Vol. 1). New York: Holt.

Janis, I. L. (1949). Problems related to the control of fear in combat. In S. A. Stouffer (Ed.), *The American soldier: Combat and its aftermath*. Princeton, NJ: Princeton University Press.

Janis, I. L. (1951). *Air war and emotional stress*. New York: McGraw-Hill.

Janis, I. L., & Mann, L. (1977). *Decision making: A psychological analysis of conflict, choice, and commitment*. New York: Free Press.

Johnston, J. H., & Cannon-Bowers, J. A. (1996). Training for stress exposure. In J. E. Driskell & E. Salas (Eds.), *Stress and human performance* (pp. 223–256). Mahwah, NJ: Erlbaum.

Johnston, J. H., Driskell, J. E., & Salas, E. (1997). Vigilant and hypervigilant decision making. *Journal of Applied Psychology, 82*, 614–622.

Jones, D. R. (1985). Self-control of psychophysiologic response to motion stress: Using biofeedback to treat airsickness. *Aviation, Space, and Environmental Medicine, 55*, 1152–1157.

Jones, J. G., & Hardy, L. (1990). *Stress and performance in sport*. Chichester, England: Wiley.

Kaempf, G. L., Klein, G., Thordsen, M. L., & Wolf, S. (1996). Decision making in complex command-and-control environments. *Human Factors, 38*, 220–231.

Keinan, G. (1987). Decision making under stress: Scanning of alternatives under controllable and uncontrollable threats. *Journal of Personality and Social Psychology, 52*, 639–644.

Keinan, G. (1988). The effects of expectations and feedback. *Journal of Applied Social Psychology, 18*, 355–373.

Keinan, G., & Friedland, N. (1996). Training effective performance under stress: Queries, dilemmas and possible solutions. In J. E. Driskell & E. Salas (Eds.), *Stress and human performance* (pp. 257–277). Mahwah, NJ: Erlbaum.

Klein, G. (1989). Recognition-primed decisions. In W. Rouse (Ed.), *Advances in man-machine systems research* (Vol. 5, pp. 47–92). Greenwich, CT: JAI Press.

Klein, G. (1996). The effect of acute stressors on decision making. In J. E. Driskell & E. Salas (Eds.), *Stress and human performance* (pp. 49–88). Mahwah, NJ: Erlbaum.

Lazarus, R. S. (1966). *Psychological stress and the coping process*. New York: McGraw-Hill.

Lazarus, R. S., & Folkman, S. (1984). *Stress, appraisal, and coping*. New York: Springer.

Locke, E. A., Frederick, E., Lee, C., & Bobko, P. (1984). Effect of self-efficacy, goals, and task strategies on task performance. *Journal of Applied Psychology, 69*, 241–251.

Mackenzie, C. F., Craig, G. R., Parr, M. J., & Horst, R. (1994). Video analysis of two emergency tracheal intubations identifies flawed decision making. *Anesthesiology, 81*, 4–12.

Markowitz, J. S., Gutterman, E. M., Link, B., & Rivera, M. (1987). Psychological responses of firefighters to a chemical fire. *Journal of Human Stress*, 84–93.

Mathews, K. E., & Canon, L. K. (1975). Environmental noise level as a determinant of helping behavior. *Journal of Personality and Social Psychology, 32*, 571–577.

Meichenbaum, D. H. (1985). *Stress inoculation training.* New York: Pergamon Press.

Meichenbaum, D. H., & Cameron, R. (1983). Stress inoculation training. In D. Meichenbaum & M. E. Jaremko (Eds.), *Stress reduction and prevention* (pp. 115–154). New York: Plenum Press.

National Transportation Safety Board. (1973). *Aircraft accident report: Eastern Airlines, Inc. L-1011, N310EA Miami Florida* (NTSB/AAR-73/14), Washington, DC.

National Transportation Safety Board. (1979). *Aircraft accident report: United Airlines Flight 173, DC-8–61, N8082U* (NTSB/AAR-79/12), Washington, DC.

Newman, J. D., & Beehr, T. (1979). Personal and organizational strategies for handling job stress: A review of research and opinion. *Personnel Psychology, 32*, 1–43.

Orasanu, J. M. (1993). Decision making in the cockpit. In E. Wiener, B. Kanki, & R. Helmreich (Eds.), *Cockpit resource management* (pp. 137–172). San Diego, CA: Academic Press.

Payne, J. W., Bettman, J. R., & Johnson, E. J. (1988). Adaptive strategy selection in decision making. *Journal of Experimental Psychology: Learning, Memory, and Cognition, 14*, 534–552.

Pennebaker, J. W., Czajka, J. A., Cropanzano, R., & Richards, B. C. (1990). Levels of thinking. *Personality and Social Psychology Bulletin, 16*, 743–757.

Perrow, C. (1984). *Normal accidents: Living with high-risk technologies.* New York: Basic Books.

Prince, C., Bowers, C. A., & Salas, E. (1994). Stress and crew performance: Challenges for aeronautical decision making training. In N. Johnston, N. McDonald, & R. Fuller (Eds.), *Aviation psychology in practice* (pp. 286–305). Hants, England: Avebury.

Rachman, S. (Ed.). (1983). Fear and courage among military bomb-disposal operators [Special issue]. *Advances in Behaviour Research and Therapy, 4*(3).

Radloff, R., & Helmreich, R. (1968). *Groups under stress: Psychological research in Sealab II.* New York: Appleton-Century-Crofts.

Regian, J. W., Shebilske, W. L., & Monk, J. M. (1992). Virtual reality: An instructional medium for visual–spatial tasks. *Journal of Communication, 42*, 136–149.

Reigeluth, C. M., & Schwartz, E. (1989). An instructional theory for the design of computer-based simulations. *Journal of Computer-Based Instruction, 16*, 1–10.

Richardson, A. (1967). Mental practice: A review and discussion (Part II). *Research Quarterly, 38*, 264–273.

Saunders, T., Driskell, J. E., Johnston, J. H., & Salas, E. (1996). The effect of stress inoculation training on anxiety and performance. *Journal of Occupational Health Psychology, 1*, 170–186.

Schendel, J. D., & Hagman, J. D. (1982). On sustaining procedural skills over a prolonged retention interval. *Journal of Applied Psychology, 67*, 605–610.

Singer, R. N., Cauraugh, J. H., Murphey, M., Chen, D., & Lidor, R. (1991). Attentional control, distractors, and motor performance. *Human Performance, 4*, 55–69.

Spelke, E., Hirst, W., & Neisser, U. (1976). Skills of divided attention. *Cognition, 4*, 215–230.

Spettell, C. M., & Liebert, R. M. (1986). Training for safety in automated person–machine systems. *American Psychologist, 41*, 545–550.

Staw, R. M., Sandelands, L. E., & Dutton, J. E. (1981). Threat-rigidity effects in organizational behavior: A multi-level analysis. *Administrative Science Quarterly, 26*, 501–524.

Streufert, S., & Streufert, S. C. (1981). *Stress and information search in complex decision making: Effects of load and time urgency* (Tech. Rep. No. 4). Arlington, VA: Office of Naval Research.

Taylor, S. E., & Clark, L. F. (1986). Does information improve adjustment to noxious medical procedures? In M. J. Saks & L. Saxe (Eds.), *Advances in applied social psychology* (Vol. 3, pp. 1–29). Hillsdale, NJ: Erlbaum.

Terris, W., & Rahhal, D. K. (1969). Generalized resistance to the effects of psychological stressors. *Journal of Personality and Social Psychology, 13*, 93–97.

Thompson, S. C. (1981). Will it hurt less if I can control it? A complex answer to a simple question. *Psychological Bulletin, 90*, 89–101.

Tryon, G. S. (1980). The measurement and treatment of test anxiety. *Review of Educational Research, 50*, 343–372.

Vineberg, R. (1965). *Human factors in tactical nuclear combat.* Alexandria, VA: Human Resources Research Office, George Washington University. (AD 463 787)

Wachtel, P. L. (1968). Anxiety, attention, and coping with threat. *Journal of Abnormal Psychology, 73*, 137–143.

Wegner, D. M., & Giuliano, T. (1980). Arousal-induced attention to self. *Journal of Personality and Social Psychology, 38*, 719–726.

Weitz, J. (1966). *Stress* (Research Paper P-251). Washington, DC: Institute for Defense Analyses. (AD 633 566)

West, L. J. (1958). Psychiatric aspects of training for honorable survival as a prisoner of war. *American Journal of Psychiatry, 115*, 329–336.

Wexley, K. N., & Latham, G. P. (1991). *Developing and training human resources in organizations* (2nd ed.). New York: HarperCollins.

Willis, M. P. (1967). Stress effects on skill. *Journal of Experimental Psychology, 74*, 460–465.

Wine, J. (1971). Test anxiety and direction of attention. *Psychological Bulletin, 76*, 92–104.

Wine, J. D. (1980). Cognitive–attentional theory of test anxiety. In I. G. Sarason (Ed.), *Test anxiety: Theory, research, and application* (pp. 349–385). Hillsdale, NJ: Erlbaum.

Worchel, S., & Yohai, S. M. L. (1979). The role of attribution in the experience of crowding. *Journal of Experimental Social Psychology, 15*, 91–104.

Wright, P. (1974). The harassed decision maker: Time pressures, distractions, and the use of evidence. *Journal of Marketing Research, 44*, 429–443.

Yamamoto, T. (1984). Human problem solving in a maze using computer graphics under an imaginary condition of "fire." *Japanese Journal of Psychology, 55*, 43–47.

Yuille, J. C., Davies, G., Gibling, F., Marxsen, D., & Porter, S. (1994). Eyewitness memory of police trainees for realistic role plays. *Journal of Applied Psychology, 79*, 931–936.

Zakay, D., & Wooler, S. (1984). Time pressure, training and decision effectiveness. *Ergonomics, 27*, 273–284.

Part III _____

Team-Level Training Strategies and Research

9

Team Coordination Training

*Daniel Serfaty, Elliot E. Entin, and
Joan H. Johnston*

In this chapter we describe a model-driven team training strategy called team adaptation and coordination training (TACT), which is aimed at enhancing team coordination performance. Team coordination strategies are trained through a dynamic adaptation of teamwork processes to changing external and internal conditions. The research is based on a multiyear effort conducted as part of the Navy Tactical Decision Making Under Stress (TADMUS) research program; one of the main research foci of TADMUS was the understanding of complex team processes and the development of team training strategies. The problem domain for these studies is tactical decision making in naval combat information centers (CIC). In the CIC, complex decision-making processes rely significantly on effective coordination by the entire CIC team.

The research reported here concerns a series of model-based experiments that simulated a five-member command team in a CIC aboard an AEGIS platform. The goal has been to investigate how teams can successfully adapt their decision-making and coordination strategies to a changing tactical environment, and how team training and structural configurations based on shared awareness of the situation can best contribute to the team's ability to maintain superior performance under a wide range of stressful operational conditions (Entin, Serfaty, & Deckert, 1994; Entin, Serfaty, Entin, & Deckert, 1993; Serfaty, Entin, & Deckert, 1993). At the inception of this research, no formal comprehensive models existed to provide hypotheses on how high-performing teams adapt their decision-making and coordination strategies to changes in situational demand.

The intent of this chapter is threefold. The first objective is to frame the results from the various studies into a single model that links the concepts of mental models, adaptive coordination, and shared knowledge into an empirically supported theoretical framework that can provide guidelines for future training and for improving the performance of military command teams. Second, we review the empirical results of the studies contributing to the development and support of the adaptive team coordination model. The third objective is to describe in some detail the measurement instruments we developed to assess team adaptation and coordination.

The study of team coordination processes and their effects on team performance outcome has shown that teams who maintain superior performance under high levels of workload and stress employ different coordination strategies from low-performing teams (Entin & Serfaty, 1990; Entin, Serfaty, Entin, & Deckert, 1993; Serfaty, Entin, & Deckert, 1993). It was found that teams exhibiting highly reliable performance possess the ability, when faced with an increasingly demanding task environment, to adapt their decision-making strategies, coordination strategies, and even their structure to maintain their performance in the presence of escalating workload and stress. We hypothesized that an important mechanism used by highly effective teams in the adaptation process is the development of a shared situational mental model of the task environment and the task itself, and a mutual mental model of interacting team members' tasks and abilities. These models are used to generate expectations about how other team members will behave (Cannon-Bowers, Salas, & Converse, 1993; McIntyre & Salas, 1995; Orasanu, 1990; Serfaty, Entin, & Volpe, 1993). Moreover, there is research evidence to show that high-performing teams make use of such models particularly when timely, error-free, and unambiguous information is at a premium, to anticipate both the developments of the situation and the needs of the other team members.

In light of these findings, we contend that *teams who develop a high level of congruence between their mental models—both situational and mutual—are able to make use of these models to anticipate the way the situation will evolve as well as the needs of the other team members.* These teams will perform consistently better under a wide range of conditions (e.g., high stress or workload). The research evidence also shows that it is this team coordination strategy of anticipating changes in the situation and needs of other team members that contributes significantly to the teams' superior performance under stress and is the reason these teams consistently perform better under a wide range of tactical conditions.

The Adaptive Team Model

In the search for effective team training strategies, research has shown that superior teams have one key quality in common: the ability to adapt to task demands. The premise of the adaptation model (Entin, Serfaty, & Deckert, 1994), illustrated in Figure 1, is that, under conditions of increasing stress, high-performing teams adapt their (a) decision-making strategy, (b) coordination strategy, and (c) organizational structure to maintain team performance at acceptable levels while keeping the perceived stress at tolerable levels.

In this model, stress is viewed as operating at two levels: as input to and output from the team processes of taskwork and teamwork. At the

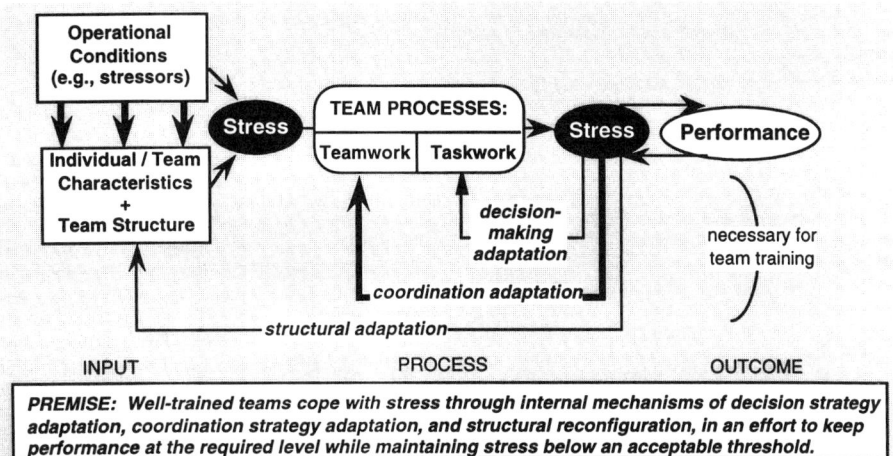

Figure 1. Theoretical framework for team adaptation.

input levels, stress is modeled as the effects of operational conditions, or stressors, on the team members and the structure. For example, the same level of task uncertainty (an external stressor) may produce different levels of input stress to different team organizational structures. At the output levels, stress is produced as a function of the type of taskwork and teamwork strategies used by the teams. Certain decision and communication strategies might require more mental effort and therefore be more stressful than others.

The basic hypothesis generated by this model is that teams have at their disposal sensing strategies and related adaptation strategies that enable them to perceive changes in stress (at both levels) and adapt their behavior in a targeted manner to cope with those changes. Rather than adopting a fixed approach to teamwork that is used under all circumstances, successful teams adapt their behavior to the demands of the situation. Hence, a critical aspect of any team training strategy is that it teach team members to be sensitive and responsive to changes in task demands (Salas & Cannon-Bowers, 1997). Most of the training research performed in this effort is aimed at the midlevel loop, targeting training and assessment of coordination and adaptation in teams.

Rasmussen (1990) suggested that flexible and dynamic work conditions are changing the concept of human error. In stable work conditions, know-how develops and "normal" ways of doings things emerge. However, complex systems require the flexible resolution of problems by individuals, with multiple degrees of freedom for action and many possible "right" answers; therefore, they introduce a requirement for continuous resolution of problems and choice among alternatives. Rasmussen (1990) argued that the human tendency to reduce degrees of freedom by reducing choices and decisions during normal operating conditions necessarily introduces occasional errors, and that *"the trick in design of reliable systems is to make sure that human actors maintain sufficient flexibility to cope with system aberrations.*" He concluded that *"dynamic shifting among alternative strategies is very important* [italics added] for skilled people as a means to resolve resource–demand conflicts met during performance" (p. 458).

Observation of teams with records of highly reliable performance suggests that the ability to adapt to a changing environment, with an associated requirement for coordination, may lie at the heart of a team's resilience to errors. LaPorte and his colleagues (LaPorte & Consolini, 1988; see also Pfeiffer, 1989; Reason, 1990) have conducted extensive observation and analysis of several types of highly reliable teams operating in real-world environments. Their observations about the common characteristics and procedures of these teams emphasize the importance of flexibility and adaptability in preventing errors in dynamic environments. The following features are characteristic of highly reliable teams:

- The team structure is *adaptive* to changes in the task environment. The reliable team has not one but several organizational structures and shifts between them as needed. LaPorte distinguishes three authority structures: routine, high-tempo, and emergency, each with a different set of characteristic practices, communication pathways, and leadership patterns. The same individuals may play completely different roles under different circumstances.
- The team maintains *open and flexible communication* lines. An adaptable team structure seems to promote the free flow of information from lowest to highest levels as well as the other way around. This is critical in situations in which lower levels in a hierarchy may have critical information that is not available to the upper levels. A number of accidents have occurred in which an aircraft's copilot or a ship's crew were aware of a problem but were reluctant to bring it to the attention of the pilot or captain (Green, 1990; Pfeiffer, 1989).
- Team members are extremely *sensitive to other members' workload and performance* in high-tempo situations. LaPorte observed that air traffic control teams are very sensitive to the overloading of any team member and, without any overt request for help, will gather around the screen of an overloaded individual to watch for danger points until the overload condition has passed.

Shared Mental Models and Team Coordination

It has been suggested by various authors (Cannon-Bowers et al., 1993; Kleinman & Serfaty, 1989; McIntyre & Salas, 1995; Orasanu, 1990) that highly effective teams have a shared mental model of the situation and of the task environment (consistent with the team's understanding of the situation) and a mutual mental model of interacting team members' tasks and abilities that generates expectations about how other team members will behave. These mental models—*situational and mutual*—are particularly useful in the absence (or scarcity) of timely, error-free, and unambiguous information. We hypothesize that teams that have developed a high level of congruence among their mental models are able to make use of these models to anticipate the way the situation will evolve as well as the needs of the other team members. These teams will perform consistently better under a wide range of conditions.

The concept of a shared mental model has proved to be a powerful mechanism for understanding the relationship between team coordination behaviors and team performance. Differences in the communication patterns of low-performance and high-performance flight crews found by Orasanu (1990) suggest that high-performance crews are more successful in developing a shared mental model of their situation. For example, the captains of high-performance crews produced more communications concerning plans and strategies than the captains of the low-performance crews. Under high-workload conditions, first officers in the high-performance crews provided more information to their captains than they did under low-workload conditions, whereas the pattern was just the opposite for the low-performance crews. Orasanu interpreted these findings as indicating that a shared mental model was developed when the captains in the high-performance crews shared their plans, allowing the first officers in those crews to provide information relevant to the plan (i.e., implicitly coordinate) under high-workload conditions.

The coordination mechanisms that support adaptation may be *explicit*, based on specific communications, or *implicit*, based on a shared mental model. Both explicit and implicit coordination modes will generate observable communication patterns; the presence *and* the absence of communication may be important. Moreover, changes in communication patterns are indicators of shifts in team coordination strategies.

Figure 2 shows an extension of Wohl's (1981) stimulus–hypothesis–option–response (SHOR) paradigm, which describes the alternative channels of implicit and explicit team coordination to support the shared mental model concept. For example, communications that provide information to a team member in the absence of requests for that information indicate an implicit coordination mechanism at work. Measures must be sensitive to changes in the team's coordination and communication patterns as they adapt their behavior to the demands of the task and the environment. Teams generally alternate between team communication modes dominated by explicit coordination and those dominated by implicit coordination. Even though a team may have a shared mental model that supports implicit coordination, some explicit exchange of information will be required to maintain that shared model as the situation changes (Orasanu & Fischer, 1992). Observation has also suggested that members of highly reliable teams are aware of the workload of other team members and *implicitly* assume some of the tasks of any individual in the team who is overloaded under stressful conditions (compensatory behavior). This dynamic reallocation of workload is observable from the team's communication patterns, and teamwork measures should be sensitive to it.

Past experimental studies (Serfaty, Entin, & Deckert, 1993; Serfaty, Entin, & Volpe, 1993) have shown a quantitative link between shared

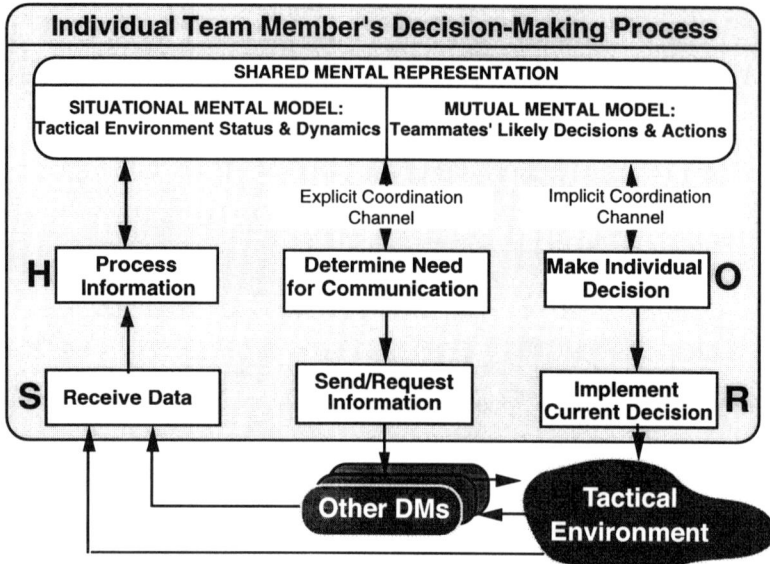

Figure 2. Extension of the SHOR paradigm to team decision-making processes. SHOR = stimulus–hypothesis–option–response paradigm; DM = decision maker.

mental models and implicit coordination strategies in a team. On the basis of the premise that a shared mental model would allow team members to coordinate their actions implicitly and to anticipate the leader's needs for information, we have defined an upward-anticipation measure, computed as the ratio of subordinates' information transfers to the leader's requests for information. Teams were observed to use a strategy that allowed them to perform accurately under high time pressure: The teams increased their use of implicit coordination based on a shared mental model as time pressure increased. The upward-anticipation ratio measure, an indicator of implicit information coordination in hierarchical teams, supports this conclusion. It was found that the anticipation ratio was almost twice as large under conditions of high time pressure as under low or moderate time pressure, indicating that subordinates were pushing information to the team leader without waiting to be asked explicitly for that information. A related experiment (Kalisetty, Kleinman, Serfaty, & Entin, 1993) established a link between *implicit coordination* and performance. A high correlation was found between the upward-anticipation ratios for six teams and their performance levels as measured by the percentage of correct target identifications. Furthermore, as hypothesized, the use of implicit coordination by the highest performing teams was *adaptive* to time pressure.

In summary, the relationship between shared mental models, team coordination, and adaptation to stress provides a theoretical basis for team training strategies aimed at improving performance through targeted enhancements to the shared mental model of the team. We argue that team coordination strategies are dominated by explicit coordination under low-workload conditions and by implicit coordination as workload increases. The transfer of information and resources in response to requests and the use of communications to coordinate action define *explicit coordination*. Anticipation of the information and resource needs of other team members through the exercise of mental models of the other decision makers or through the exercise of a common mental representation of the situation defines *implicit coordination*.

Training Teams to Be Adaptive

The two key questions addressed by this research were the following: Can teams be trained to use the strategies of communication and coordination employed by high-performing teams under stressful conditions? Moreover, can teams be taught to recognize the signs of increasing stress so they can institute efficient procedures to adapt their coordination strategies effectively and thereby mitigate some of the debilitating effects of high stress?

Earlier studies by Serfaty et al. (1993), Entin et al. (1993), Orasanu (1990), and others afforded us the opportunity to examine the strategies employed by high- and low-performing teams. Guided in part by these observations, the team training literature (e.g., Salas, Cannon-Bowers, & Johnston, 1997), the results of our experiments, and the team adaptation model, we designed the TACT team training strategy and an experiment to assess the effectiveness of the training procedure (Entin et al., 1994). TACT was developed to train team members to adapt their coordination strategies to take account of changes in workload or stress. The TACT procedure focuses on the intermediate feedback loop of the diagram in Figure 1 (i.e., coordination adaptation) by training the CIC teams on adaptive coordination skills. The TACT experiment was designed to demonstrate that a team training package that combines adaptability skills, coordination skills, and the exercise of a specific shared situation-assessment procedure would be effective in improving the team performance. It is therefore important to be able to frame the study findings within the shared mental model concept and its relationship with team coordination.

Method

Participants

We drew 60 active-duty military officers from two training sites to serve as participants. Twelve CIC anti-air warfare (AAW) teams of five individuals each were formed, and the team members allocated themselves to the five simulated watchstations (see Figure 3): tactical action officer (TAO),

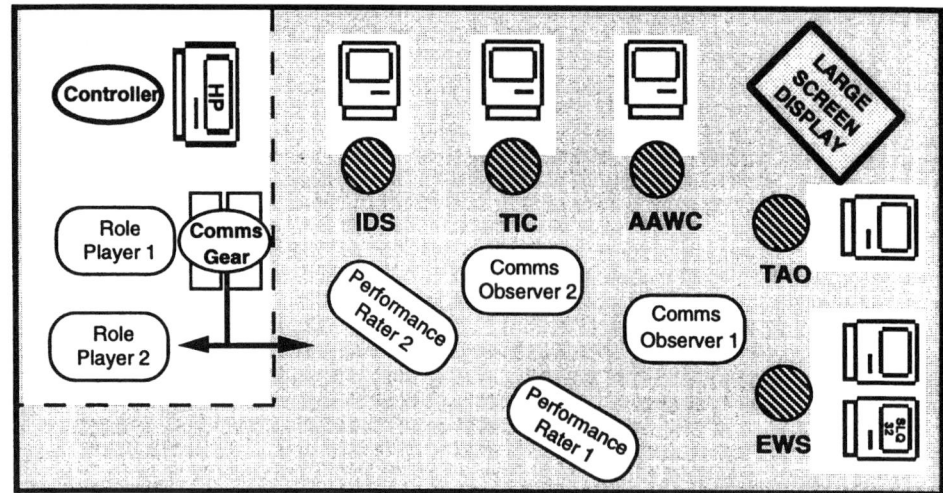

Figure 3. Team members' assignment to watchstations and simulation environment layout. IDS = identification supervisor; TIC = tactical information coordinator; AAWC = anti-air warfare coordinator; TAO = tactical action officer; EWS = electronic warfare supervisor; COMMS = communications.

identification supervisor (IDS), tactical information coordinator (TIC), anti-air warfare coordinator (AAWC), and electronic warfare supervisor (EWS). Four teams were randomly assigned to either a control condition, a training condition, or a training plus periodic situation-assessment briefings by TAO condition.

Independent Variables and Design

The design manipulated three independent variables. The training had three levels: a control condition in which no team adaptation and coordination training occurred, a team adaptation and coordination training (TACT) condition, and a condition including TACT plus periodic situation-assessment updates by the TAO (TACT+). The stress factor varied workload-induced stress over two levels: high and low stress. Two of four scenarios developed by the Naval Air Warfare Center Training System Division (NAWCTSD) presented a high-workload/ambiguity environment, whereas the remaining two scenarios presented a low-workload/ambiguity environment. Pretest and posttest were the two levels of the last factor. The three independent factors were crossed and organized into a pretest–posttest control group experimental design, as depicted in Figure 4.

Measurement Instruments

Team performance outcome measures. To assess the mission performance outcome of each team, the team performance outcome measure (Entin et al., 1994) was used. This measure is composed of 12 behaviorally

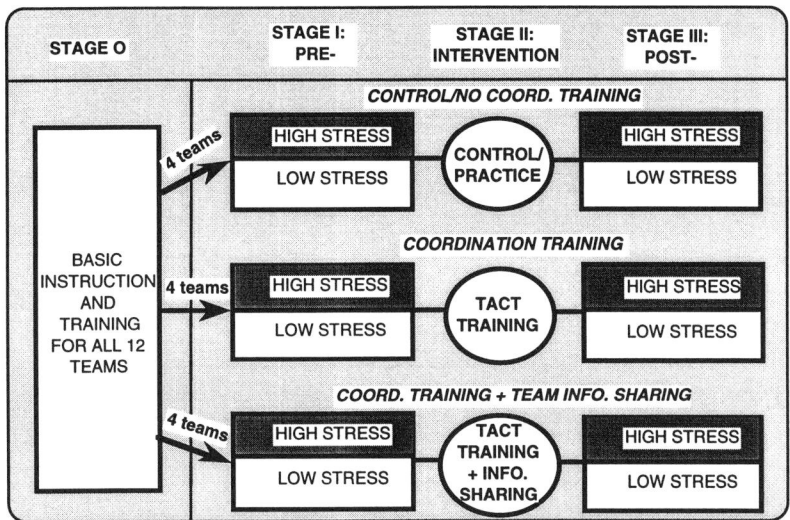

Figure 4. Experimental design. TACT = team adaptation and coordination training.

anchored items and is based on the Anti-Air Warfare Team Performance Index and the Behavior Observation Booklet (Johnston, Smith-Jentsch, & Cannon-Bowers, 1997). Examples of the items are shown in Figure 5. The instrument is designed to be completed by a trained subject-matter expert like a naval officer at the end of each trial.

Entin et al. (1994) showed that the team performance outcome measure enjoys high reliability (coefficient alpha = .97) and high construct validity. The latter was determined factor analytically (see Nunnally, 1967). A principal-components factor analysis with varimax rotation computed for the 12 items of the performance-assessment measure extracted a single factor that accounted for more than 73% of the variance. Moreover, all the items of the performance outcome measure correlated at or above .74 with the extracted performance factor. This implied that the items composing the performance-assessment instrument were tapping one common factor, and clearly that factor was team performance.

Teamwork process measures. To assess team processes, the Teamwork Observational Form was employed. This instrument consists of 15 behaviorally anchored items designed to assess the six dimensions of teamwork (i.e., team orientation, communication behavior, monitoring behavior, feedback behavior, backup behavior, and coordination behavior; McIntyre & Salas, 1995). The teamwork items were adapted from teamwork assessment instruments used by Serfaty, Entin, and Deckert (1993) and Entin et al. (1993) that in turn were based on the Anti-Air Warfare Team Observation Measure (ATOM; Johnston et al., 1997). A sample of items from the Teamwork Observational Form is depicted in Figure 6.

The reliability and validity analyses of the teamwork measure were

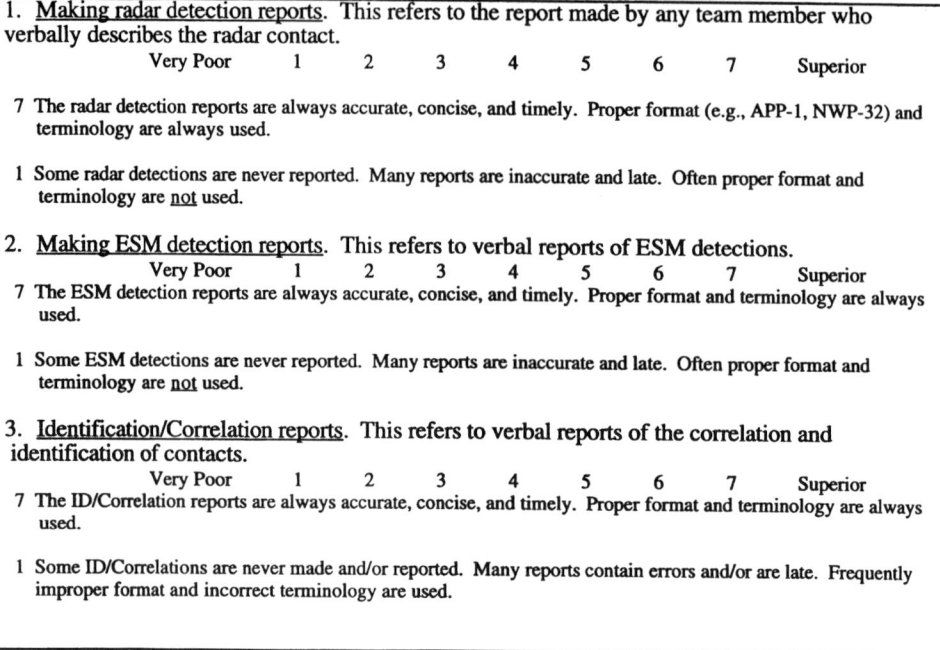

1. <u>Making radar detection reports</u>. This refers to the report made by any team member who verbally describes the radar contact.

Very Poor   1   2   3   4   5   6   7   Superior

7 The radar detection reports are always accurate, concise, and timely. Proper format (e.g., APP-1, NWP-32) and terminology are always used.

1 Some radar detections are never reported. Many reports are inaccurate and late. Often proper format and terminology are <u>not</u> used.

2. <u>Making ESM detection reports</u>. This refers to verbal reports of ESM detections.

Very Poor   1   2   3   4   5   6   7   Superior

7 The ESM detection reports are always accurate, concise, and timely. Proper format and terminology are always used.

1 Some ESM detections are never reported. Many reports are inaccurate and late. Often proper format and terminology are <u>not</u> used.

3. <u>Identification/Correlation reports</u>. This refers to verbal reports of the correlation and identification of contacts.

Very Poor   1   2   3   4   5   6   7   Superior

7 The ID/Correlation reports are always accurate, concise, and timely. Proper format and terminology are always used.

1 Some ID/Correlations are never made and/or reported. Many reports contain errors and/or are late. Frequently improper format and incorrect terminology are used.

Figure 5.   Items from the Team Performance Outcome Measure.

carried out by Entin et al. (1994) in a manner similar to that for the team performance measure. The reported reliability was high (coefficient alpha = .98) and a principal-components factor analysis extracted two factors that account for 86% of the variance. On the basis of the highest loading items, the first factor dealt with giving and receiving assistance, and the second factor embraced communication and anticipation. Each of the items loaded at or above .70 with one of the two factors, indicating good congruence. From a theoretical perspective, the two common factors of giving and receiving assistance and communication and anticipation define the construct of teamwork. It appeared that the teamwork instrument demonstrated high construct validity.

Team communication and coordination measures. The core of the communication and coordination assessment was composed of two communication-recording forms, or matrices, used to record the communication of the TAO and team in real time. Figure 7 shows that the two matrices are laid out in a similar fashion. The columns of the matrices are the message destinations (e.g., TAO to TIC, TAO to all, TIC to TAO, IDS to team). The rows of the matrices are the type (e.g., requests, transfers) and content (e.g., information, action and task) of communication messages. Two psychologists with previous experience coding verbal behavior worked with videotaped scenarios of helicopter pilots to practice and calibrate their mutual coding. After approximately 16 hours of practice and discussion, the two coders attained an 85% agreement level. A coder work-

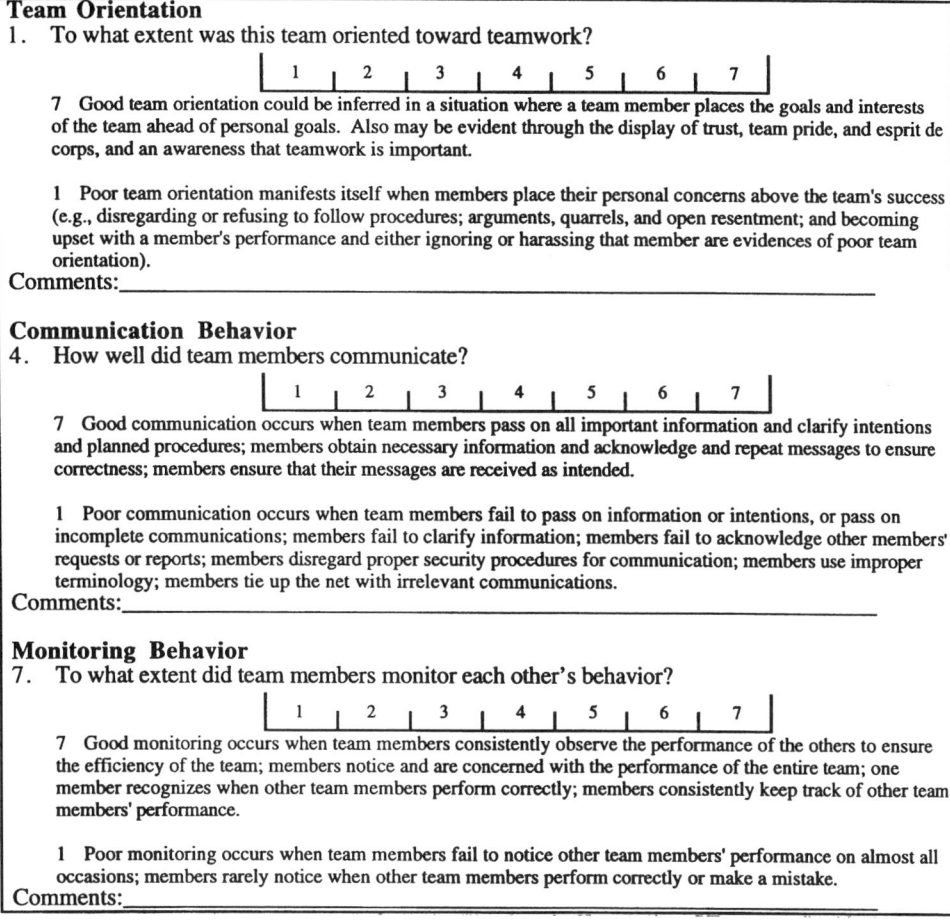

Team Orientation

1. To what extent was this team oriented toward teamwork?

| 1 | 2 | 3 | 4 | 5 | 6 | 7 |

7   Good team orientation could be inferred in a situation where a team member places the goals and interests of the team ahead of personal goals. Also may be evident through the display of trust, team pride, and esprit de corps, and an awareness that teamwork is important.

1   Poor team orientation manifests itself when members place their personal concerns above the team's success (e.g., disregarding or refusing to follow procedures; arguments, quarrels, and open resentment; and becoming upset with a member's performance and either ignoring or harassing that member are evidences of poor team orientation).

Comments:_____

Communication Behavior

4. How well did team members communicate?

| 1 | 2 | 3 | 4 | 5 | 6 | 7 |

7   Good communication occurs when team members pass on all important information and clarify intentions and planned procedures; members obtain necessary information and acknowledge and repeat messages to ensure correctness; members ensure that their messages are received as intended.

1   Poor communication occurs when team members fail to pass on information or intentions, or pass on incomplete communications; members fail to clarify information; members fail to acknowledge other members' requests or reports; members disregard proper security procedures for communication; members use improper terminology; members tie up the net with irrelevant communications.

Comments:_____

Monitoring Behavior

7. To what extent did team members monitor each other's behavior?

| 1 | 2 | 3 | 4 | 5 | 6 | 7 |

7   Good monitoring occurs when team members consistently observe the performance of the others to ensure the efficiency of the team; members notice and are concerned with the performance of the entire team; one member recognizes when other team members perform correctly; members consistently keep track of other team members' performance.

1   Poor monitoring occurs when team members fail to notice other team members' performance on almost all occasions; members rarely notice when other team members perform correctly or make a mistake.

Comments:_____

Figure 6.   Selected items from the Teamwork Observational Form.

ing with either matrix recorded a tally mark for each utterance denoting a particular message type and content, and from whom and to whom that message was directed. The various communication and coordination measures were derived from the raw communication matrices by summing appropriate rows, columns, or both.

Entin et al. (1994) noted that the form of the communication and coordination measures was quite different from that of the more traditional measures of performance outcome or teamwork. The communication and coordination measures were not composed of items or rating scales, but were based on summations of rows and columns of the matrices yielding amounts, rates, and ratios. The amounts, rates, and ratios constitute the various communication and coordination variables, and because the variables were not of the same form, the same reliability and validity analyses performed on the preceding dependent variable could not be performed here.

TACT EXPERIMENT (TADMUS)
CIC TEAM COMMUNICATION RECORDING FORM: **TAO**
TEAM #____ SITE ____ DATE ____ OBSERVER ____ SCEN #____ PERIOD 1 2 3

TYPE & CONTENT		TAO to:					
		TIC	IDS	AAWC	EWS	All	Out
REQUESTS	INFOR-MATION						
	ACTION & TASK						
	PROBLEM SOLVING & PLANNING						
TRANSFERS	INFOR-MATION						
	ACTION & TASK						
	PROBLEM SOLVING & PLANNING						
ACKNOWLEDGMENTS							

TACT EXPERIMENT (TADMUS)
CIC TEAM COMMUNICATION RECORDING FORM: **TEAM**
TEAM #____ SITE ____ DATE ____ OBSERVER ____ SCEN #____ PERIOD 1 2 3

TYPE & CONTENT		TIC to:		IDS to:		AAWC to:		EWS to:		Team to:
		TAO	Team	TAO	Team	TAO	Team	TAO	Team	Out
REQUESTS	INFOR-MATION									
	ACTION & TASK									
	PROBLEM SOLVING & PLANNING									
TRANSFERS	INFOR-MATION									
	ACTION & TASK									
	PROBLEM SOLVING & PLANNING									
ACKNOWLEDGMENTS										

Figure 7. Team communications recording matrices. TAO = tactical action officer; TIC = tactical information coordinator; IDS = identification supervisor; AAWC = anti-air warfare coordinator; EWS = electronic warfare supervisor.

The **Pre-Mission Questionnaire** consisting of 20 items was devised to assess:

- the extent to which team members could anticipate one another's needs
  - —e.g., "Team members should be able to anticipate each other's information needs during the mission."
- the confidence team members had in one another
  - —e.g., "How much confidence do you place in the ability of the other members of your team to accomplish this mission?"
- the extent to which team members believe they should monitor other team members
  - —e.g., "To what extent should team members monitor other team members for signs of stress?"
- the extent to which team members believe they should readjust to changes in workload
  - —e.g., "To what extent should team members change their work strategy in response to high stress/workload?"
- attitudes about the team atmosphere
  - —e.g., "A team member's decision-making ability is as good in stressful situations as it is in non-stressful conditions."

Figure 8. Themes and sample items in the Pre-Mission Questionnaire.

Team self-report and workload measures. Whereas the teamwork and communication matrix instruments assessed various team processes using trained observers, the next two questionnaires employed self-report procedures to assess team processes directly from the participants. These descriptive measures were taken to supplement the observer-based measures described in the first three categories. The model and related hypotheses were tested using the objective observer-based measures only. The first of the self-report measures, the Pre-Mission Questionnaire, is composed of 20 items that were designed to assess attitudes about team and task dynamics, another window into team processes. Figure 8 delineates the specific themes and provides sample item stems. Each of the attitudinal items was accompanied by a 7-point scale that allowed participants to express their agreement or disagreement with the item's subject.

The overall measure showed moderate internal consistency (coefficient alpha = .55), which Entin et al. (1994) considered understandable given the multifactor composition of the questionnaire's design. In this case, we did not conduct the factor analysis to test the hypothesis of a single underlying construct, but instead to verify the existence of the several factors the questionnaire was designed to assess. Six factors accounting for 60% of the variance were extracted by a principal-components factor analysis. The factors were named according to their highest loading items. Each of the items correlated at or above .46 with one of the six factors. Factors 1 through 6, respectively, were named: Confidence, Stress Awareness, Error Awareness and Correction, Effect of Workload Stress, Experience and Effect of Stress, and Understanding and Anticipating Needs. These underlying factors appeared to show a reasonable correspondence to the themes of the Pre-Mission Questionnaire.

Whereas the Pre-Mission Questionnaire inquired about team members' attitudes and expectations concerning team processes, the Post-Mission Questionnaire was derived to assess participants' observations as to whether certain team processes occurred and to what extent they occurred. These team processes were anticipation, confidence, stress awareness, and monitoring. Each item was accompanied by a 7-point scale on which participants indicated the extent to which each process referred to in the item occurred. The measure was reliable (coefficient alpha = .88) and appeared to demonstrate high construct validity. Two factors were extracted by a principal-components factor analysis: Factor 1, named (by the highest loading items) Team Coordination, accounted for 54% of the variance, and Factor 2, named Confidence, accounted for an additional 14% of the variance. All items correlated .64 or above with one of the two factors (Entin et al., 1994).

Finally, administered as part of the Post-Mission Questionnaire were the six items of the Task Load Index (TLX; Hart & Staveland, 1988) that assessed subjective workload. The TLX is a self-report measure of workload that elicits a rating on six dimensions (mental demand, physical demand, temporal demand, performance, effort, and frustration). Designed at the National Aeronautics and Space Administration (NASA), the TLX has shown high validity and reliability (Lysaght et al., 1989).

Procedure

At the onset of each experimental session, the five-person teams were given a brief overview of the TADMUS project, refresher instruction on the Decision Making Evaluation Facility for Tactical Teams (DEFTT) simulator, instruction on the watchstations, and three practice scenarios on the DEFTT simulator. Before each data collection scenario, the teams received a mission brief delineating their mission, goals, potential threats, and rules of engagement. The TAOs were always afforded time to brief their teams if they so desired. A typical tactical scenario ran between 25 and 30 minutes.

The training intervention was composed of four components—instruction, demonstration, practice, and feedback—and occurred in three phases (Salas, Cannon-Bowers, & Johnston, 1997). During Phase 1, teams were taught how to identify signs and symptoms of stress in themselves, other team members, the team, and the environment, and they were instructed on the adaptive use of five general team coordination strategies to adapt to increases in workload and stress. During Phase 2, participants viewed three pairs of specially prepared videotaped vignettes demonstrating a team using good and poor strategies, as well as indications of heightened stress affecting team behavior. In Phase 3, teams were given the opportunity to practice what they had learned by completing two 12-minute training scenarios and receiving and providing process feedback. They also viewed a videotape featuring a retired Navy flag officer reviewing what had been presented in training. Training for TACT+ was identical to that for TACT with one notable addition. During TACT+ training, the TAOs were given specific instructions on how to give brief (about 30 seconds) periodic situation-assessment updates every 3 minutes to the rest of the

CIC team. Team members were taught to interpret the information contained in the TAO's briefs and to use this information effectively. The teams in the control condition received a sham training procedure that included the reading of two articles and the opportunity to practice on the two training scenarios. The control intervention ran about as long as the TACT and TACT+ trainings (about 2 hours).

Before and following the training intervention, each team performed one high-stress and one low-stress scenario. Stress presentation was counterbalanced within and between teams. As a manipulation check to show that the stress manipulation was effective, an analysis of the TLX Workload Scale (Hart & Staveland, 1988), included as part of the Post-Mission Questionnaire, was performed. The analysis revealed significantly higher workload scores in the high- than in the low-stress scenarios, $F(1, 54) = 19.16, p < .01$, empirically supporting the workload-induced stress manipulation.

To assess teamwork and performance, two naval officers from each experimental site were trained for about 4 hours to use the Team Performance Outcome Measure and Teamwork Observational Form. The agreement between the two naval officers in each trial was found to be high (coefficient alpha = .79). The two observers coding team communications sat just behind the team and were connected by earphones to the communication network used by the team. One observer coded the verbal interactions of the TAO, and the other observer coded the verbal interaction among the remaining team members.

Results

Team Training and Outcome Performance

An average outcome performance measure was derived by taking the mean of the 12 items. Analysis of the average performance measure showed clearly that the effect of the training intervention was significant. As illustrated by Figure 9, the teams receiving TACT and TACT+ performed significantly better than the control teams, $F(2, 9) = 4.91, p < .04$. In addition, as Figure 10 shows, teams receiving either training performed better after than before training, $F(1, 9) = 5.95, p < .05$, whereas the teams in the control condition showed about the same level of performance pre- and posttraining.

Stress and Performance

Teams that received the training were apparently able to reduce their coordination and communication overheads, thereby having more time to devote to the anti-air war task, which resulted in better performance. The training intervention also helped teams deal with stress. As shown in Figure 11, teams performed better under high stress after the training inter-

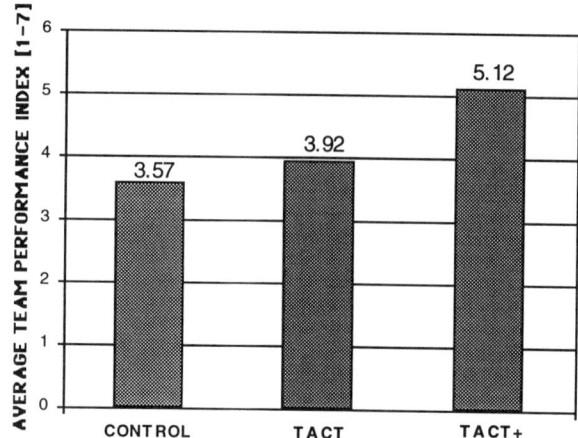

Figure 9. Outcome performance for the two experimental conditions and the control condition.

ventions than they did under low stress before the training. This result leads us to believe that the training was effective in making the team more resilient to stress under both low- and high-stress scenarios.

Training and Teamwork

The original teamwork skill measures (i.e., team orientation, communication behavior, monitoring behavior, feedback behavior, backup behavior, and coordination behavior) were averaged and analyzed. Results of the teamwork skill measure closely paralleled the performance findings. The teamwork means for the three groups depicted in Figure 12 differed significantly, $F(2, 9) = 5.48$, $p < .03$. These findings support the argument that training improves a team's teamwork skills, which in turn leads to better performance. Further analysis also showed that teamwork in the

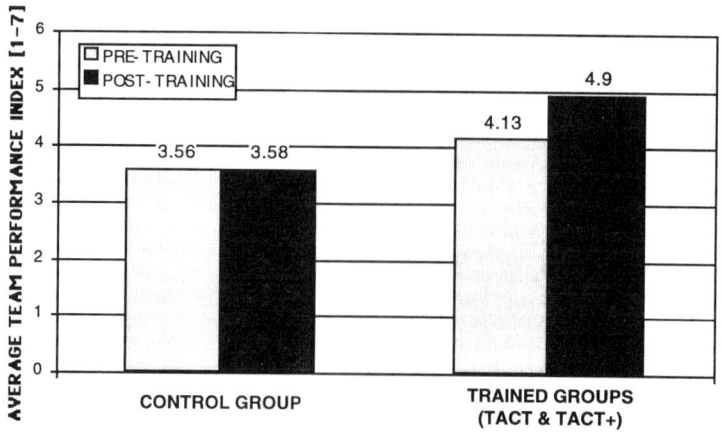

Figure 10. Team performance pre- and post-training.

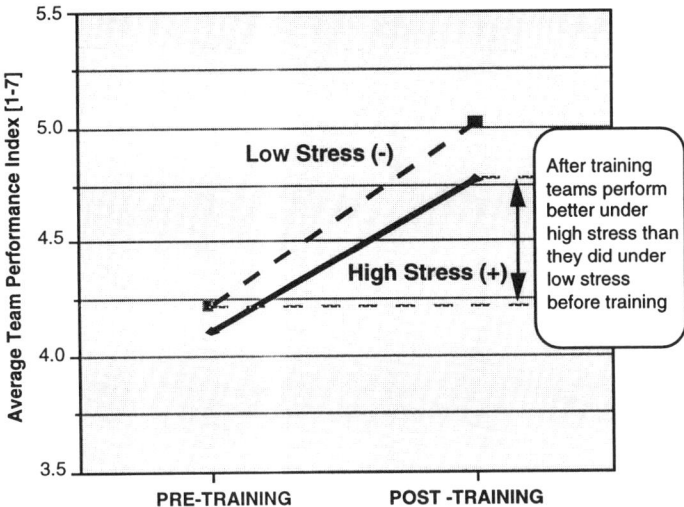

Figure 11. Performance of teams receiving training in high- and low-stress scenarios.

two experimental conditions receiving training improved significantly after training.

Training and Team Communications

The overall communication rate indicated that teams were quite busy sending messages; on the average, teams sent 7.6 messages per minute. The breakdown of the message traffic for the five team members is shown in Figure 13. As expected, the TAO had the highest message rate. The next highest rate, that of the AAWC, was roughly half that of the TAO. The message rates for the remaining three team members were lower and about equal to each other. The general direction of message flow in the teams is depicted in Figure 14. *Downward communication* refers to messages sent from the TAO to various members of the team; *upward com-*

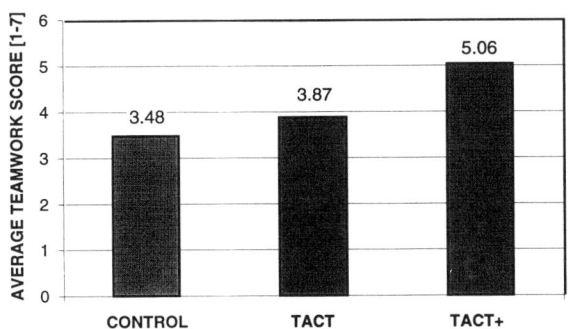

Figure 12. Teamwork score for the two experimental conditions and control condition.

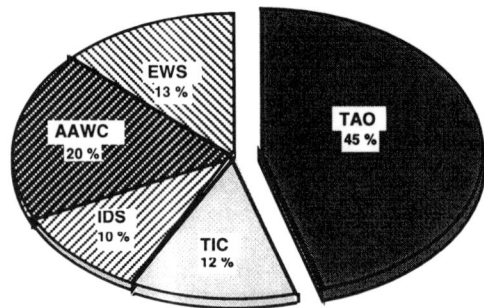

Figure 13.   Distribution of communications. TAO = tactical action officer; TIC = tactical information coordinator; IDS = identification supervisor; AAWC = anti-air warfare coordinator; EWS = electronic warfare supervisor.

munication refers to messages sent from various team members to the TAO; *lateral communication* refers to messages sent among the team members other than the TAO; and *outward communication* refers to messages sent by the TAO to the "outside world."

Several communication measures were derived from the two recording matrices and analyzed. Examination of the upward communication rate revealed that in the low-stress scenarios, team members communicated less with the TAO after than before training, whereas in the high-stress conditions, team members communicated more with the TAO after than before training, $F(1, 9) = 5.87$, $p < .04$. In low-stress scenarios, upward communication was the same in the control and TACT conditions and somewhat lower in the TACT+ condition. Under high stress, the upward communication rate was greatest in the TACT condition and lowest in the TACT+ condition. The TAO's periodic situation briefs to the team in the TACT+ condition seemed to reduce the team members' need to communicate with the TAO.

Concerning the TAO's communication (downward communication rate), we found that the TAOs communicated less with their teams after training, $F(1, 9) = 4.38$, $p < .07$.[1] This finding was due entirely to the decrease in downward communication rate in the low-stress scenarios, however, because downward communication was the same pre- and posttraining in the high-stress conditions, $t(9) = 1.88$, $p < .05$, one-tailed. Further analysis showed that the downward communication rate was the same pre- and posttraining in the control condition, decreased posttraining in the TACT condition, and increased slightly in the TACT+ condition posttraining, $F(2, 9) = 6.19$, $p < .025$. The latter findings might be the result of subordinate team members' tendency to communicate more with TAOs in the TACT conditions, as noted previously.

The amount of communication subordinate team members had among themselves was called the *lateral communication rate*. Subordinate team

---

[1]Given the nascent aspects of this work, we decided to relax the standard alpha levels to avoid Type II errors.

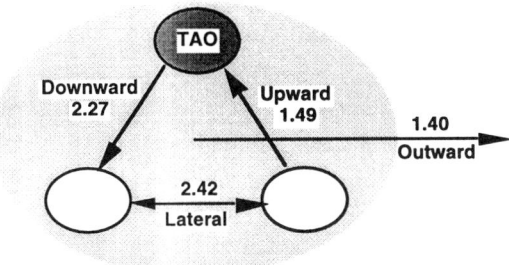

Figure 14.   Direction of communications. TAO = tactical action coordinator.

members communicated more among themselves in the high- than low-stress scenarios, $F(1, 9) = 6.15$, $p < .04$. We also found that the lateral communication rate was somewhat lower after than before training for control teams, increased somewhat after training for TACT teams, and decreased substantially after training for TACT+ teams, $F(2, 9) = 3.81$, $p < .065$. The marginally significant three-way interaction, $F(2, 9) = 3.57$, $p < .075$, sheds further light on these findings. The decrease in lateral communication observed for the control teams after training was due to the lower lateral communication rate in the low-stress scenario. Although TACT teams showed a mild increase in lateral communication after training, the pattern of results was far more involved when stress was considered. TACT subordinate team members communicated more in the high- than low-stress scenarios before training, whereas subsequent to training, the pattern was reversed: Lateral communication fell in the high- versus the low-stress scenarios. Perhaps these results indicate that the TACT subordinate team members shifted to a more adaptive strategy under high stress, posttraining. The subordinate team members in the TACT+ condition also may have shifted to a more adaptive strategy posttraining, given the drop in the lateral communication rate, but the strategy they chose still obliged them to communicate more among themselves in the high-stress scenarios than did the TACT team members.

Training and Adaptive Coordination

The team communication results showed that communication patterns were altered in ways predicted and taught by the TACT training procedure. For example, changes in the communication patterns were dominated by a stronger upward information push and stronger anticipatory behavior for the TACT and the TACT+ teams, supporting the hypothesis that implicit team coordination mechanisms are essential to effective adaptation to stress.

Several anticipation ratios were computed to track further the anticipatory behavior that was central to the TACT procedure. Overall, we found a significant increase in several category-based anticipation ratios (e.g., upward information) as a result of the training. The overall team

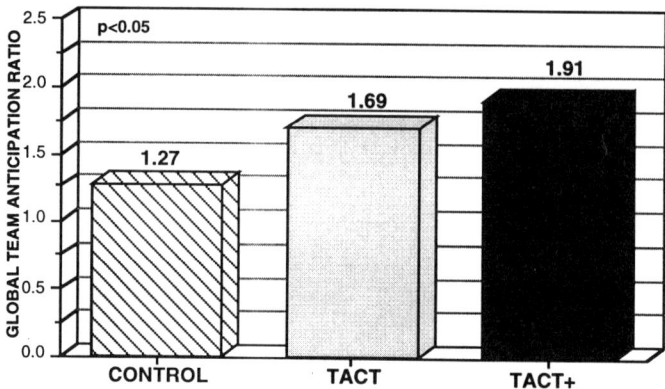

Figure 15. Training and anticipatory behavior. TACT = team adaptation and coordination training.

anticipation ratio (shown in Figure 15) was computed as the ratio of transfers to transfers plus requests across all categories of communication. Both the TACT and TACT+ interventions produced significantly higher rates of anticipatory behavior than found in the practice-only control group. We did not find significant differences in anticipatory behavior between low- and high-stress conditions, however. The lack of evidence of differential adaptation to stress might be due to the relatively small stress differentiation present in the scenarios. In other words, the high-stress scenarios might not have induced stress in the teams to a degree warranting changes in anticipatory behavior (see Figure 1). However, as discussed earlier, other differences in communication patterns owing to stress levels were detected. This may suggest a hierarchical model of coordination adaptation in which patterns of communications change gradually as stress increases. This intriguing hypothesis could be tested in further studies.

Team Coordination and Shared Mental Models

The TACT objectives were designed to enhance team performance through team behavior improvement. The TACT strategies were focused on team coordination and adaptation skills (teamwork), not on tactical decision skills (taskwork). The taskwork performance of the team was only indirectly affected by the improved teamwork skills. Team members were instructed on how to improve the way they dealt with the overhead required to keep the team organized and well coordinated under varying conditions of stress. Teams were trained to develop a collection of team coordination strategies and to understand which strategy works best under a particular circumstance. The focus was always on teamwork processes, however, and on how to make changes to adapt to stress. We believe these are universal team skills that transcend individual proficiencies.

The training approach in TACT is based on behavioral modeling; that

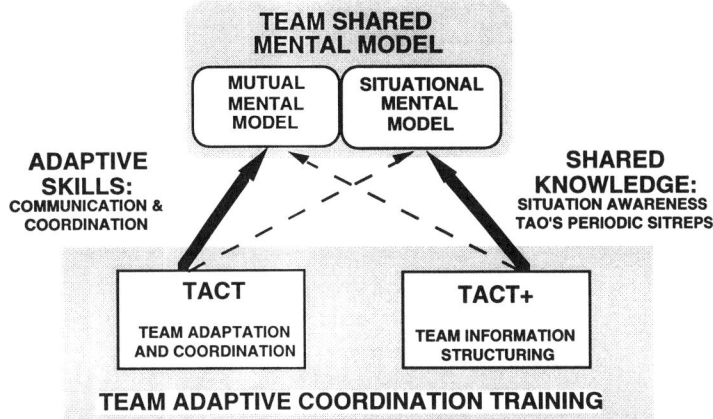

Figure 16. Mental models and team training. TAO = tactical action officer; TACT = team adaptation and coordination training.

is, teams were taught how to develop skills dealing with coordination behaviors (Bowers, Salas, Prince, & Brannick, 1992; Cannon-Bowers, Salas, & Converse, 1990). TACT+ further enhanced the skill-based training by the periodic sharing of the TAO's situation assessment, which was added to enhance the congruence of the team members' mental models (Figure 16).

The data showed that both adaptive skills and shared knowledge affect the mental models developed by a team. Adaptive team communication and coordination skills have their primary impact on the team's mutual mental models. Acquired through formal team training and honed by practice, the adaptive communication and coordination skills foster various teamwork skills, such as monitoring and backup behavior, that provide the observational opportunities necessary to learn other team members' needs, proclivities, and tasks. This in-process team learning in turn has the effect of maintaining highly congruent mutual mental models that allow team members to shift from explicit to implicit communication modes and to anticipate the needs of other team members. Such behaviors increase team performance and efficiency, particularly when normal operations are replaced by highly stressful conditions.

Periodic situation-assessment updates by the team leader work through another mechanism. Periodic updates solidify and validate the common knowledge as well as the situational mental model shared by team members. Stress and high workload distract team members and can alter their understanding of the tactical environment. Periodic updates refocus the team and bring forward critical elements of the current and future tactical situation. The data obtained in the TACT study supported, to a certain extent, this Team Training × Behavior × Cognition interaction model.

Lessons Learned

Lesson 1: Adaptive Teams Are High Performers

With the research described in this chapter, we have attempted to link several cognitive science concepts in team performance and training, such as shared knowledge, mental models, and team coordination, in a common theoretical framework supported by empirical findings. Observations of teams and individuals operating in complex and stressful environments reveal that superior performance is often associated with a high degree of adaptability. Flexibility in decision making and teamwork processes is achieved through successful strategies of adaptation to stress. Variations in the level of stress generated by environmental stressors such as task uncertainty and time pressure can be sensed by the performing entities, and they stimulate changes in behavior, strategy, and even organizational structure.

Lesson 2: Team Adaptive Coordination Is Trainable

For naval command teams operating in a shipboard CIC, a rich set of adaptation mechanisms, such as team coordination strategies and group situational updates, are available to cope with the changing levels of stress. The TACT procedure was conceived as a model-based strategy to train command teams to cope successfully with stressful operational conditions by using coordination and communication strategies based on a multilevel model of adaptation to stress. These strategies include both team behavior modeling (adaptive coordination strategies) and team information structuring (periodic situational updates). Significant improvements in team performance and teamwork skills were obtained for teams who received TACT training in a simulation-based experiment.

Lesson 3: Significant Improvements in Team Performance Are Achievable With a Relatively Simple Training Intervention

All the teamwork skill improvements were achieved following a relatively short training period (about 2 hours). We employed a rather classic approach to training and adhered rigidly to the prescribed principles of training: instruction, demonstration, practice, and feedback. Although our focus was on communication and coordination skills, we were able to induce significant improvements in the *tactical performance* of the teams, which occurred because of intervention-induced improvements to teamwork process skills.

Lesson 4: Shared Knowledge Is Necessary, but Not Sufficient

For optimal performance, one must also develop team coordination strategies that operate on, update, and maintain the team's shared mental models. The experimental findings on adaptive team training and performance were framed by a common theoretical premise: Teams can be trained to develop a mental model of their common task that enables them to use the team structure to maintain team coordination and performance under a wide range of conditions. Effective team training depends on two key factors: the ability to help teams develop shared mental models, and the ability to train teams in using these mental models to adapt their coordination strategies to accommodate situational demands. Techniques for development of the team's mutual mental models, such as cross-training or exposure to each other's jobs (Duncan, Burns, & Rouse, 1994; Volpe, Cannon-Bowers, Salas, & Spector, 1996), may not be sufficient by themselves to enable the team to maintain performance under stress. Team members need to learn how to exercise these shared mental models through specific training of coordination strategies.

Lesson 5: To Maximize Performance, One Should Combine Knowledge Training With Team Skills Training (Hypothesis)

We hypothesize that further strengthening of the shared mental model will have a significant positive effect on the team and therefore on team performance. Enhancements to the shared mental model can be achieved through integration of cross-training or interpositional training (Salas & Cannon-Bowers, 1997), which familiarizes team members with their teammates' tasks, functions, and informational needs in a dynamic fashion, with current skill-based training components. When all team members consistently share a common awareness and understanding of the situation, the correct tasks are accomplished in a timely manner; in short, teams perform well. The explanation of these findings and the use of the proposed framework are important for the design and implementation of future team and distributed team training systems in the military. The integration of the TACT (and TACT+) component with enhanced mental model training could produce a training package superior to existing ones—one that will achieve greater levels of CIC team performance. Future team training procedures will need to include an integrated approach to team training. Skill-based training (coordination and adaptation), sound communication procedures (through information structure), and knowledge-based training (mental models) are all necessary ingredients for fostering superior team performance.

Lesson 6: Model-Based Team Training Provides a Practical Mechanism for Explaining Process Changes and Performance Improvements

Successful team training procedures derived from a theoretical basis are showing promise. Although significant progress has been made, there are several questions that need to be explored further. Among them are the following: (a) How does one develop and maintain accurate team mutual mental models? Training individual team members to perform other members' tasks may provide a possible method but not necessarily a practical one. (b) In the TACT program the participants were collocated. Are there other coordination skills beyond those developed under the TACT program that are especially important for distributed training? (c) The feedback component of the TACT program is promising, but it demands further research. Can specific behaviors be isolated, recognized, and tagged during practice sessions, and can they be fed back to team members?

Conclusion

This research was designed to apply theory-based principles of team adaptation and coordination in a simple team training strategy. The training produced significantly positive effects on team performance. A further direction for this research could be to integrate and test training procedures to support the development of principles of team training for coordination and adaptation to stress. In particular, emphasis should be placed on training CIC teams to acquire a set of appropriate coordination strategies and communication procedures to apply to a shared knowledge of the tactical situation that will support teamwork under conditions of stress. Training principles include the acquisition of a common situational awareness; a recognition of stressful conditions that may affect performance; and the use of adaptive strategies necessary to maintain performance at the required level, minimize the occurrence of errors, and hold individual and team stress below an acceptable threshold. The fact that these training procedures have emerged from a strong theoretical basis, yet have maintained a relative simplicity of application, could make them ideal candidates for onboard training systems.

References

Bowers, C. A., Salas, E., Prince, C., & Brannick, M. (1992). Games teams play: A methodology for investigating team coordination and performance. *Behavior Methods, Instruments and Computers, 24*, 503–506.

Cannon-Bowers, J. A., Salas, E., & Converse, S. (1990). Cognitive psychology and team training: Training shared mental models of complex systems. *Human Factors Society Bulletin, 33*(12), 1–4.

Cannon-Bowers, J. A., Salas, E., & Converse, S. (1993). Shared mental models in expert team decision making. In N. J. Castellan, Jr. (Ed.), *Current issues in individual and group decision making* (pp. 221–246). Hillsdale, NJ: Erlbaum.

Duncan, P. C., Burns, J. J., & Rouse, W. B. (1994). *A mental model-based approach to training teams working with complex systems* (tech. rep.). Norcross, GA: Search Technology.

Entin, E. E., & Serfaty, D. (1990). *Information gathering and decision making under stress* (TR-454). Burlington, MA: ALPHATECH.

Entin, E. E., Serfaty, D., & Deckert, J. C. (1994). *Team adaptation and coordination training* (TR-648-1). Burlington, MA: ALPHATECH.

Entin, E. E., Serfaty, D., Entin, J. K., & Deckert, J. C. (1993). *CHIPS: Coordination in hierarchical information processing structures* (TR-598). Burlington, MA: ALPHATECH.

Green, R. (1990). Human error on the flight deck. In D. E. Broadbent, J. Reason, & A.

Baddely (Eds.), *Human factors in hazardous situations* (pp. 503–512). Oxford, England: Clarendon Press.

Hart, S. G., & Staveland, L. (1988). Development of NASA-TLX (Task Load Index): Results of empirical and theoretical research. In P. A. Hancock & N. Meshkati (Eds.), *Human mental workload* (pp. 139–183). Amsterdam: Elsevier.

Johnston, J. A., Smith-Jentsch, K. A., & Cannon-Bowers, J. A. (1997). Performance measurement tools for enhancing team decision making. In M. T. Brannick, E. Salas, & C. Prince (Eds.), *Team performance assessment and measurement: Theory, method, and application*. Hillsdale, NJ: Erlbaum.

Kalisetty, S., Kleinman, D. L., Serfaty, D., & Entin, E. E. (1993). Coordination in hierarchical information processing structures (CHIPS). *Proceedings of the 1993 JDL Command and Control Research Symposium*, Fort McNair, Washington, DC.

Kleinman, D. L., & Serfaty, D. (1989). Team performance assessment in distributed decision-making. In J. Gibson et al. (Eds.), *Proceedings of the Symposium on Interactive Networked Simulation for Training*, Orlando, FL.

LaPorte, T. R., & Consolini, P. M. (1988). *Working in practice but not in theory: Theoretical challenges of high reliability organizations*. Annual meeting of the American Political Science Association, Washington, DC.

Lysaght, R. J., Hill, S. G., Dick, A. O., Aamondon, B. D., Linton, P. M., Wierwille, W. W., Zaklas, A. L., Bittner, A. C., & Wherry, R. J. (1989). *Operator workload: Comprehensive review and evaluation of operator workload methodologies* (TR-851). Alexandria, VA: U.S. Army Research Institute for the Behavioral and Social Sciences.

McIntyre, R. M., & Salas, E. (1995). Team performance in complex environments: What we have learned so far. In R. Guzzo & E. Salas (Eds.), *Team effectiveness and decision making in organizations*. San Francisco: Jossey Bass.

Orasanu, J. M. (1990). *Shared mental models and crew decision making* (CSL Report 46). Princeton, NJ: Cognitive Science Laboratory, Princeton University.

Orasanu, J. M., & Fischer, U. (1992). Team cognition in the cockpit: Linguistic control of shared problem solving. In *Proceedings of the 14th annual conference of the Cognitive Science Society*. Hillsdale, NJ: Erlbaum.

Pfeiffer, J. (1989, July). The secret of life at the limits: Cogs become big wheels. *Smithsonian*.

Rasmussen, J. (1990). Human error and the problem of causality in analysis of accidents. In D. E. Broadbent, J. Reason, & A. Baddely (Eds.), *Human factors in hazardous situations* (pp. 449–460). Oxford, England: Clarendon Press.

Reason, J. (1990). The contribution of latent human failures to the breakdown of complex systems. In D. E. Broadbent, J. Reason, & A. Baddely (Eds.), *Human factors in hazardous situations* (pp. 475–484). Oxford, England: Clarendon Press.

Salas, E., & Cannon-Bowers, J. A. (1997). Methods, tools, and strategies for team training. In M. A. Quinones & A. Ehrenstein (Eds.), *Training for a rapidly changing workplace: Applications of psychological research* (pp. 249–279). Washington, DC: American Psychological Association.

Salas, E., Cannon-Bowers, J. A., & Johnston, J. H. (1997). How can you turn a team of experts into an expert team? Emerging training strategies. In C. Zsambok & G. Klein (Eds.), *Naturalistic decision making* (pp. 359–370). Hillsdale, NJ: Erlbaum.

Serfaty, D., Entin, E. E., & Deckert, J. C. (1993). *Team adaptation to stress in decision making and coordination with implications for CIC team training* (TR-564, Vol. 1–2). Burlington, MA: ALPHATECH.

Serfaty, D., Entin, E. E., & Volpe, C. E. (1993). Adaptation to stress in team decision-making and coordination. In *Proceedings of the Human Factors and Ergonomics Society 37th annual meeting*. Santa Monica, CA: Human Factors Society.

Volpe, C. E., Cannon-Bowers, J., Salas, E., & Spector, P. E. (1996). Walking in each other's shoes: The impact of cross-training on team functioning. *Human Factors 38*, 87–100.

Wohl, J. (1981). Force management decision requirements for Air Force tactical command and control. *IEEE Transactions on Systems, Man, and Cybernetics 11*, 610–639.

10

Training Team Leaders to Facilitate Team Learning and Performance

Scott I. Tannenbaum, Kimberly A. Smith-Jentsch, and Scott J. Behson

Organizations have increasingly turned to the use of work teams as a common structural building block (Guzzo & Salas, 1995). Teams are valued not only for their potential performance capabilities but also for their capacity to promote learning among team members (Katzenbach & Smith, 1993). They are used in a wide range of military and civilian environments, including many stressful environments such as those examined in the TADMUS project.

Given the prevalence of teams, researchers and practitioners have sought to devise mechanisms to develop, improve, and maintain effective team performance. Research has been devoted specifically to investigating the effectiveness of theoretically based training programs for teams (see Salas & Cannon-Bowers, 1997; Salas, Cannon-Bowers, & Johnston, 1997).

One means of improving team effectiveness is to elevate the task proficiency of individual team members through enhanced selection or individual training (Tannenbaum, Salas, & Cannon-Bowers, 1996). Alternatively, because team members are by definition interdependent, teams can enhance their effectiveness by improving the way team members interact with each other (Salas, Dickinson, Converse, & Tannenbaum, 1992). Team processes, such as communication, coordination, shared situational awareness, compensatory behavior, and decision making can contribute to team effectiveness (Cannon-Bowers, Tannenbaum, Salas, & Volpe, 1995). A variety of structured interventions are available that may potentially enhance team interactions. For example, cross-training, team mental model training, team building, and team coordination training have all been implemented with some success (Salas et al., 1997; Tannenbaum et al., 1996).

Yet team development should not be limited to structured interventions. Each time a team engages in an activity together (whether that activity is a team training event or simply the normal performance of team

---

The views expressed herein are those of the authors and do not reflect the official position of the organizations with which they are affiliated.

responsibilities), the potential exists for team members to learn about their interactions and to enhance their subsequent performance. In fact, because few newly formed teams immediately adopt maximally effective interaction strategies, it is essential that team members learn from their experiences. In highly stressful environments, it is important that they learn to improve their interactions prior to peak stress events, as these events can have a debilitating effect on team interactions and team decision making (Cannon-Bowers & Salas, chap. 2, this volume).

Unfortunately, not all teams demonstrate the propensity to learn, nor do teams always improve through experience (Hackman, 1990). Teams need to be *prepared* to learn from their experiences. For example, does the team spend time before and after key events to ensure that team learning occurs?

Team leaders can contribute to team effectiveness in many ways (e.g., Komaki, Desselles, & Bowman, 1989), but one key role they can play is as a facilitator of on-going team development (Kozlowski, Gully, Salas, & Cannon-Bowers, 1996). Therefore, this stream of the TADMUS project focuses on training team leaders to facilitate team discussions before and after team activities, to maximize learning from those experiences, and to enhance subsequent interactions and performance. The effort draws on existing theoretical and empirical foundations, incorporates field observations of team leaders during team training, and culminates in an experimental study of team leader training.

This chapter describes the effort and its implications for team effectiveness. First, we briefly discuss the role of the team leader as a performance/learning facilitator. Next, we present a cyclical model of team learning with a focus on "post-action reviews." On the basis of the existing literature and field observations of team leaders during training, we then identify several critical team leader behaviors for supporting team learning. Afterward, we briefly describe an experimental study in which we trained team leaders to demonstrate critical team leader behaviors and facilitate effective pre-briefs and post-action reviews. The chapter culminates with some lessons learned for enhancing team performance.

Team Leader as Team Performance Facilitator

In traditional structures the team leader often acts as a supervisor; engaging in *command-and-control* activities; providing explicit directions to team members; and making most, if not all, team decisions. Recently, there has been a shift toward team leaders serving in a different capacity. In many organizations, the team leader's role is shifting toward more of a coach or facilitator (Caminiti, 1995). In this role, the team leader spends somewhat more of his or her time coaching individuals and facilitating team performance. It is important to note that this is not a dichotomous distinction. Team leaders are not strictly supervisors or facilitators; supervisors often facilitate and facilitators often supervise. Rather, the issue is the amount of time spent serving as facilitator versus supervisor. Several factors influence the evolution of the team leader's role.

Rapidly unfolding situations and distribution of tasks across team members are becoming more common. Yet traditional command-and-control activities are unlikely to have a direct effect on team outcomes in these circumstances (Adelman, Zirk, Lehner, Moffett, & Hall, 1986). Research suggests that in complex situations, effective team decision making tends to be characterized by implicit, flexible, and adaptive coordination (e.g., load sharing and mutual performance monitoring) in the absence of explicit directives (Adelman et al., 1986; Fleishman & Zaccaro, 1992). This calls for team leaders who work with and prepare their team for future events by learning from past events, rather than assuming that the team can always be given direct task instructions as is the case in the traditional supervisory role.

Increased organizational expectations for continuous learning and improvement are driving team leaders to spend more time working with their team to identify ways to develop themselves and improve team performance. This is more consistent with the facilitator role than the traditional supervisor role. A recently developed model of leadership and team effectiveness (Kozlowski, Gully, McHugh, Salas, & Cannon-Bowers, 1996) stresses this developmental role of the team leader. Kozlowski, Gully, McHugh, Salas, and Cannon-Bowers hold that the role that leaders play in the ongoing learning and development of team members is "an overarching process that extends throughout the lifecycle of the team" (p. 259) and that this process provides a foundation for all future team interactions.

Simultaneously, training researchers have acknowledged that training cannot serve as the sole source of learning and performance enhancement (Robinson & Robinson, 1995). Recent research reveals that individuals typically attribute less than 10% of their competency development to formal training (Tannenbaum, 1997). Experiential learning is important as is learning from supervisors and peers. Team leaders "function as an extension of the more formal instructional system" (Kozlowski, Gully, McHugh, Salas, & Cannon-Bowers, 1996, p. 262). Team leaders stimulate learning by being a direct source of information as well as by creating an effective learning environment, supporting more formal training experiences, and facilitating and encouraging team discussions.

Expert facilitators help teams agree on goals, roles, and procedures; ensure that all team members contribute; discourage disruptive behaviors; manage conflict; guide teams' decision-making processes, communicate clearly with all team members; and observe and accurately interpret group dynamics (Burns, 1995). They need to know how to conduct efficient, effective, highly participative meetings (Isgar, Ranney, & Grinnell, 1994).

The transition from supervisor to team leader is not trivial. It requires a shift in leadership attitudes as well as behaviors and thus represents a long, often difficult process (Ray, Hines, & Wilcox, 1995). In many work settings, such as the military and other hierarchical organizations where legitimate authority and the chain of command are integral parts of the culture, the transition from supervisor to facilitator can be particularly difficult. This transition may threaten an individual's identity, as team leaders are required to use new skills and change familiar behavior patterns (Steckler & Fondas, 1995). In addition, the transition from supervisor to facilitator may be perceived by some team leaders as a sign of weakness or lack of authority (Caminiti, 1995). Finally, even team leaders who have successfully adopted their facilitator role have a tendency to lapse into their command and control behavior in stressful situations (Isgar et al., 1994).

As arduous as it may be to make the transition from supervisor to facilitator, such a shift in leadership roles is essential for team learning. Formal training in teamwork skills is not always feasible or sufficient. It is frequently the case that team members' teamwork skills must be developed in the actual performance environment. Team members will learn much about how to interact as part of a team through actual team interaction led by the team leader. Therefore, the primary responsibility for the development of integrated task and teamwork skills necessarily falls to the team leader (Kozlowski, Gully, McHugh, Salas, & Cannon-Bowers, 1996). One possible way to initiate or support the role transition is to train team leaders how best to encourage and facilitate team learning.

Team Learning Cycle

Although in theory, each team activity may afford the opportunity for learning, realistically it is often difficult to learn while engaging in task activities. This is particularly true for activities that require high levels of attention. While performing such tasks, individuals often allocate most of their cognitive resources to the task itself (Kanfer & Ackerman, 1989). Individuals may also dedicate some cognitive resources to *self*-monitoring, learning, and correction (i.e., meta-cognition). However, that leaves little attentional resources available for learning about and correcting *team* processes during task performance.

To optimize team learning, teams need to prepare before engaging in task performance (i.e., pre-brief) and then provide feedback to one another after engaging in task performance (i.e., post-action reviews). This team learning cycle is depicted in Figure 1 and can pertain to either a team in training or in an on-going work environment.

The term *pre-brief* refers to a scheduled team meeting prior to engaging in a specific team activity. A pre-brief can provide an opportunity for the team to confirm its performance strategies, clarify team members' roles and expectations, discuss anticipated events, and focus on key performance issues. Moreover, it has been demonstrated that the climate established by team leaders in a pre-brief can quickly set the stage for subsequent team interactions (Smith-Jentsch, Salas, & Brannick, 1994).

Effective teams use pre-briefs to better articulate their plans and strategies (Orasanu, 1990). In addition, in some circumstances, teams can set both performance goals (what we want to accomplish) and learning-

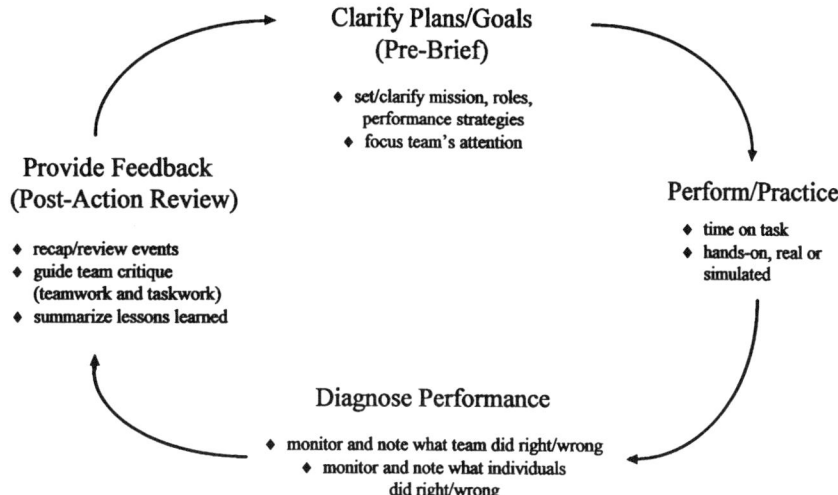

Figure 1.   Team learning cycle.

oriented goals (what we want to learn or improve) during the pre-brief sessions (Dweck, 1986).

Pre-briefs are essential for the setting of collective goals and strategies. These shared goals and strategies can help foster subsequent coordination and communication (Kozlowski, Gully, Salas, & Cannon-Bowers, 1996). Pre-briefs give teams the opportunity to understand the linkages between their individual roles and develop contingency plans (e.g., "when should I do X?"), thus allowing team members to better predict and anticipate each other's actions. By setting clear role-behavior expectations before an actual performance episode, teams can reduce role conflict and ambiguity. These role difficulties have been shown to increase stress, frustration, and tension and to lower satisfaction, motivation, and performance (Jackson & Schuler, 1985; Schaubrock, Ganster, Sime, & Ditman, 1993). Avoiding undue role stress is important because groups tend to deal with stress by becoming more rigid. For example, groups under stress tend to constrict their information processing through reduced communication and narrower attention (Burgess, Riddle, Hall, & Salas, 1992).

Pre-briefing can also help direct attentional resources toward team processes in need of improvement by reflecting on the team's previous performances. As noted earlier, individuals have a limited amount of cognitive resources (Kanfer & Ackerman, 1989). An effective pre-brief should help focus attention and help set the stage for follow-up discussions during subsequent post-action reviews.

The next stage of the cycle refers to the *performance* of team activities. In a training setting, this is usually an opportunity to practice trained skills by performing a role play, work sample, or team simulation. Experiences that require trainees to actively produce the knowledge, skill, or technique that is being trained can be an effective part of an instructional strategy (Campbell, 1988). Continuous opportunities to practice and perform tasks can also help lead to the development of error-correction strategies (Ivancic & Hesketh, 1995).

In the work environment, *performance* refers to a team's effort to complete any definable team activity (e.g., flying an aircraft from point A to B, assembling an electronics product, or delivering a team sales presentation to a client). As we noted earlier, it is often difficult to diagnose and correct one's performance while engaging in an activity. For teams operating in complex task environments, this is further complicated because often no team member, including the team leader, can observe everything that the other team members have done correctly or incorrectly (Orasanu & Salas, 1993). In those settings, systematic learning about and correction of team processes while performing is almost impossible.

As the team performs, team members implicitly or explicitly develop their own view of what they, the team, and other team members did well and poorly. To effectively *diagnose* their team's performance, leaders and team members must hold accurate mental models of their task, equipment, and teammates (Cannon-Bowers, Salas, & Converse, 1993), as well as the manner in which critical teamwork processes manifest themselves in the context of their team task. If the team conducted a thorough prebrief, and if the team shares consistent mental models (Cannon-Bowers et al., 1993), the individual diagnoses should converge. However, in practice, because team members may observe different aspects of the team's performance and view even common events through their own unique perceptual lens, it would be unusual for all team members to reach identical conclusions. Therefore, these need to be discussed during a post-action review.

Post-action review refers to the systematic process of sharing observations and interpretations about team functioning (i.e., processes) and performance. It is a means of providing collective feedback to team members and represents an activity often guided by an instructor during team training. However, in many circumstances, particularly on the job or in embedded training, a post-action review can be led by the leader of the team as well.

Feedback is central to the learning process (Ilgen, Fisher, & Taylor, 1979) and group-level feedback can improve team performance (Pritchard, Jones, Roth, Stebing, & Ekeberg, 1988). Feedback has both informational and motivational properties (Ilgen et al., 1979). According to Swezey and Salas (1992), one guideline for effective team training is that it "should include systematic procedures for providing feedback information for the trainees while they are learning team skills" (p. 238). This is equally relevant for improving team performance in situ as it is in a training environment, because performing a task without appropriate feedback can inhibit or even interfere with on-going learning (Komaki, Heinzmann, & Lawson, 1980).

Further, it is often advantageous to get feedback from multiple sources, for example, from both supervisors and peers, to get a more accurate view of multiple facets of performance. For this to happen, the team leader must create a team environment that fosters participation in the feedback process and, therefore, must be able and willing to probe, ask questions, confront ambiguities, and offer recommendations. A post-action review should enable teams to clarify misperceptions, refine strategies, and begin to develop subsequent plans (Smith-Jentsch et al., 1994).

Team Leader Behaviors During Pre-Briefs and Post-Action Reviews

Team leaders play a central role in facilitating or hindering team improvement (Yukl, 1989). As noted earlier, team leaders are increasingly asked not only to guide their team and to provide individual coaching and training, but also to promote team learning and assist the team in learning *as a team*.

One way that a team leader can promote team learning is by supporting the learning cycle described earlier—facilitating effective pre-briefs and post-action reviews. The feedback, leadership, and training literatures suggest a number of specific behaviors that team leaders can exhibit during briefings to enhance team learning. To supplement this literature, we observed team leaders conducting pre-briefs and post-action reviews in a variety of naval team training environments including both shore-based and shipboard simulations. The briefings we observed ranged from initial team training given by commanding officers to their precommissioned ship units in a high-fidelity simulator, to integrated team training provided by shipboard instructors underway. These briefings yielded many examples of leader behaviors that seemed to accelerate team learning, as well as those that appeared to squelch dialogue and team learning.

Collectively, our observations, along with related research findings from several diverse literatures, provide a basis for positing the types of leader behaviors that facilitate team learning. The following section summarizes our observations and describes eight team leader behaviors that we propose should be exhibited during pre-briefs and post-action reviews.

1. Provide a self-critique early in the post-action review. This is particularly important in complex, dynamic team environments in which the team leader cannot observe all team behaviors. An effective briefing environment must be one in which team members feel comfortable admitting when they were confused or when they made mistakes during task performance (Senge, 1990). Team members will take cues from and modify their behaviors on the basis of the team leader's behaviors and not simply what the leader says (Weiss, 1977). Therefore, the team leader must model effective behaviors during briefings that his or her team can emulate, such as acknowledging if they were confused or made a mistake at some point during team performance. This is particularly important as the commission of, admission of, and subsequent analysis of errors made during training sessions are beneficial to the transfer of trained behaviors to the job (Ivancic & Hesketh, 1995; Schmidt & Bjork, 1992).

Some of the team leaders we observed were very quick to acknowledge their own shortcomings following a team training exercise. In these cases we found that team members generally followed suit, owning up to their role in coordination breakdowns. This was particularly true when the team leader was a commanding officer. Conversely, a few team debriefs degenerated into finger pointing and accusations for teams whose leaders were unwilling to critically examine their own performance.

Consistent with previous research (Smith-Jentsch et al., 1994), the team leaders we observed seemed to quickly establish a climate that was either open and forgiving or divided and demeaning. Admitting mistakes early in the post-action review was a clear and strong sign that the leader was interested in fostering personal growth and not placing blame.

2. *Accept feedback and ideas from others.* In addition to self-critique, peer feedback is important for team learning. Thus, to optimize learning in a post-action review, team members must be willing to accept feedback on their performance that is offered by others. To encourage this behavior, the team leader should model the acceptance of feedback and ideas from team members. Failure to accept feedback or, even worse, openly discrediting individuals who offer feedback has two negative repercussions. First, individuals may stop providing feedback to the team leader, isolating the team leader from potentially valuable input. Second, team members learn that it is acceptable to reject or ignore input from others, decreasing the chance that feedback will lead to learning. In contrast, teams that learn to value and endorse dissent are more likely to avoid complacency and stagnation (Nahavandi & Aranda, 1994).

The impact of a leader's response to feedback from team members has been demonstrated experimentally in the context of a flight crew pre-brief (Smith-Jentsch et al., 1994). A confederate crew member was scripted to offer a viewpoint that opposed that of the captain in a pre-flight brief. Although in all cases the confederate's suggestion was incorrect, half the participants observed the captain carefully consider the input, thank the confederate for speaking up, and explain why his suggestion would not work. The other half of the participants observed the captain reject the input outright and demean the confederate offering such a suggestion. Results indicated that participants who observed the confederate receive punishment for offering feedback were less likely to confront ambiguities, correct errors made by the captain, and offer suggestions during a simulated flight than those participants who had observed their captain positively reinforce the confederate crew member.

In our observations of pre-briefs and post-action reviews, we had the opportunity to witness team leaders who fit into both categories. For example, those leaders who were open to feedback on their own performance could be seen to benefit from input regarding how well they communicated their priorities to the team and how often team members required situation updates. On the other hand, leaders who were not open to feedback often expressed frustration with the lack of progress their teams appeared to be showing. It did not occur to these leaders that they may be part of the problem. Moreover, it was clear that no one on the team was interested in being the one to point this out.

Unique developmental feedback can come from multiple sources in a team briefing. Ideally, team members will learn how they can support one another and leaders will learn how to better lead their teams by listening to and accepting input from other team members. Therefore, it is critical that leaders demonstrate a willingness to accept feedback on their own performance.

3. Avoid person-oriented feedback; focus on task-focused feedback. The way in which team members offer input to one another can determine in part whether it will be accepted. Amason, Thompson, Hochwarter, and Harrison (1995) distinguished between two types of conflict in groups: affective and cognitive. Affective conflict is seen as dysfunctional for groups and is centered around personal issues and emotions, whereas cognitive conflict is seen as potentially beneficial for groups and is based on ideas and strategies. Amason et al. (1995) observed that effective teams tend to experience cognitive conflict, but not affective conflict. Implied in this observation is that, as a team leader attempting to promote effectiveness, one should promote free exchange of ideas, including disagreements, but should try to eliminate personal friction.

A major source of interpersonal strain is destructive, person-focused feedback in which evaluative judgments are made about the individual instead of her or his ideas, behavior, or performance (Anderson & Rodin, 1989). Person-focused feedback can lead to frustration, aggression, and decreases in group morale and cohesion. In addition, such person-focused feedback can have a chilling effect on future group communication, which will also adversely affect team learning. This is not to say that negative feedback should never be used. As long as negative feedback is handled constructively, it does not typically lead to negative consequences (Baron, 1988). In short, feedback provided to individual team members should be focused on the individual's actions and behaviors and not on the person.

Unfortunately, some of the team leaders we observed used personal attacks as a means of shaming team members into improving their performance. These leaders later explained to us that they felt that public embarrassment was one of the most powerful tools they had to motivate team members to improve. This attitude assumes that lack of motivation is the root of all performance problems. Although the personally focused feedback we heard (e.g., "EWS is really a talker, he loves to hear the sound of his own voice") may have motivated some to change their ways, it rarely included direction to the team regarding the specific behaviors that needed to change (e.g., how and when it is appropriate for EWS to make situation reports). Moreover, the recipients of person-oriented comments were often so busy defending themselves that they failed to recognize the potential value of the underlying point attempting to be made.

Performance feedback is most useful when it is stated in terms of behaviors that need to change instead of as a personal attack. When leaders provide behavior-based rather than person-oriented feedback they facilitate acceptance of their comments as well as encourage team members to state their concerns in a similar fashion.

4. Provide specific, constructive suggestions when providing feedback. Although they should not be the only source of feedback, team leaders are perhaps the most valuable single source of team performance feedback. Individuals respond more positively to specific feedback than to vague feedback (Ilgen, Mitchell, & Fredrickson, 1981; Liden & Mitchell, 1985). This is particularly true when providing negative feedback (Anderson & Rodin, 1989; Baron, 1988). In addition, specific feedback provides direction to the recipients regarding how they can subsequently correct or improve their performance, thus fulfilling both a motivational and a cuing function (Nadler, 1979). In short, the provision of specific, constructive feedback is an effective way for team leaders to help identify and summarize specific areas for improvement for both the team as a whole and for individual members.

Many of the leaders we observed provided specific examples of performance problems and offered solution-oriented suggestions to their teams on how they could improve. However, a significant number of team leaders offered vague broad brush impressions, such as "communications were sloppy," without identifying specific instances or preferred approaches. These leaders tended to argue that "once a performance problem is noted the solution will be obvious." This attitude assumes that all team performance problems are due to a lack of effort or motivation and not to a lack of knowledge or skill.

To maximize team learning, leaders should provide specific examples of performance deficiencies as well as solutions. This is likely to help ensure team members both understand and accept goals for improvement.

5. Encourage active team member participation during briefings and reviews and not simply state one's own observations and interpretations of the team's performance. Team leaders should guide and facilitate team debriefings rather than "lecture." Lecturing or one-way communication may be a less effective method of adult learning than sharing experiences and observations when complex responses (such as those necessary for effective team functioning) are required (Goldstein, 1993). In addition, a team leader who lectures is forcing her or his interpretation of team events onto the group. As expert as this observation and opinion may be, it only represents one perspective. Because most team situations are complex and dynamic, it is important to gather and integrate as many relevant points of view as possible when debriefing or critiquing performance (Rouse, Cannon-Bowers, & Salas, 1992). Using a predominantly one-way communication strategy is less compatible with the role of team leader as facilitator and is more in line with the role of team leader as supervisor.

The team leaders we observed varied significantly in the extent to which they engaged team members in the post-action review. The majority of shipboard briefs were predominantly one-way communications of problem areas from the leader to team members. This is due in part to the time constraints that exist for shipboard training. Team training exercises are often scheduled back to back with little time in between to conduct a thorough debriefing. In this environment, both leaders and team members are motivated to keep the briefs short so they can catch up on some much needed sleep before the next exercise. Many team leaders who conducted post-action reviews in a lecture-type fashion complained that the blank looks they received from their team members left them wondering how much learning had occurred.

A second group of leaders attempted to solicit team member input; however, they missed the mark by asking questions in a way that did not guide or motivate team members toward responding in a meaningful way. For example, asking the team to critique themselves after the team leader had already expressed his or her assessment rarely elicited new information. Moreover, general closed-ended questions, like "how were communications?" elicited answers like "great" or "not so good" but rarely specific problems or solutions.

Other team leaders adopted a more effective probing style, posing specific thought-provoking questions to the team before offering their own assessment of "what went wrong." This type of questioning approach led team members to describe specific misunderstandings and offer unique solutions, often bringing to light problems of which the leader was unaware.

These observations are consistent with findings from an earlier study that indicated that an open-ended questioning approach significantly increases team members' willingness to confront ambiguities, ask questions, and offer potential solutions (Smith-Jentsch et al., 1994). In addition, it has been shown that team member critique can increase commitment to subsequent team decisions as long as team members perceive that their input is given fair consideration (Korsgaard, Schweiger, & Sapienza, 1995).

Team member participation in the post-action review is critical to accurately diagnosing team problems, assessing team member understanding of lessons learned, and encouraging commitment toward goals for team improvement. Team leaders can solicit meaningful participation from team members by asking them to critique their individual and team performance using specific open-ended questions before offering team leader opinions.

6. *Guide briefings to include discussions of "teamwork" processes as well as "task work."* Although the provision of process feedback is not widely implemented or universally supported in the literature (e.g., McLeod, Liker, & Lobel, 1992), there is evidence attesting to its importance. The success of many team-building and process consultation interventions is predicated on the assumption that group process can have an impact on group task performance (Tannenbaum, Beard, & Salas, 1992). Moreover, there is some evidence to suggest that an emphasis on the mastery of underlying processes rather than on performance outcomes may facilitate retention and generalization of training (Kozlowski, Gully, Salas, & Cannon-Bowers, 1996). This has been described as the difference between training teams to make the right decision in a given scenario versus training them to "make the decision right" using processes that will on average lead to more correct decisions (Cannon-Bowers & Salas, 1997).

When discussing processes that contribute to team outcomes, there is growing evidence that one can distinguish between two relatively independent tracks; task work skills and teamwork skills (McIntyre & Salas, 1995). Technical skills that are specific to the type of task performed by a team, such as "stick and rudder" skills for a pilot, are referred to as task work skills. Teamwork skills are differentiated from task work skills in that they involve coordination with other team members that can be important for a variety of different team tasks, such as leadership. A number of researchers have demonstrated that teamwork skills contribute a significant amount of unique variance in team performance above and beyond that accounted for by the individual task work abilities of team members (Smith-Jentsch, Johnston, and Payne, chap. 4, this volume; Stout, Salas, & Carson, 1994). However, team members typically receive little structured feedback on the quality of their team *interactions*. This may be because team leaders are traditionally more experienced at assessing task work. Because it is generally easier to make objective assessments of task work skills than it is of teamwork skills, leaders may feel more confident providing task-work-related feedback to teams. Similarly, team members may feel more comfortable discussing task work issues than teamwork issues.

The majority of the team briefings we observed emphasized team outcomes and task work skills to a much greater degree than teamwork processes. When teamwork processes were mentioned in a post-action review, team leaders most often spoke in very general terms (e.g., "coordination was smooth" or "communication broke down"). The team leaders who appeared to be most effective at helping their teams improve tended to guide their teams toward discussing how teamwork had an impact on performance at specific points in an exercise. This approach seemed to make teamwork concepts more tangible for their teams and the resulting discussions more productive.

There is increasing evidence that feedback on teamwork skills is as important as feedback on task work skills. Team members do not always understand the specific teamwork behaviors related to the performance of their team. Therefore, team leaders should guide their teams in reviewing the role that teamwork processes, in addition to task work processes, play in determining performance outcomes.

7. Refer to prior pre-briefs and team performance when conducting subsequent debriefs. This provides a sense of continuity, reliability, and consistency, all of which are desirable attributes of a feedback source (Herold, Liden, & Leatherwood, 1987). In addition, reiterating areas for improvement on the basis of previous performance helps focus team members on critical issues and may help avoid recurring mistakes. As we illustrated in Figure 1, goals established in post-action reviews should be reviewed as part of the next pre-brief. This helps to remind team members of lessons learned from previous exercises and allows them to plan how these lessons generalize to a new context or scenario.

Continuing with the learning cycle, topics of discussion in the post-action review should reflect the goals or criteria set in the pre-brief. This continuity is important for solidifying stimulus–response linkages. Moreover, team members are more likely to accept feedback in a post-action review if it is based on criteria established and agreed on earlier in the pre-brief.

In our observations of post-action reviews, we noted examples of leaders who were very effective at weaving a consistent thread that bound pre-briefs, performance, and post-action reviews. Teams briefed by these leaders appeared to be more focused and motivated than those briefed by leaders who failed to make consistent links between briefs.

In other instances, the topics discussed in the pre-briefs we observed bore little resemblance to those addressed in the subsequent post-action review. Teams that received these disjointed briefings appeared to be frustrated, and some complained that they were being expected to "hit a moving target," as their leader's criteria for success seemed to be an elusive goal.

Team learning is expected to be enhanced when connections among pre-briefs, performance, and post-action reviews are made explicit for team members. Team leaders play a critical role in establishing these linkages by referring to previous goals set and lessons learned and by drawing similarities and distinctions that aid team members in generalizing performance strategies.

8. Vocalize satisfaction when individual team members or the team as a whole demonstrates improvements. Voicing satisfaction with improvements should serve as a reinforcing mechanism and enhance motivation to learn. Moreover, positive reinforcement should enhance individual self-efficacy (Gist & Mitchell, 1992) and team or collective efficacy (Travillian, Baker, & Cannon-Bowers, 1992), which contributes to improved performance. Finally, by making improvement a priority and referring to it often, a team leader is communicating her or his dedication to continual learning and improvement.

Some of the team leaders we observed focused almost exclusively on what the team did wrong, whereas others made time for noting that their team had corrected problems from a previous exercise. Highlighting team improvement is not only motivating to team members, it is also necessary in order to teach them what is correct. As one team leader put it "if you don't tell them what they are doing right, they might try to fix what isn't broke." When a team leader fails to point out to his or her team that they have corrected a performance problem, team members are left to wonder whether (a) their leader didn't notice the problem this time, (b) their leader didn't mention it because he or she is tired of repeating himself or herself, or (c) they have improved sufficiently so that it is no longer a problem. Thus, it is important for team leaders to tell the team what they should keep doing right as well as what they should stop doing wrong. Voicing satisfaction with team improvements provides team members with important information as well as motivation to grow and to learn as a team.

TADMUS Experiment

Given the potentially important role of briefings for team learning, and the central role team leaders play in conducting briefings, if team leaders can be taught to conduct effective briefs, team effectiveness should be enhanced. Moreover, briefing skills are an example of a "transportable" team competency, one that can be used in different settings with different team members (Cannon-Bowers et al., 1995). This is important because it suggests that the advantages of acquiring effective briefing skills should carry over to future team settings. This points to the potential on-going utility of training leaders to conduct effective briefs.

There is a growing literature demonstrating the efficacy of team leader training, particularly when based on behavior modeling principles (Burke & Day, 1986). However, we know of no research that has specifically examined the effects of training leaders how to conduct effective pre-briefs and debriefs. Thus, as part of the TADMUS effort, we designed a training program that incorporates the principles of effective briefings noted above and tested its effectiveness in enhancing subsequent team performance. In particular, we wanted to examine the following three research questions:

1. Can team leaders be trained to conduct more effective pre-briefs and post-action reviews? Will they exhibit appropriate briefing behaviors?
2. Will team members act differently during pre-briefs and post-action reviews if their team leaders exhibit appropriate briefing behaviors?
3. Do enhancements in leader and team briefing behaviors affect team performance?

Method

Participants. Seventy naval officers were recruited to receive team training in groups of 14 5-person teams. Within each team, the officer with the most relevant shipboard experience was selected to be the team leader. Each team leader was randomly assigned to receive either training for briefing skills or control training. Leaders were trained individually because only one team could be run on each day. Because briefing skills are a "transportable" team competency, these skills can be trained at the individual level (Cannon-Bowers et al., 1995).

Briefing skills training. Team leaders in the experimental condition received 2 hours of training on effective pre-briefing and debriefing skills. The training followed a behavior modeling format. To ensure the relevance of the program, content and examples for the training were based on the observation work and subject matter expert input. The learning objectives for trainees were (a) to distinguish between effective and ineffective briefing behaviors and (b) to become skilled at using a set of some of the effective briefing behaviors described earlier.

First, the trainer discussed the team learning cycle, effective briefing behaviors, and how the behaviors are related to team learning. Second, examples of effective and ineffective debriefs were shown on videotape. Following each videotape debrief, team leaders were asked to critique the model's debriefing skills, using the effective briefing behaviors as criteria. Third, team leaders viewed a 15-minute videotape of a team performing and were asked to imagine that they were the team leader. At the conclusion of the tape, the team leader conducted a practice debrief with two scripted, role-playing team members. Finally, the team leader critiqued themselves on their use of the critical briefing behaviors and received feedback from the instructor.

Control training. Team leaders in the control condition received two hours of computer-based training on the buttonology necessary to navigate through menus used to obtain information from the console of an AEGIS combat information center (CIC). This same training was given to all other team members in both conditions while their leaders were trained in a separate room.

Team task. Each team participated in three 30-minute performance scenarios administered through a computer-based simulation of the CIC of a navy warship. Five networked computers were used to simulate interdependent work stations within the CIC. Each of the performance scenarios contained three "key events." These events were specifically designed to require coordination and cooperation between the team members in order to arrive at an accurate team decision. Additional ambiguity and workload demands were included in subsequent scenarios to provide teams with graduated exposure to the stressful conditions that exist in a CIC (Johnston, Smith-Jentsch, & Cannon-Bowers, 1996).

Each team member had access to different information through the consoles and was linked to other team members through a communication network. Role players played outside parties and followed scripts to ensure consistency around key events.

The type of team in this study is an example of what Sundstrom, DeMeuse, and Futrell (1990) referred to as an "action team." They worked on a structured but dynamic task, had common goals but specialized individual tasks, and had differentiated roles and distributed expertise; in addition, coordinated patterns of interdependency were specified.

Procedure. Each five-person team was randomly assigned to one of the two experimental conditions. The team's participation lasted one 8-hour day. The team leader received 2 hours of training in a separate room regardless of condition. During this time the remaining four teammates received the same computer-based training given to leaders in the control condition.

When the team members reconvened they received 1 hour of instruction on how to operate their consoles before participating in two practice sessions. The first practice session involved individual practice on their console and lasted approximately 10 minutes. The second session involved team interaction and lasted approximately 15 minutes. Teams were permitted to debrief after the second practice session. After a lunch break, teams performed a series of increasingly difficult performance scenarios. Team leaders were asked to lead briefings before and after each scenario. All team leaders were given a summary of the three key events after each scenario. They were told to take as much time as necessary to conduct the briefings. The three scenarios, with their respective pre-briefs and post-action reviews, lasted approximately 4 hours.

Measures

Briefing behaviors. Two condition-blind raters independently evaluated leader and team member behaviors from transcripts of the pre-briefs and after-action reviews. Using a standardized coding sheet, coders rated 3 leader behaviors (e.g., reminded team of areas to focus on) and 3 team behaviors (e.g., asked questions) for each pre-brief and 17 team leader behaviors (e.g., provided suggestions) and 10 team behaviors (e.g., critiqued self) for each after action review. Consensus ratings were used in the analysis of training on leader and teammate briefing behaviors.

Teamwork behaviors. Teams were evaluated on several teamwork behaviors, six of which were pertinent to the current study. These were providing backup, stating priorities, providing situation updates, using proper phraseology, passing relevant information before having to be asked, and exploiting all available sources of information. Two condition-blind subject matter experts individually rated these behaviors and then collectively assigned a consensus rating on each behavior. The ratings of teamwork behaviors used in this study were developed as part of a larger TADMUS performance measurement effort and have been linked to performance outcomes (see Smith-Jentsch, Johnston, & Payne, chap. 4, this volume).

Team performance outcomes. The critical performance outcome for the team task involved the prompt and accurate identification of important aircraft that were linked to scenario events. Behaviorally anchored rating scales designed to evaluate the accuracy and timeliness of the teams' situation assessment were developed jointly by researchers and subject matter experts as part of the overall TADMUS performance measurement effort (see Smith-Jentsch, Johnston, & Payne, chap. 4, this volume). Two condition-blind subject matter experts assigned team performance scores for each scenario event after viewing videotapes of the team's performance. Consensus ratings were used in all analyses involving team performance.

Results

Detailed results of the analyses are available in Tannenbaum, Smith-Jentsch, and Cannon-Bowers (1998). In this section we report general results for each of the three guiding research questions.

Research Question 1: Can team leaders be trained to conduct more effective pre-briefs and post-action reviews? Will they exhibit appropriate briefing behaviors? Despite the small sample size and limited power, trained team leaders exhibited significantly more of the effective briefing behaviors than their control group counterparts. These differences were more apparent in the post-action reviews than in the pre-briefs partly because the information provided in pre-briefs is generally much more structured (e.g., geopolitical information and rules of engagement). Although there were no significant differences in their attempts to involve others during pre-briefs (most attempted at least somewhat), or provide situational updates (most did), trained team leaders were significantly more likely to remind their team members to concentrate on key areas for improvement during the next performance scenario. This should have helped direct the team's attention and resources to those areas in greatest need of improvement.

Post-action reviews conducted by trained team leaders looked distinctly different from other post-action reviews. Trained leaders were significantly more likely to ask the team members to critique themselves (encouraging their involvement). They were also more likely to guide the team to consider teamwork behaviors (e.g., error correction and backup). They provided much more positive feedback, requested feedback more frequently, probed much more, and were more likely to critique themselves as well. Finally, they were significantly more likely to recap and review events and summarize lessons learned. All the other indicators were in the hypothesized direction but were nonsignificant (e.g., providing specific suggestions and expressing satisfaction with improvement). On the basis of these results, it is clear that team leaders can be trained to exhibit effective briefing behaviors, particularly during post-action reviews.

Research Question 2: Will team members act differently during pre-briefs and post-action reviews if their team leaders exhibit appropriate briefing behaviors? Team members exhibited no significant differences on the three pre-brief behaviors—asking questions, offering suggestions and opinions, or clarifying the situation. This is not surprising considering team leaders in both conditions appeared to involve their teams similarly during pre-briefs. Trained team leaders were more likely to remind the team to focus on teamwork-related issues, but this did not appear to affect the team's behavior in the pre-briefs.

On the other hand, teams with trained leaders exhibited several critical differences during post-action reviews. They were significantly more likely to engage in discussions about teamwork behaviors. They were also more likely to critique themselves and offer suggestions to others. Although a few other behaviors failed to reach significance (e.g., critiquing the team leader), it appears clear that the team leaders were able to create an environment wherein team members were more comfortable to admit concerns, provide feedback to others, and discuss processes as well as outcomes.

Research Question 3: Do enhancements in leader and team briefing behaviors affect team performance? Teams with trained leaders demonstrated more effective teamwork behaviors during their.performance runs than did those with untrained leaders. In particular, they stated team priorities explicitly, provided more frequent situation updates, and exploited more of the sources of information available to them in assessing key events. Means for the remaining teamwork behaviors were not significantly different but were all in the hypothesized direction except for backup behavior. It has been demonstrated that backup behavior is less consistent across events when the need for backup is not controlled for in a team (Smith-Jentsch, Johnston, & Payne, chap. 4, this volume).

Finally, teams with trained leaders achieved superior performance outcomes. Specifically, these teams' assessment of key scenario events were more accurate and timely than were those made by teams that were briefed by untrained leaders. Overall, it appears that "better" briefings, and the application of the team learning cycle, results in better teamwork and enhanced team performance outcomes.

Lessons Learned: Implications for Enhancing Team Effectiveness

On the basis of the literature, our observations of teams in a variety of team training environments, and the empirical study that we conducted, we have compiled the following lessons learned about team learning:

1. Teams learn and perform better when they effectively prepare for (pre-brief) and subsequently compare perceptions about (post-action review) their performance—that is, when the team learning cycle works.
2. Team leaders that are effective at encouraging team learning demonstrate a different set of behaviors during briefings than do team leaders who are less effective.
3. Team leaders do not necessarily possess the skills required for conducting effective briefings—for example, they tend to overrely on one-way communications.
4. Teams do not naturally conduct effective briefings—for example, they tend to gravitate toward discussing outcomes and task work skills to the exclusion of teamwork skills.
5. Team leaders cannot see everything the team does during task performance—team members can provide unique information about individual and team effectiveness if they believe it is appropriate and acceptable.
6. Team members learn about appropriate behaviors during briefings by observing their team leader—for example, if the team leader critiques himself, team members are also more likely to admit mistakes.
7. Team leaders can be trained to demonstrate more effective briefing

behaviors—for example, to probe more and to guide the team to consider teamwork behaviors.

8. Teams whose team leaders have been trained to conduct effective briefs will prepare better (pre-brief), compare their perceptions of performance better (post-action review), and will perform more effectively than teams with untrained team leaders.

Implications

The results of our investigations and the lessons identified have numerous implications for enhancing team effectiveness. This is particularly true given several recent trends in both industry and the military.

One trend that we discussed earlier is the increasingly common expectation for team leaders to serve as coach/facilitators. Team leaders are being called on to develop team members both individually and collectively and to create an environment wherein team learning can occur. When team leaders can facilitate the learning cycle and encourage team members to contribute their own unique perspective on team performance, teams are more likely to learn.

However, our observations confirm the experience of Caminiti (1995) and others that the transition to this new role does not come naturally to all team leaders. Team leaders do not necessarily possess the skills they need, and simply "empowering" teams to discuss their own performance will not necessarily result in effective learning. In particular, our observations suggest that not all teams naturally know which aspects of performance to discuss (teamwork as well as task work), or how to discuss them in a constructive manner (behavior based and solution oriented vs. vague or accusatory). There is a clear need to prepare team leaders to perform their new role effectively. Providing team leaders with the type of training described in this chapter can help support the role transition.

A second prevalent trend is the increasing use of nontraditional training methods (Bassi, Benson, & Cheney, 1996). There is beginning to be a shift away from traditional instructor-led classroom training. We are starting to see a greater use of action learning methods and a greater reliance on semistructured, experiential learning (Chao, 1997). For example, some organizations are coupling initial training with on-going sessions to discuss how the training is being used, including what has been working and what obstacles need to be overcome. This approach to learning relies on individuals who can facilitate post-action reviews. As this approach proliferates, organizations will not be able to rely on instructors to facilitate these sessions. Instead, they will need a cadre of trained team leaders who can handle this as part of their on-going team development responsibilities.

There is a parallel trend in the naval environment. The push to reduce shore-based training expenditures necessitates a return to times when the workplace was the primary classroom. Such an approach would return the large number of years now being spent in the classroom to functional time spent in the fleet. It has been recognized that "this is not simply a change to existing procedure; it is an entirely new culture," one where "the senior enlisted are returned to the deckplates, where they are best suited, maintaining their proficiency while they serve as mentors to junior sailors" (Tonning, 1997, p. 64). In the future, there is likely to be an increased reliance on shipboard learning. In general, there is an increasing expectation that every time a team performs together, whether by simulation or in normal duty, its members should be able to learn and enhance their performance. Unfortunately, our observations suggest that this does not necessarily occur. Moreover, shipboard training poses additional problems of time constraints, space constraints, and fatigue. For these reasons it is imperative that team leaders (and shipboard instructors) be equipped with the skills and structure they need to maximize limited debriefing times.

A third trend is that more individuals are being asked to serve as team leaders for multiple teams—either concurrently or sequentially. In industry it is increasingly common for individuals to be asked to lead a function or departmental team as well as a project or cross-functional team. In the military, team leaders routinely rotate to new assignments. These trends can make it difficult to assemble intact teams for training, particularly when leadership demands are coupled with a dynamic team membership. In that type of environment, how can teams enhance their teamwork skills?

When it is difficult to assemble an intact team, it may still be feasible to train the team leader to facilitate team learning. The leader behaviors that foster enhanced team learning are applicable in most team settings. Conducting effective pre-briefs and post-action reviews is a transportable skill. This means that trained leaders should be able to apply these skills to the multiple teams they are leading, or carry it with them to their next team assignment.

Future Research Needs

Although this program of research has some immediate implications for preparing teams, some additional research questions remain. One assumption of the learning cycle is that pre-briefs and post-action reviews are as applicable to on-going work assignments as they are to team training environments. Although the theory would support that contention, future research should examine the efficacy of training team leaders to facilitate team briefs in a nontraining context. There is a need to address the following questions. How can team leaders best structure periodic pre-briefs and post-action reviews as part of the normal work environment? Will this have a similar effect on team learning as it did in the team training environment? A new program of research that has grown out of the TADMUS program will begin to address these questions and others regarding how to maintain a continuous learning environment at sea (Smith-Jentsch, Cannon-Bowers, & Salas, 1997).

One potential advantage of training team leaders to conduct effective briefings is the generalizability of these skills to subsequent leadership assignments. If this occurs, then the impact of this type of intervention is on-going and far reaching. Although the literature suggests that training of this type should generalize to subsequent assignments, future research should empirically examine this assumption.

Another advantage of the training is its impact on team members. This research demonstrates that team leader behaviors during briefings influence team member behaviors during briefings. Yet some team members will eventually assume team leader responsibilities. A few pertinent questions for future research are, Will team members apply what they learned from observing their trained team leader to the teams they eventually lead? Are the effects of modeling strong enough to extend the benefits of the training beyond the current leader to future leaders? Will allowing team members to occasionally facilitate team debriefs help them develop and practice their briefing skills?

The type of training described in this chapter can serve as a stand-alone intervention. However, it may be most powerful when used to supplement other team interventions. For example, trained team leaders should be better able to conduct post-action reviews to support team simulations, collective training, team building interventions, and action learning events. Moreover, as shipboard technology begins to incorporate more built-in performance feedback mechanisms, team leaders may be able to supplement those with highly focused team debriefs. Future research should examine the usefulness of team leader training for enhancing the effectiveness of other team interventions.

Summary and Conclusion

This chapter described a stream of research within the broader TADMUS effort. On the basis of existing research and observations of actual teams, we developed a model of how the team learning cycle can work. We isolated specific leader behaviors that should optimize the benefits of team pre-briefs and post-action reviews. These served as the foundation of a behavior-modeling-based training program for team leaders. We then empirically tested the training with naval team leaders.

The empirical study revealed that a relatively inexpensive, 2-hour intervention can have a powerful effect on team behaviors and performance. Future applications of the principles identified in this stream of research have the potential to help enhance the quality of team decision making in both the military and private sectors—a major objective of the TADMUS program.

References

Adelman, L., Zirk, D. A., Lehner, O. E., Moffett, R. J., & Hall, R. (1986). Distributed tactical decision-making: Conceptual framework and empirical results. *IEEE Transactions on Systems, Man, and Cybernetics, 16*, 794–805.

Amason, A. C., Thompson, K. R., Hochwarter, W. A., & Harrison, A. W. (1995). Conflict: An important dimension in successful management teams. *Organizational Dynamics, 24,* 20–35.

Anderson, S., & Rodin, J. (1989). Is bad news always bad? Cue and feedback effects on intrinsic motivation. *Journal of Applied Social Psychology, 19,* 449–467.

Baron, R. A. (1988). Negative effects of destructive criticism: Impact on conflict, self-efficacy, and task performance. *Journal of Applied Psychology, 73,* 199–207.

Bassi, L., Benson, G., & Cheney, S. (1996). The top ten trends. *Training and Development, 50,* 28–42.

Burgess, K. A., Riddle, D. L., Hall, J. K., & Salas, E. (1992). *Principles of team leadership under stress.* Paper presented at 38th annual meeting of the Southeastern Psychological Association, Knoxville, TN.

Burke, M. J., & Day, R. R. (1986). A cumulative study of the effects of managerial training. *Journal of Applied Psychology, 71,* 232–245.

Burns, G. (1995). The secrets of team facilitation. *Training and Development, 49,* 46–52.

Caminiti, S. (1995, Feb. 20). What team leaders need to know. *Fortune, 131,* pp. 93–100.

Campbell, D. J. (1988). Training design for performance improvement. In J. P. Campbell & R. J. Campbell (Eds.), *Productivity in organizations: Frontiers of industrial and organizational psychology* (pp. 177–215). San Francisco: Jossey-Bass.

Cannon-Bowers, J. A., & Salas, E. (1997). A framework for developing team performance measures in training. In M. T. Brannick, E. Salas, & C. Prince (Eds.), *Team performance assessment and measurement: Theory, methods, and applications* (pp. 45–62). Hillsdale, NJ: Erlbaum.

Cannon-Bowers, J. A., Salas, E., & Converse, S. A. (1993). Shared mental models in expert team decision making. In N. J. Castellan, Jr. (Ed.), *Individual and group decision making: Current issues* (pp. 221–246). Hillsdale, NJ: Erlbaum.

Cannon-Bowers, J. A., Tannenbaum, S. I., Salas, E., & Volpe, C. E. (1995). Defining team competencies and establishing team training requirements. In R. Guzzo & E. Salas (Eds.), *Team effectiveness and decision making in organizations* (pp. 333–380). San Francisco: Jossey-Bass.

Chao, J. T. (1997). Unstructured training and development: The role of organizational socialization. In J. K. Ford & Associates (Eds.), *Improving training effectiveness in work organizations.* Hillsdale, NJ: Erlbaum.

Dweck, C. S. (1986). Motivational processes affecting learning. *American Psychologist, 41,* 1040–1048.

Fleishman, E. A., & Zaccaro, S. J. (1992). Toward a taxonomy of team performance functions. In R. W. Swezey & E. Salas (Eds.), *Teams: Their training and performance* (pp. 31–56). Norwood, NJ: Ablex.

Gist, M. E., & Mitchell, T. R. (1992). Self-efficacy: A theoretical analysis of its determinants and malleability. *Academy of Management Review, 17,* 183–211.

Goldstein, I. L. (1993). *Training in organizations.* Pacific Grove, CA: Brooks/Cole.

Guzzo, R. A., & Salas, E. (Eds.). (1995). *Team effectiveness and decision making in organizations.* San Francisco: Jossey-Bass.

Hackman, R. J. (1990). *Groups that work (and those that don't).* San Francisco: Jossey-Bass.

Herold, D. M., Liden, R. C., & Leatherwood, M. L. (1987). Using multiple attributes to assess sources of performance feedback. *Academy of Management Journal, 30,* 826–835.

Ilgen, D. R., Fisher, C. D., & Taylor, M. S. (1979). Consequences of individual feedback on behavior in organizations. *Journal of Applied Psychology, 64,* 349–371.

Ilgen, D. R., Mitchell, T. R., & Fredrickson, J. W. (1981). Poor performers: Supervisors' and subordinates' responses. *Organizational Behavior and Human Performance, 27,* 386–410.

Isgar, T., Ranney, J., & Grinnell, S. (1994). Team leaders: The key to quality. *Training and Development, 48,* 45–47.

Ivancic, K., & Hesketh, B. (1995). Making the best of errors during training. *Training Research Journal, 1,* 103–125.

Jackson, S. E., & Schuler, R. S. (1985). A meta-analysis and conceptual critique of research on role ambiguity and role conflict in work settings. *Organizational Behavior and Human Decision Processes, 36,* 16–78.

Johnston, J. H., Smith-Jentsch, K. A., & Cannon-Bowers, J. A. (1996). Performance measurement tools for enhancing team decision making. To appear in M. T. Brannick, E. Salas, & C. Prince (Eds.), *Team performance assessment and measurement: Theory, method, and application* (pp. 311–327). Hillsdale, NJ: Erlbaum.

Kanfer, R., & Ackerman, P. L. (1989). Motivation and cognitive abilities: An integrative/aptitude–treatment interaction approach to skill acquisition. *Journal of Applied Psychology, 74,* 657–690.

Katzenbach, J. R., & Smith, D. K. (1993). *The wisdom of teams: Creating the high-performance organization.* Boston: HBR Press.

Komaki, J. L., Desselles, M. L., & Bowman, E. D. (1989). Definitely not a breeze: Extending an operant model of effective supervision to teams. *Journal of Applied Psychology, 74,* 522–529.

Komaki, J. L., Heinzmann, A. T., & Lawson, L. (1980). Effect of training and feedback: Component analysis of a behavioral safety program. *Journal of Applied Psychology, 65,* 261–270.

Korsgaard, A. M., Schweiger, D. M., & Sapienza, H. J. (1995). Building commitment, attachment, and trust in decision-making teams: The role of procedural justice. *Academy of Management Journal, 38,* 60–84.

Kozlowski, S. W. J., Gully, S. M., McHugh, P. P., Salas, E., & Cannon-Bowers, J. A. (1996). A dynamic theory of leadership and team effectiveness: Developmental and task contingent leader roles. In G. R. Ferris (Ed.), *Research in personnel and human resources management* (Vol. 14, pp. 253–305). Greenwich, CT: JAI Press.

Kozlowski, S. W. J., Gully, S. M., Salas, E., & Cannon-Bowers, J. A. (1996). Team leadership and development: Theory, principles, and guidelines for training leaders and teams. In M. Beyerlein, S. Beyerlein, & D. Johnson (Eds.), *Advances in interdisciplinary studies of work teams: Team leadership* (Vol. 3, pp. 253–292). Greenwich, CT: JAI Press.

Liden, R. C., & Mitchell, T. R. (1985). Reactions to feedback: The role of attributions. *Academy of Management Journal, 28,* 291–308.

McIntyre, R. M., & Salas, E. (1995). Measuring and managing for team performance: Emerging principles from complex environments. In R. Guzzo & E. Salas (Eds.), *Team effectiveness and decision making in organizations* (pp. 149–203). San Francisco: Jossey-Bass.

McLeod, P. L., Liker, J. K., & Lobel, S. A. (1992). Process feedback in task groups: An application of goal setting. *Journal of Applied Behavioral Science, 28,* 15–41.

Nadler, A. (1979). The effects of feedback on task group behavior: A review of the experimental research. *Organizational Behavior and Human Performance, 23,* 309–338.

Nahavandi, A., & Aranda, E. (1994). Restructuring teams for the re-engineered organization. *Academy of Management Executive, 8,* 58–68.

Orasanu, J. (1990). *Shared mental models and crew decision making.* Paper presented at the 12th annual conference of the Cognitive Science Society, Cambridge, MA.

Orasanu, J., & Salas, E. (1993). Team decision making in complex environments. In G. Klein, J. Orasanu, R. Calderwood, & C. E. Zsambok (Eds.), *Decision making in action: Models and methods* (pp. 327–345). Norwood, NJ: Ablex.

Pritchard, R. D., Jones, S. D., Roth, P. L., Stebing, K. K., & Ekeberg, S. E. (1988). Effects of group feedback, goal setting, and incentives, on organizational productivity. *Journal of Applied Psychology, 73,* 337–358.

Ray, R. G., Hines, J., & Wilcox, D. (1995). Training internal facilitators. *Training and Development, 48,* 45–48.

Robinson, D. G., & Robinson, J. C. (1995). *Performance consulting: Moving beyond training.* San Francisco: Berrett-Koehler.

Rouse, W. B., Cannon-Bowers, J. A., & Salas, E. (1992). The role of mental models in team performance in complex systems. *IEEE Transactions on Systems, Man, and Cybernetics, 22,* 1296–1308.

Salas, E., & Cannon-Bowers, J. A. (1997). Methods, tools, and strategies for team training. In M. A. Quinones & A. Ehrenstein (Eds.), *Training for a rapidly changing workplace: Applications of psychological research* (pp. 249–279). Washington, DC: American Psychological Association.

Salas, E., Cannon-Bowers, J. A., & Johnston, J. H. (1997). How can you turn a team of experts into an expert team? Emerging training strategies. In C. Zsambok & G. Klein (Eds.), *Naturalistic decision making* (pp. 359–370). Hillsdale, NJ: Erlbaum.

Salas, E., Dickinson, T. L., Converse, S. A., & Tannenbaum, S. I. (1992). Toward an understanding of team performance and training. In R.W. Swezey & E. Salas (Eds.), *Teams: Their training and performance* (pp. 3–29). Norwood, NJ: Ablex.

Schaubrock, J., Ganster, D. C., Sime, W. E., & Ditman, D. (1993). A field experiment testing supervisory role clarification. *Personnel Psychology, 46,* 1–25.

Schmidt, R., & Bjork, R. (1992). New conceptualizations of practice: Common principles in three paradigms suggest new concepts for training. *Psychological Science, 3,* 207–217.

Senge, P. M. (1990). *The fifth discipline: The art and practice of the learning organization.* New York: Doubleday.

Smith-Jentsch, K. A., Cannon-Bowers, J. A., & Salas, E. (1997). New wine in old bottles: A framework for theoretically-based OJT. In K. A. Smith-Jentsch (chair), *Toward a continuous learning environment.* Symposium presented at the 12th annual conference of the Society for Industrial/Organizational Psychology, St. Louis, MO.

Smith-Jentsch, K. A., Salas, E., & Brannick, M. T. (1994). Leadership style as a predictor of teamwork behavior: Setting the stage by managing team climate. In K. Nilan (chair), *Understanding teams and the nature of teamwork.* Symposium presented at the 9th annual conference of the Society for Industrial/Organizational Psychology, Nashville, TN.

Steckler, N., & Fondas, N. (1995). Building team leader effectiveness: A diagnostic tool. *Organizational Dynamics, 23,* 20–35.

Stout, R. J., Salas, E., & Carson, R. (1994). Individual task proficiency and team process: What's important for team functioning. *Military Psychology, 6,* 177–192.

Sundstrom, E., DeMeuse, K. P., & Futrell, D. (1990). Work teams: Applications and effectiveness. *American Psychologist, 45,* 120–133.

Swezey, R. W., & Salas, E. (1992). Guidelines for use in team-training development. In R. W. Swezey & E. Salas (Eds.), *Teams: Their training and performance* (pp. 219–245). Norwood, NJ: Ablex.

Tannenbaum, S. I. (1997). Enhancing continuous learning: Diagnostic findings from multiple companies. *Human Resource Management, 36,* 437–452.

Tannenbaum, S. I., Beard, R. L., & Salas, E. (1992). Team building and its influence on team effectiveness: An examination of conceptual and empirical developments. In K. Kelley (Ed.), *Issues, theory, and research in industrial/organizational psychology* (pp. 117–153). Amsterdam: Elsevier.

Tannenbaum, S. I., Salas, E., & Cannon-Bowers, J. A. (1996). Promoting team effectiveness. In M. West (Ed.), *Handbook of work group psychology* (pp. 503–529). Sussex, England: Wiley & Sons.

Tannenbaum, S. I., Smith-Jentsch, K. A., & Cannon-Bowers, J. A. (1998). *Enhancing team effectiveness through team leader training.* Manuscript in preparation.

Tonning, J. (1997, Feb.). Training tomorrow's Navy. *Naval Institute Proceedings, 123,* 62–64.

Travillian, K., Baker, C. V., & Cannon-Bowers, J. A. (1992). *Correlates of self and collective efficacy with team functioning.* Paper presented at the 38th annual meeting of the Southeastern Psychological Association, Knoxville, TN.

Weiss, H. M. (1977). Subordinate imitation of supervisor behavior: The role of modeling in organizational socialization. *Organizational Behavior and Human Performance, 19,* 89–105.

Yukl, G. (1989). Managerial leadership: A review of theory and research. *Journal of Management, 15,* 251–289.

11

Team Dimensional Training: A Strategy for Guided Team Self-Correction

*Kimberly A. Smith-Jentsch, Rhonda L. Zeisig,
Bruce Acton, and James A. McPherson*

The current trend toward downsizing, in industry as well as the military, has required that teams learn to manage and train themselves. Team members must take responsibility for evaluating their own performance, diagnosing root causes of performance problems, identifying solutions, and planning for future tasks. Effective teams have a natural tendency to reflect on past performance together and "self-correct" following a critical event. This critical event may be an important presentation for a marketing group, a play-off game for a basketball team, or a combat exercise for a Navy ship team. Such after-action reviews have been likened to a "replay at the bar" in which team members discuss what happened and why, resolve miscommunications, and share expectations (Blickensderfer, Cannon-Bowers, & Salas, 1997a).

Although some teams have a natural tendency to engage in effective "team self-correction," many other teams are unsuccessful at using this mechanism for a variety of reasons. First, teams often are not provided with sufficient time or a sanctioned forum for team self-correction. Second, team discussions may be unfocused and fragmented. Third, team members may not possess the diagnostic skills needed to evaluate their own performance or the communication skills necessary to provide feedback to one another in a constructive way (i.e., assertively, not aggressively). Fourth, the climate within a team may not reinforce open communication and participative decision making.

Researchers have begun to suggest interventions to help teams avoid these pitfalls, and, thereby, maximize the potential benefits of team self-correction sessions (Blickensderfer et al., 1997a; chapter 10, this volume).

We thank the Battle Force Tactical Training program, ATGPAC, ATGLANT, and the Surface Warfare Officer's School for their support. We also recognize the staff of the five Navy warships who have participated in TDT demonstrations: USS *Hue City*, USS *Constellation*, USS *Yorktown*, USS *The Sullivans*, and USS *Gladiator*. We greatly appreciate the efforts of the many shipboard instructors who contributed to the refinement of TDT.

For example, Blickensderfer, Cannon-Bowers, and Salas (1997b) demonstrated that feedback skills training can be used to prepare team members to use team self-correction sessions more effectively. Specifically, team members who received such training were more likely to have consistent expectations about team task strategies after a team debriefing session. This chapter describes another strategy for enhancing teams' ability to self-correct, Team Dimensional Training (TDT). TDT incorporates *guided team self-correction* to develop team members' teamwork-related knowledge and skills.

Guided Team Self-Correction

Guided team self-correction refers to the use of a facilitator who (a) keeps the team's discussion focused, (b) establishes a positive climate, (c) encourages and reinforces active participation, (d) models effective feedback skills, and (e) coaches team members in stating their feedback in a constructive manner (Smith-Jentsch, Payne, & Johnston, 1996).

Guidance provided by the facilitator is designed to allow teams to reap the greatest possible return on investment when engaging in team self-correction. Essentially, this is done by helping the team to determine *what* specific topics they should be discussing and *how* they should be discussing them. Guided team self-correction can be used to improve team performance by fostering various types of shared knowledge among team members (e.g., expectations, teamwork processes, teammate-specific preferences). On the basis of the specific focus of a guided team self-correction session, a debriefing guide is prepared that outlines specific questions that the facilitator can use to stimulate a productive team discussion.

TDT uses guided team self-correction as a way of helping teams to develop shared, accurate knowledge about the components of teamwork and to accelerate their mastery of targeted teamwork skills. This chapter (a) provides a detailed description of TDT, (b) presents guidelines for the use of TDT in conjunction with exercise-based team training, (c) summarizes the training effectiveness data collected thus far, and (d) offers directions for future research.

Mental Models of Teamwork

For teams to "self-correct" coordination breakdowns effectively, team members must accurately determine which specific teamwork processes they use well and which they do not. Moreover, all team members must agree with this diagnosis and commit themselves to an agreed-on solution. Because team members are likely to have had different experiences before joining a team, their knowledge and opinions about the nature of teamwork may differ substantially. Team members may find it difficult to assess their performance collectively because they are working from different frames of reference, or "mental models."

The term *mental model* has been used to refer to knowledge structures, or cognitive representations, that humans use to organize new information; to describe, explain, and predict events (Rouse & Morris, 1986); and to guide their interactions with others (Gentner & Stevens, 1983; Rumelhart & Ortony, 1977). Rentsch, Heffner, and Duffy (1990) found that individuals' mental models of teamwork differed as a function of their previous experience in team settings. Specifically, those with greater team experience described teamwork using fewer, more abstract dimensions and represented their knowledge more consistently.

When team members share accurate mental models of the teamwork processes that influence their performance, they should be better able to (a) uncover performance trends and diagnose deficiencies, (b) focus their practice appropriately on specific goals, and (c) generalize the lessons they learn to new situations. It is important that team trainers and team leaders facilitate the development of accurate and shared mental models of teamwork among team members. TDT was designed to accomplish this objective for teams on board Navy warships. A second objective for TDT was to accelerate improvement on four teamwork skill dimensions that have been linked to team performance through the Tactical Decision Making Under Stress (TADMUS) program (see chapter 4, this volume). Finally, TDT was designed to reduce the workload of shipboard instructors by providing them with tools and guidelines for conducting exercise-based team training.

A Strategy for Guided Team Self-Correction

In the U. S. Navy, the trend toward substituting shore-based training with increased shipboard training (i.e., training inside the lifelines) has placed an added burden on overworked and often undertrained shipboard instructors. These instructors are frequently assigned positions on shipboard training teams (SBTTs) on the basis of their technical expertise as watchstanders and not their experience or skill as trainers. Therefore, there is a great demand for tools and strategies that can support SBTTs in providing high-quality shipboard training.

One component of a shipboard instructor's task for which he or she may be particularly unprepared is the evaluation and training of teamwork skills. It has been demonstrated that teamwork skills play a significant role in ensuring effective team performance above and beyond individual team members' technical, or "taskwork," skills (Stout, Salas, & Carson, 1994; chapter 4, this volume). However, traditionally, little guidance has been provided to SBTTs on how to evaluate and train teamwork. The result has been that the components of teamwork are generally less understood by instructors, are considered less objective and more difficult to evaluate, and therefore are less emphasized in training. We discovered this pattern during our observations of postexercise prebriefs and debriefs conducted on board a number of Navy ships (see chapter 10, this volume).

There is currently an encouraging movement in the Navy, however, toward "objective-based" or "event-based" training. Event-based assessment and training, or EBAT (Dwyer, Fowlkes, Oser, Salas, & Lane, 1997; Dwyer, Oser, Salas, & Fowlkes, in press), involves linking together training objectives, exercise events, performance criteria, and developmental feedback strategies. This systematic approach is taking hold across the Navy as well as other organizations (e.g., the airline industry) and is expected to improve greatly the quality of exercise-based training. TDT is a concrete example of how an EBAT approach can be applied to objective-based training for teamwork skills. Instructors (a) systematically collect data on specific teamwork behaviors, (b) organize these data into a format that is structured according to a data-driven model of teamwork, and (c) facilitate guided team self-correction, which is based on that teamwork model.

A Model of Teamwork

The model of teamwork on which TDT is based was derived from analyses of performance data from 100 combat information center (CIC) teams. Four critical teamwork dimensions, defined by a set of specific behaviors, were identified (see chapter 4, this volume). The first dimension, information exchange, involves passing relevant pieces of information before having to be asked, providing periodic "big picture" situation updates, and seeking out information from all available resources. The second dimension, communication, includes using proper phraseology and complete standard reporting procedures; speaking in a clear, intelligible tone of voice; and avoiding excess chatter. Supporting behavior, the third dimension, includes promptly correcting team errors and providing and requesting backup or assistance when needed. Finally, team initiative/leadership includes providing guidance to team members, as well as stating clear priorities for the team.

The behaviors within each of the four dimensions were found to be relatively consistent across exercise events, whereas sets of behaviors defining different dimensions were relatively independent. Furthermore, results indicated that experienced teams could be discriminated from inexperienced teams on the basis of their performance on the four dimensions. Thus, the data-driven model of teamwork identified through TADMUS research was adopted as the learning objective for TDT. This model of teamwork provides instructors with a frame of reference from which to (a) focus team members' attention during an exercise prebrief, (b) observe the team's performance during an exercise, (c) diagnose the team's strengths and weaknesses after an exercise, and (d) guide the team through a self-critique of their performance (see Figure 1).

To date, more than 400 shipboard instructors have been trained to use the TDT method. These instructors have demonstrated TDT at several shore-based training facilities, including Surface Warfare Officer's School, afloat training groups, and the AEGIS Training Center, as well as and on five Navy warships (i.e., USS *Hue City*, USS *Constellation*—twice, USS

Team Dimensional Training
Prebriefing - Performance - Debriefing Cycle

PREBRIEF
- Clarify mission
- Focus team's attention on the four teamwork dimensions
- Remind team of specific goals from previous TDT

DEBRIEF
- TDT facilitator(s) sets the stage and recaps key exercise events
- TDT facilitator(s) guides team critique of the four teamwork dimensions
- Lessons learned are summarized and goals for improvement are set

PERFORM / OBSERVE
- Time on task; real or simulated
- Instructors record positive and negative examples under the four teamwork dimensions

DIAGNOSE PERFORMANCE
- After exercise, instructors come to consensus on categorization of selected examples
- Instructors transfer 1-2 best examples of each teamwork dimension to the debriefing guide
- Strengths and goals for improvement are identified for each dimension

Figure 1.   Teamwork cycle. TDT = Team Dimensional Training.

Yorktown, USS *The Sullivans*—twice, and USS *Gladiator*). The following sections describe each of the components of the TDT process in detail and present guidelines for using this strategy that were developed on the basis of these experiences.

The Prebrief

The TDT cycle begins with an exercise prebrief. During this prebrief, instructors (a) define the TDT teamwork dimensions and their component behaviors, (b) focus the team's attention on these objectives as the members perform an exercise, (c) present an outline of how the TDT cycle will progress, and (d) establish a positive learning climate. The following sections describe each of these components and offer guidelines on how to optimize the benefits of the TDT prebrief.

Defining terms. It is during the TDT prebrief that team members are first introduced to the TDT dimensions and their component behaviors. This declarative knowledge about the TDT model of teamwork provides team members with a common vocabulary with which to discuss their performance. A TDT prebriefing guide is provided to instructors to aid them in defining TDT terms (see Figure 2). However, it is important that instructors become familiar with these definitions before prebriefing so that they do not appear simply to be reading from the guide.

Focusing the team's attention. Previous research has demonstrated that trainees who focused on the "mastery" of process-related goals were

TEAM DIMENSIONAL TRAINING
PREBRIEF

This exercise will give you the opportunity to work on improving the **processes** your team uses to accomplish its objectives. In particular, I'd like you to focus on four dimensions of teamwork; **information exchange, communication delivery, supporting behavior,** and **team initiative & leadership**. Before we start the exercise, let's go over the specific components of these four dimensions.

INFORMATION EXCHANGE

The first dimension is **information exchange**. Effective information exchange allows the team to develop and maintain a shared situation awareness.

The components of **information exchange** are:
- **Utilizing all available sources of information**
- **Passing information to the appropriate persons without having to be asked**
- **Providing periodic situation updates which summarize the big picture**

COMMUNICATION

The second dimension is **communication**. While information exchange deals with what is passed to whom, this dimension involves **how** that information is delivered.

The components of **communication delivery** are:
- **Proper phraseology**
- **Completeness of standard reports**
- **Brevity/ Avoiding excess chatter**
- **Clarity/ Avoiding inaudible comms**

SUPPORTING BEHAVIOR

The third dimension is **supporting behavior**. This involves compensating for one another in order to achieve team objectives.

The components of **supporting behavior** are:
- **Monitoring and correcting team errors**
- **Providing and requesting backup or assistance to balance workload**

INITIATIVE/LEADERSHIP

The fourth dimension is **initiative/ leadership**. Anyone on the team can demonstrate initiative/leadership.

The components of **initiative/leadership** are:
- **Providing guidance or suggestions to team members**
- **Stating clear and appropriate priorities**

After the exercise, you will be asked to critique your own performance on these four dimensions. I will facilitate this process of team self-correction using a strategy called Team Dimensional Training, or TDT.

For TDT to be effective, all team members need to contribute to the process of identifying problems and coming up with solutions to those problems. I encourage everyone to take advantage of this opportunity to ask questions, state concerns, clarify issues, and offer suggestions on how your team can improve.

Figure 2.   Prebriefing guide. TDT = Team Dimensional Training.

better able to generalize what they learned to a new task than those who focused on outcome-related performance goals (Kozlowski, Gully, Smith, Nason, & Brown, 1995). Therefore, it is expected that a focus on understanding and applying teamwork processes should help team members to generalize the lessons learned from team exercises to novel situations. For example, if a team member knows that an important component of teamwork involves monitoring signs of stress in others and offering backup or assistance when needed, he or she can consciously look for opportunities to apply this knowledge in a variety of novel situations.

The learning objectives for a TDT evolution are to master the four teamwork dimensions and their component behaviors. This focus on mastery of team-level processes rather than maximization of exercise-specific outcomes may depart significantly from what instructors and teams are familiar with. Therefore, it is important that this distinction be clearly stated to the team during the TDT prebrief. After presenting information that situates the team in a particular exercise (e.g., rules of engagement, mission goals, geopolitical data), TDT instructors should stress that the specific purpose of the exercise is to develop teamwork, not individual task training. In addition, if the team previously participated in a TDT evolution, the members should be reminded of the specific goals for improvement that were set during the previous TDT debrief.

Presenting an outline of the TDT cycle. TDT is based on the concept that team members and instructors should be partners in the learning process. This message is conveyed to team members in the prebrief, in which they are informed about how the TDT process will unfold. Instructors briefly outline the steps that will follow for the team, including requirements for their participation.

Establishing a positive learning climate. It has been demonstrated that perceptions of team climate are established quickly and early in a team's development and that leader briefing behavior has a powerful effect on such perceptions (Smith-Jentsch, Salas, & Baker, 1996). Therefore, it is important that instructors use the TDT prebrief to convey genuine interest in team member input and growth. Specific prebriefing behaviors that are most effective in this regard include asking for questions or comments as well as encouraging and reinforcing participation (see chapter 10, this volume).

Guidelines for prebriefing the team before a TDT exercise:

> *Guideline 1.* During the prebrief, emphasize that team members' focus should be on the mastery of teamwork processes rather than performance outcomes.
> *Guideline 2.* Provide the team with a common vocabulary by defining teamwork terms.
> *Guideline 3.* Provide the team with an advanced organizer that describes the guided team self-correction process, highlighting requirements for team members' participation.
> *Guideline 4.* Convey genuine interest in team member input and growth; establish a positive learning climate from the start!

Exercises

TDT exercises must be planned and conducted to elicit a sufficient number of concrete teamwork examples. The data collected by instructors during these exercises will later serve as a springboard for team discussion, a mechanism for conveying distinctions among behavioral categories, and a way of convincing team members that behaviors within the four teamwork dimensions do, in fact, contribute to their success or failure as a team. The following sections describe lessons learned regarding how to plan and conduct a TDT exercise as well as how to collect performance samples systematically.

Exercise planning. When planning to conduct a TDT evolution, instructors must determine which team members they want to participate and then select or create an exercise that contains events that are expected to strain coordination among those team members. Such exercises need not last longer than 30 minutes. In our experience, a well-prepared 30-minute exercise provides more than enough examples to be discussed in a TDT debrief. In fact, such brief exercises are preferred when using TDT because they leave more time for a thorough debrief and also because team members have difficulty remembering details from longer exercises. The most important feature of a TDT exercise, regardless of length, is that it include preplanned situations that place heavy coordination demands on a select group of team members.

Two points are important when selecting team members to participate in a TDT evolution. First, instructors should attempt to ensure that the chosen exercise requires coordination among all the selected team members and that all team members who are key to this coordination are included in the debrief. At times, TDT is used with only a subset of a larger team that has participated in a shipwide exercise. In these cases, some team members who participated in the exercise but had little coordination with others may not be included in the TDT debrief. We have found that when these team members are included in the TDT prebrief and debrief, they often feel frustrated because they do not have any input to contribute. In a shipboard environment, these trainees may also resent the fact that they are being kept from other tasks (or much needed sleep) for no reason. Conversely, it is difficult for team members to piece together facts and resolve breakdowns when key team members are not included in the TDT debrief. Therefore, when planning a TDT evolution, instructors should carefully consider which team members are appropriate to include given the nature of exercise events.

A second consideration when selecting team members to include in a TDT evolution is team size. Team size tends to be inversely related to the amount of participation of any one team member. Larger team debriefs require greater skill on the part of instructors to draw out reluctant contributors from the team. However, experienced TDT facilitators have successfully obtained participation from teams of 5 to 22 members.

Exercise conduct. Generally speaking, instructors avoid communicating with, or coaching, watchstanders during a TDT exercise. Two exceptions to this rule are (a) when safety is an issue and (b) when a particular mistake would take the exercise in a direction that would prevent other training objectives from being met. In this way, team members are given the opportunity to make mistakes and witness the effects of those mistakes firsthand. These mistakes become learning points that are later discussed in the debrief. Shipboard instructors distinguish this "assessment mode" from the "training mode," during which feedback and suggestions are provided to watchstanders as they perform an exercise.

Data collection. During a TDT exercise, instructors are responsible for recording in detail at least one example of each behavior under their assigned dimension. However, observations of behaviors falling under one of the other dimensions are also noted if an instructor is unsure whether the individual responsible for that dimension has missed it.

Ideally, each instructor is asked to focus on observing one of the four dimensions. This procedure of assigning one instructor per dimension is designed to ensure that a sufficient number of examples are collected for each teamwork behavior. When multiple instructors each try to observe all four dimensions, it is possible that one or more of the teamwork behaviors will be overlooked. It is also easier for instructors to focus on behaviors within only one dimension. There are times when it is not possible to have four instructors participate in a TDT evolution. In such cases, an instructor who is highly experienced as a TDT facilitator may be able to handle all four dimensions. However, an inexperienced TDT facilitator is likely to have difficulty.

Each performance example is recorded on the TDT worksheet (see Figure 3). This worksheet provides a timeline and information about pre-

SUPPORTING BEHAVIOR

RECORD OBSERVATIONS OF:

GROUND TRUTH	± TEAM ERRORS CAUGHT AND CORRECTED	± BACKUP BEHAVIOR
BEGIN EVENT ONE		
00-00 DETECT TACAIR 7021 076/54 VICINITY BUSHEHR	Track Sup made identifications for the ID Operator while he was having trouble shooting the link	
00-30 DETECT APQ-120 FROM TACAIR 7021 B-076		
01-00		
01-30 LOST CONTACT ON TACAIR 7021 AS IT DESCENDS FOR TOUCH AND GO		
02-00 COMAIR 7031 T/O KHARK 026/38	TAS Operator thought SWC ordered track 2501 to be covered with birds, but ordered track 2510. No one noticed the error – wrong track was engaged.	
02-30 REGAIN TACAIR 7021 T/O BUSHEHR 073/52		
03-00 COMAIR 7051 C/C 180	Surface Supervisor notified the Surface Tracker that he was on the wrong IVCS net.	

Figure 3. Worksheet.

scripted exercise events to aid instructors in anticipating situations when teamwork may break down among various team members. In this way, they can position themselves accordingly and ensure that they are monitoring the correct communication net at various points in the exercise.

Instructors are told to record "representative samples" of teamwork strengths and weaknesses, rather than list comprehensively every error that occurred in an exercise. These performance samples should be recorded in detail so they can later serve as concrete illustrations of TDT learning objectives. Furthermore, they should be representative of the team's behavior throughout the exercise. It is particularly useful to record instances when the team's teamwork skills (or lack thereof) clearly contributed to effective or ineffective team outcomes at key exercise events.

Initially, instructors tend to record only examples of performance problems and not examples of effective teamwork. However, when they get to the point of trying to prepare their TDT debrief, they quickly realize that they have incomplete information in their worksheets. Another common mistake is the tendency of instructors to focus their attention almost exclusively on the formal leader of the team, particularly when assigned to observe behaviors within the team initiative/leadership dimension. It should be noted that any team member can demonstrate initiative/ leadership or any of the other dimensions. If the examples collected by instructors (positive or negative) during an exercise are limited to one or two team members, those team members may feel singled out and other team members may feel ignored. Therefore, instructors need to learn to seek out performance samples actively from multiple team members.

Guidelines for planning and conducting team exercises, and collecting data to stimulate guided team self-correction:

> *Guideline 5.* Use relatively brief (30–60 minutes) team exercises to generate concrete teamwork examples.
> *Guideline 6.* Embed two to three trigger events into the team exercise that are expected to strain coordination among specific team members.
> *Guideline 7.* Whenever possible, assign one instructor or facilitator to observe each higher order teamwork dimension.
> *Guideline 8.* Allow team errors to unfold naturally; avoid interfering with team coordination (or lack thereof) during the exercise.
> *Guideline 9.* Seek out and record two to three detailed examples of each type of teamwork behavior that are representative of the team's performance throughout the exercise.
> *Guideline 10.* Actively seek both positive and negative teamwork examples.
> *Guideline 11.* Be sure that the examples collected do not focus exclusively on a few teammates (e.g., the formal leader).

Consensus Meeting

Immediately following a TDT exercise, instructors meet separately from trainees (a) to clarify, classify, and prioritize their observations and (b) to diagnose strengths and weaknesses and identify specific goals for improvement. This information is used to prepare a debrief that is organized around the TDT model of teamwork. The following sections describe the TDT consensus meeting and present guidelines designed to facilitate the rapid and accurate preparation of a TDT debrief.

Data reduction, clarification and synthesis. After a TDT exercise, instructors who have been involved in data collection select one to two examples of each component behavior (both positive and negative) in the TDT model. Selecting appropriate examples involves an interactive dialogue in which instructors request and offer clarification as well as discuss the classification of samples. Only the clearest and most significant examples fitting each category are chosen. Instructors strive to eliminate redundancy so that multiple instructors do not "beat a dead horse" when debriefing their assigned dimension.

The TDT debriefing guide provides space to record these examples within an outline that is organized around the four teamwork dimensions (see Figure 4). Preparation of a TDT debrief is accelerated when instructors focus their data synthesis by scanning their notes for specific types of

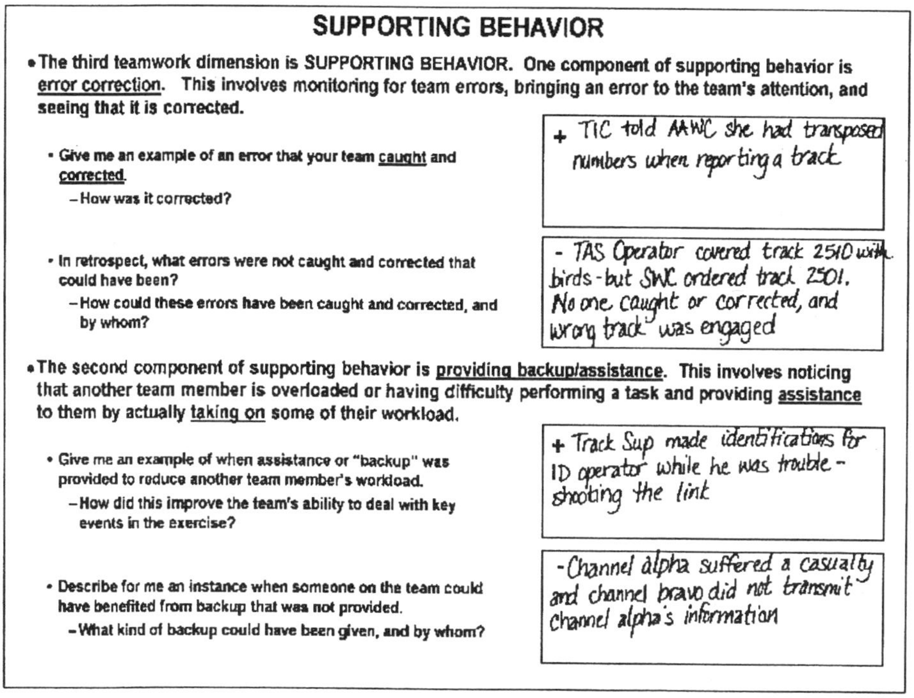

SUPPORTING BEHAVIOR

• The third teamwork dimension is SUPPORTING BEHAVIOR. One component of supporting behavior is error correction. This involves monitoring for team errors, bringing an error to the team's attention, and seeing that it is corrected.

 • Give me an example of an error that your team caught and corrected.
  – How was it corrected?

 + TIC told AAWC she had transposed numbers when reporting a track

 • In retrospect, what errors were not caught and corrected that could have been?
  – How could these errors have been caught and corrected, and by whom?

 − TAS Operator covered track 2510 with birds – but SWC ordered track 2501. No one caught or corrected, and wrong track was engaged

• The second component of supporting behavior is providing backup/assistance. This involves noticing that another team member is overloaded or having difficulty performing a task and providing assistance to them by actually taking on some of their workload.

 • Give me an example of when assistance or "backup" was provided to reduce another team member's workload.
  – How did this improve the team's ability to deal with key events in the exercise?

 + Track Sup made identifications for ID operator while he was trouble-shooting the link

 • Describe for me an instance when someone on the team could have benefited from backup that was not provided.
  – What kind of backup could have been given, and by whom?

 − Channel alpha suffered a casualty and channel bravo did not transmit channel alpha's information

Figure 4.   Debriefing guide.

examples, moving through the guide box by box rather than examining all their notes chronologically. Instructors usually learn this after their first experience in a TDT consensus meeting. Generally, the first consensus meeting attempted by a group of instructors lasts 1–1.5 hours. However, this average time drops to 15 minutes after three to four experiences using TDT. The TDT debriefing guide is designed to facilitate a rapid turnaround time for debrief preparation. This is desirable from both a practical and a learning standpoint because feedback is considered most effective the closer it is in time to performance.

Diagnosis of strengths and goals for improvement. On the basis of the discussion of performance samples gathered during a TDT exercise, instructors diagnose trends in the team's performance. This diagnosis is summarized using the teamwork summary tables, an example of which is shown in Figure 5. A strength and a specific goal for improvement are agreed on among the instructors for each of the four dimensions. This summary is designed to focus the team's attention on a manageable number of specific, attainable goals for improving teamwork processes on subsequent exercises.

It is important to note that the identified strengths and goals must be specific but short. As shown in Figure 5, for example, a team's strength under communications may be the consistent use of proper phraseology. Conversely, the goal for improvement on this dimension may be for team

TEAMWORK SUMMARY

INFORMATION EXCHANGE		SUPPORTING BEHAVIOR	
STRENGTH	Passing critical information to teammates without being asked	STRENGTH	Assisted each other well by sharing workload during stressful events
GOAL FOR IMPROVEMENT	Need to use all sources of information - not just one	GOAL FOR IMPROVEMENT	Need to monitor each other's actions better and correct mistakes or errors

COMMUNICATION		INITIATIVE / LEADERSHIP	
STRENGTH	Phraseology	STRENGTH	Provided constructive and appropriate suggestions to team mates
GOAL FOR IMPROVEMENT	Need to reduce excess chatter on the nets	GOAL FOR IMPROVEMENT	Priorities need to be explicitly stated

Figure 5. Summary table.

members to formulate a message in their minds before "keying their mike" so they do not tie up communications with unnecessary "uhh, umms" (i.e., improve brevity). The teamwork summary should not rehash everything that was recorded in the previous sections of the debriefing guide. This is a common mistake that instructors make when initially learning to use TDT.

Guidelines for synthesizing data from a TDT exercise and preparing a guide for team self-correction:

> *Guideline 12.* Review exercise notes for one to two representative samples of teamwork that have been recorded in sufficient detail and are *clear* examples (both positive and negative) of each component behavior in the TDT model.
> *Guideline 13.* Record detailed descriptions of the selected examples within the outline provided in the TDT debriefing guide (organized around the TDT model of teamwork).

The Debrief

The TDT debrief uses guided team self-correction to enhance teamwork mental models and performance. This process involves (a) setting the stage for team participation; (b) soliciting examples of teamwork from the previous exercise that are consistent with the TDT model; (c) managing the resulting team discussion; and (d) helping the team to diagnose strengths and weaknesses, identify solutions, and establish goals for improvement. The following sections describe these components of the TDT debrief and present guidelines for facilitators.

Setting the stage. Instructors begin a TDT debrief by setting the stage for team participation. This involves providing team members with the ability and motivation to engage themselves in the team critique. First, team members are reminded of the exercise objectives by again being given definitions of the four teamwork dimensions and their component behaviors; the purpose is to reinforce their declarative knowledge about teamwork and establish a common vocabulary for the subsequent debrief.

Second, team members are given an advanced organizer for how the debrief will proceed. Specifically, team members are told that they will be led through a self-critique of their performance, which flows dimension by dimension and links teamwork processes to key event outcomes. Third, an instructor briefly summarizes the outcome of about two or three exercise events to help refresh the team's memory and to clarify *what* happened before asking team members to discuss *why* it happened. Visual aids such as exercise timelines or a compressed video replay of events (e.g., bird's-eye view) are useful in clarifying outcomes and triggering team members' recall.

The final step in setting the stage involves encouraging team members to talk openly about their observations, ask questions, and offer suggestions. Eye contact with each team member is critical for encouraging active participation in the debrief. When instructors fail to make eye contact with all team members, they send a message that not everyone is expected to have useful input. Instructors have a tendency to focus their eye contact on team leaders and other high-ranking team members. This problem is magnified when higher ranking team members position themselves front and center. We have found that placing the lowest ranking team members, and others who may be reluctant to speak up, front and center at the start of the debrief increases the amount of eye contact they receive and subsequently the amount of input they offer. It is also useful to speak to the leader ahead of time to explain that the goal of the debrief is to get all team members to contribute. This advance notice can be useful in preventing the team leader from inadvertently monopolizing the team discussion.

Soliciting examples that are consistent with the TDT model. Once the stage has been set, instructors begin the process of guided team self-correction. Using the TDT debriefing guide, instructors ask the team to provide concrete examples of specific teamwork behaviors (e.g., backup, error correction) organized by dimension (e.g., supporting behavior). Ideally, a separate instructor facilitates this process for each dimension. This helps to reinforce the multilevel model of teamwork, in which subcomponents are organized under four overarching dimensions. It also tends to keep the debrief moving and adds variety to the debrief. However, this arrangement is often not possible, and it is not a prerequisite for conducting TDT.

The team is always asked to provide an example of their behavior that fits a particular category before the instructor provides his or her own observations. This step is important to engage the team in the learning process. Team members are much less likely to offer an example if the instructor has already provided one, especially an example of negative teamwork. In fact, the instructor's notes should be discussed only if the team does not provide an appropriate example or an error not noted by the team is so significant that it warrants additional discussion. The instructor's notes can be thought of as "hip-pocket items"—items used to stimulate discussion and clarify distinctions among behavioral categories. As a team becomes more familiar with the TDT process, the instructor will have to talk less and less.

One common mistake that instructors make is to give team members insufficient time to think of a response before offering their own input. Instructors need to remember that team members generally do not have the same level of understanding of the TDT dimensions as they do. Moreover, team members do not have the detailed information that instructors have in their debriefing guides. Therefore, it may take 10–15 seconds of silence before a team member has extracted an example from memory and determined that it fits the category in question. If an instructor appears too eager to inject his or her own input, team members may begin to doubt whether the instructor is genuinely interested in their observations. Therefore, instructors should consciously pause and make eye contact with all team members after each question is asked. We have found that adherence to this simple recommendation has dramatic effects in terms of team member involvement in a TDT debrief.

Instructors should ask for an example of each type of behavior, whether they noticed any examples themselves. Usually team members are aware of many more performance problems than instructors can observe. If a question is skipped simply because the instructor did not see any examples of a particular behavior, he or she could be missing an opportunity to correct a problem uncovered by the team. Similarly, many instructors have initially made the mistake of prefacing a question with a comment such as, "I didn't notice any instances of improper phraseology —did anybody else?" This type of question is unlikely to encourage team members to point out ineffective behaviors that were missed by the instructor.

To remind instructors to ask for a team critique first, probes are listed in the debriefing guide above the spaces provided for instructor comments. These questions are specific, open-ended, and behavioral in nature. For instance, "Give me an example of a time when backup was needed, but wasn't provided." This approach was found to be most effective in eliciting concrete examples from team members (see chapter 10, this volume). In contrast, general questions like "How were communications?" were found to generate answers like "Good" or "Sloppy," which are not useful in diagnosing specific problems or identifying solutions.

For each component behavior, instructors ask first for a positive example (e.g., effective backup) and then for a negative example (e.g., failure to provide backup). This is important because generalization of training has been shown to be enhanced by both effective and ineffective models (Baldwin, 1992). Discussion of both positive and negative examples of the behavioral categories is expected to strengthen team members' understanding of those categories and to improve their team skills by highlighting not only what to stop doing but also what to keep doing. Moreover, an overemphasis on weaknesses without reinforcing strengths can hurt collective efficacy and team morale.

For every ineffective example noted, the instructor asks follow-up questions to get team members to identify negative outcomes that either resulted or could have resulted (e.g., "How did/could have this instance of improper phraseology slow the team's response to events or lead to confusion?"). Finally, the team is asked to identify corrective action (e.g., "What would the appropriate phraseology have been?"). Ideally, someone on the team will provide an acceptable solution. However, if no one does, the instructor should offer a solution to the problem before moving on to the next topic.

Managing the team discussion. In a TDT debrief, the instructor's role is that of a facilitator. This means that he or she must keep the discussion focused and reinforce participation. In addition, the instructor should

coach team members in linking concrete examples to appropriate components in the TDT model and in stating their concerns to one another in a
way that is behavior based, specific, solution oriented, and nonthreatening. The TDT debriefing guide provides a structured outline for team self-
correction. It is up to the instructor to ensure that the team discussion
does not stray into areas beyond the scope of this outline. When this is
allowed to happen, the debrief becomes lengthy, and team members lose
perspective and become restless. In addition, an instructor needs to limit
the number of "appropriate" examples discussed within a behavioral category. Generally speaking, two to three examples of each behavioral category are sufficient to make the point. One common mistake that instructors make is to offer their own observations after the team has already
provided a sufficient number of examples. However, instructors quickly
learn that doing this unnecessarily lengthens the debrief and defeats the
purpose of getting the team to "self-correct." Whereas it is important to
limit the number of examples discussed for each component behavior, it is
equally important to avoid making team members feel that the instructor
is rushing them through the debrief or cutting them off when they attempt
to contribute. This involves walking a fine line and is one of the more
difficult aspects of facilitating a TDT debrief.

An instructor's behavior should send the message that he or she is
genuinely interested in the team's input. Team members should be positively reinforced by the instructor after offering input, particularly when
a team member highlights one of his or her own mistakes. Because the
goal of the debrief is to diagnose and correct team problems, correctly
identifying one's own mistakes should be rewarded rather than punished.
If a team member is ridiculed for admitting a mistake during the debrief,
it is unlikely that he or she or others on the team will be willing to critique
themselves in the future. This defeats the purpose of guided team self-
correction, which is to maximize the resources available to identify and
correct team coordination breakdowns.

Team members are generally more willing to accept and strive toward
goals for improvement when they participate in their own self-critique.
Getting team members to talk about their own role in teamwork breakdowns is a critical component of the learning process. Instructors generally
find this point easy to understand in theory; however, in practice they
often find it difficult to provide positive reinforcement of the admission of
errors. In such cases, instructors should focus on the fact that a team
member was able to extract a concrete example from his or her personal
experience and correctly link it to a category within the TDT model of
teamwork. The most successful instructors learn to react to such admis-
sions of guilt with responses such as the following: "That's exactly the kind
of thing I'm looking for when I refer to incomplete reports" or "That's a
great example of improper phraseology." Often, when a team member ad-
mits to making a mistake, he or she has some idea, in retrospect, of what
could have been done differently. In such cases, giving the team member
the opportunity to identify his or her own corrective behavior fosters pride
and commitment to improvement. Of course, it is the instructor's role to

ensure that the solutions the team ultimately agrees on are reasonable before moving on.

Another aspect of managing team discussions can be difficult: dealing with team members' suggestions that are inappropriate or stated in a way that is nonconstructive. Not every solution identified by the team will be reasonable. It is the job of the instructor to explain why a solution is incorrect without "shooting it down."

In addition, instructors who facilitate guided team self-correction must learn how to handle vague or threatening comments among team members. Often, when a team member provides feedback to another team member, a valid point gets lost in the delivery. Therefore, it is the role of the instructor to get that team member to restate his or her feedback in a way that is behavior based and solution oriented, rather than personal. For example, a team member may state, "All the information I got from Joe was garbage." To this, an instructor might ask, "What exactly was it that you needed from Joe that was not provided?" If an instructor cannot draw out a more constructive comment from the team member, the instructor may attempt to interpret the comment and restate it him- or herself more constructively. In this way, the instructor not only clarifies the issue at hand, but also models effective feedback skills and brings the conversation back to the teamwork model. For example, the instructor might say, "So it sounds like the lesson here is that the team needs more frequent situation updates on track emitters."

The team discussions during a TDT debrief not only correct team problems, but also develop team members' procedural knowledge regarding the TDT model of teamwork. It is the instructor's role to use the concrete examples brought forth to discuss distinctions among behavioral categories. For example, trainees need to understand why excess chatter, clarity, completeness of reports, and phraseology are all subcomponents of communication and not of information exchange. Suggested dialogue is provided to aid instructors in explaining these distinctions. For example, within the TDT model of teamwork, *information exchange* includes *when* and *what* to communicate, whereas the *communication* dimension involves communication *style*, or *how* one delivers a message. A shared understanding of these distinctions is important to avoid misunderstandings when team members use specific TDT terms. As instructors become familiar with the TDT model, they learn to explain distinctions among the behaviors without having to read from the guide.

Another way in which instructors facilitate team members' procedural knowledge regarding the TDT model is to use TDT terms consistently to paraphrase or discuss examples offered by the team. When team members report performance instances that fit well within a category in the TDT model, instructors can reinforce the connection by stating, for example, "So, in this case backup provided from Tom helped you to deal with the large number of track IDs you had piling up." After a series of experiences with TDT, team members begin to pick up on the terminology and use it to express themselves to their teammates. With experience, team members come to share a common understanding of the TDT model and accept TDT terms as an agreed on vocabulary they can use to communicate. This makes TDT debriefs more effective and more efficient. Like the learning curve in instructor consensus meetings, the learning curve for teams to become familiar with the debriefing process is relatively steep. Whereas the first TDT debrief a team participates in may last in excess of an hour, this time can be cut in half after approximately three to four experiences.

Facilitating diagnosis and goal setting. The final step of the guided team self-correction process involves diagnosing strengths and weaknesses and setting specific goals for improvement. Recommended goals are recorded by instructors during the consensus meeting; however, these goals may change as a function of what is learned during the debrief. Team members often bring to light problems that were missed by instructors. It is important that both team members and instructors agree on the most significant goals for improvement under each of the four dimensions. These goals then feed subsequent TDT prebriefs.

Guidelines for facilitating a TDT debrief:

> *Guideline 14.* Position low-ranking team members front and center, with leaders and other higher ranking team members on the ends, before beginning a TDT debrief.
>
> *Guideline 15.* Recap key exercise events at the start of a TDT debrief; use visual aids, when available, to trigger team members' recall.
>
> *Guideline 16.* Keep the team's discussion focused by following the structure of the TDT debriefing guide; don't allow the team to get too far off track.
>
> *Guideline 17.* Ask the team to describe concrete examples of each specific type of teamwork behavior, organized by higher order dimension in the TDT debriefing guide.
>
> *Guideline 18.* Make key features that differentiate among teamwork categories explicit when asking for concrete examples.
>
> *Guideline 19.* Pause and make eye contact with all team members after asking for a particular type of example.
>
> *Guideline 20.* Reserve one's own observations of teamwork for instances when team member input is not forthcoming or a distinction between categories needs to be clarified.
>
> *Guideline 21.* Reinforce linkages between concrete examples and components of teamwork by paraphrasing team member input using model-specific terminology.
>
> *Guideline 22.* Follow up descriptions of effective and ineffective teamwork by asking the team how each of these influenced or could have influenced the outcome of key events.
>
> *Guideline 23.* Model effective feedback skills using specific, behavior-based, solution-oriented examples.

Guideline 24. Reinforce team member participation, especially self-critique.

Guideline 25. Guide team members in stating their feedback constructively.

Guideline 26. Guide the team in coming to consensus on a specific goal for improvement under each higher order teamwork dimension. Highlight these at the end of the session.

Does Team Dimensional Training Change Mental Models of Teamwork?

The previous sections of this chapter have described the components of TDT. This strategy was designed to foster mental models of teamwork that are consistent with the data-driven model of teamwork identified through TADMUS research. Next, we summarize a validation study that was conducted to examine the impact of TDT on instructors' mental models (Smith-Jentsch, Campbell, Ricci, & Harrison, 1997).

A 1-day instructor-training workshop was developed to teach shipboard training teams (SBTTs) to use the TDT method. This workshop evolved over a period of 3 years on the basis of feedback from the more than 400 instructors who participated in it. During this period, class sizes ranged from 8 to 35 participants, with most of the workshops including 24 participants. Approximately two thirds of those trained to use TDT were members of one of the afloat training groups (ATGPAC, ATGLANT). These instructors are responsible for ensuring that SBTTs are capable of providing effective training to those on their own ships. SBTT members and instructors from various shore-based facilities (e.g., Surface Warfare Officers School made up the remaining TDT workshop participants.

Instructor-Training Workshop

The TDT instructor-training workshop involved approximately 4 hours of classroom training, followed by hands-on practice using TDT. The practice portion of the workshop lasted between 4 and 8 hours depending on class size. An attempt was made to allow each instructor to practice using TDT at least once; however, this was not always possible. Therefore, we were able to examine the added impact of the hands-on portion of the workshop by comparing the performance of instructors who had an opportunity to practice and receive feedback with that of those who did not.

Classroom. The first half of the instructor-training workshop presented the TDT model of teamwork. First, the four teamwork dimensions and their behavioral components were defined. This declarative knowledge formed the basis for a class discussion regarding the relationships and distinguishing features among the behavioral categories. Next, the TDT process for guiding teams toward evaluating and correcting deficiencies in these areas was described.

Active practice and feedback. The second half of the TDT instructor-training workshop involved hands-on practice using the TDT method to prebrief, evaluate, and debrief a team in conjunction with a series of two to four simulation-based training exercises. This portion of the training was designed to foster generalization, discrimination, and strengthening of instructors' teamwork schemas, or mental models. Specifically, instructors were given the opportunity to practice applying the TDT model of teamwork by extracting behavioral examples from novel situations (exercises) and classifying them appropriately. Generalization and discrimination were particularly sharpened during the postexercise consensus meeting, when instructors critically analyzed the categorization and prioritization of one another's observed examples. This procedural teamwork knowledge was strengthened by repeated practice using TDT in a series of exercises.

In some cases, trainees from the class have taken turns playing the roles of team members participating in the exercise, allowing them to experience TDT from both sides. Alternatively, actual team members can be used. When this is the case, it is important to explain to the team that

---

Exhibit 1.   Managing the Team Discussion

SAY: The last area that I would like to discuss is your training team's ability to handle team member comments. In order for team dimensional training to work, instructors need to stroke team members for admitting mistakes or confusion, explain why solutions are rejected if they must be, and restate or paraphrase team member feedback to make it more constructive.

ASK:
- Were team members positively reinforced for admitting mistakes or confusion?
- Do you recall any examples of an instructor missing an opportunity to reinforce a team member positively for admitting a mistake or negatively reinforced him or her?

ASK:
- When team members offered suggestions that had to be rejected, did the debriefers explain why?
- Do you recall any examples of an instructor rejecting a suggestion without explaining why?
- Can you think of any examples of a team member providing feedback to another in a way that was not specific enough or stated in an insulting or accusatory manner?
- If so, did the debriefer make sure that he or she or the team member restated the comment in a more constructive way?
- Were there any nonconstructive team member comments that were never restated?

the evolution is designed to allow their instructors to learn a new strategy for conducting training. The team should also be asked to be patient with instructors as they sharpen their ability to use TDT effectively. Finally, as soon as possible after instructors have practiced facilitating a TDT debrief, they should be given an opportunity to discuss their performance as a training team and to set goals for improvement on the next evolution. Using a guide similar to the TDT debriefing guide, the workshop instructor facilitates this "training team self-correction" session (see Exhibit 1), which is considered to be critical in helping instructors to fine-tune their use of TDT.

Summary of Results

To examine the impact of the TDT instructor workshop on mental models of teamwork, data were collected from 27 combat systems instructors and 12 seamanship instructors. Instructors' mental models of teamwork were evaluated using a card-sorting task. Examples of teamwork behaviors that fit each of the behavioral categories within TDT were listed on a set of 26 note cards (e.g., "AAWC notified TAO when a helo, which had been performing touch and goes, suddenly landed."). Immediately before and following the TDT workshop, instructors were asked to sort these examples into categories that were meaningful to them. The contents of each instructor's piles were evaluated for consistency with the TDT model of teamwork. This resulted in pre- and posttraining scores reflecting similarity to the TDT model for each instructor.

Results indicated that instructors' mental models of teamwork became significantly more similar to the TDT model following 1 day of instructor training (Smith-Jentsch et al., 1997). This analysis was performed separately for combat systems and seamanship instructors because, although the dimensions remained the same, behavioral examples to be sorted into those dimensions differed across the two domains. Results indicated that the mental models of both groups of instructors significantly changed in the appropriate direction.

Because not all of the instructors who participated in the workshop received hands-on practice using TDT, we were able to examine the added impact of this portion of the training on the development of mental models. Results showed that although mental models of teamwork were modestly affected by the classroom training alone, instructors who received the hands-on practice showed twice as much change. This difference was statistically significant, suggesting that hands-on practice added substantially to the procedural knowledge needed to apply the TDT model (Smith-Jentsch et al., 1997).

Does Team Dimensional Training Affect Team Performance?

Obtaining empirical validation of TDT-related performance improvement in a shipboard training environment has been challenging. Although seven shipboard demonstrations have been conducted, many factors varied across these demonstrations. For example, the experience level of the teams trained using TDT, the size of those teams, and the type of teams varied with almost every demonstration. In addition, the exercises used for these demonstrations differed in length and difficulty. Finally, the instructors on site were not condition-blind and were just learning to use the TDT method. Given these factors, it was necessary to collect ratings from an independent set of subject matter experts who assessed performance post hoc from audio recordings.

To illustrate the impact of TDT on team performance, we describe the results from one such case study. In this instance, one precommissioned CIC team had participated in TDT as part of a developmental test of the Battleforce Tactical Training (BFTT) system. Their performance on four consecutive TDT evolutions was evaluated from audiotape by two experienced evaluators who had no knowledge of what was discussed during any of the associated debriefing sessions. These evaluators assessed the team on each of the component behaviors that make up the four TDT dimensions using numerical rating scales developed as part of TADMUS (see chapter 4, this volume).

Performance ratings were plotted for each of the TDT behaviors over the course of the four exercises and compared against team goals set during the preceding team debriefs. Inspection of performance trends indicated that improvement on specific behaviors did not occur uniformly. In fact, performance improvement on each teamwork behavior could be directly linked to the particular TDT debrief in which that behavior was set as a goal. For example, no situation updates were given nor were priorities ever stated in the first exercise. Both of these problems were diagnosed in the first team debrief. However, the team determined that their most important goal for the second exercise was to provide more situation updates. Improvement on this target behavior was demonstrated in the next exercise and continued in the third and fourth exercises. In the second TDT debrief, the team decided to focus on stating priorities more explicitly as a goal. Consequently, performance ratings indicated improvement on this teamwork behavior during the third exercise, which was maintained during the fourth exercise. In sum, the observed trends from this case study indicated that the team accurately diagnosed specific teamwork-related problems, demonstrated immediate improvement on targeted goals, and maintained this level of performance during new exercises by generalizing lessons learned.

We have demonstrated previously that teams who effectively use the component behaviors within the TDT model of teamwork consistently make more accurate situation assessments (see chapter 4, this volume); TDT was designed to improve team performance through the mastery of those component behaviors. To investigate further the impact of TDT on performance, the experimental team's overall performance was plotted against the average overall performance of a group of five comparable teams in our database that had participated in the same exercises without TDT. Results indicated that the average performance rating received by the control group dropped as subsequent exercises became more difficult. In contrast, although the experimental team's performance was almost identical to that of the proxy control group on the first exercise, their overall performance improved on subsequent exercises despite the increased difficulty. These results, together with the data collected on instructor mental models and numerous anecdotal reports from shipboard instructors who have used TDT, suggest that it is a promising new strategy for improving team performance.

In fact, on the basis of positive experiences with this team training strategy, the afloat training group ATGPAC expressed its commitment to pursue TDT in the following Navy message (COMAFLOATRAGRUPAC, 1996) issued to other training commands:

> As we continue to migrate training on board our ships, we must provide shipboard training teams the tools to ensure their success. This methodology is one of these tools. . . . While this concept was originally designed to support combat systems training teams . . . [TDT] is equally applicable to all shipboard training teams. . . . our goal is to have all Pacific fleet Afloat Training Group shipboard trainers equipped with the fundamental techniques of this process.

Since this message was sent, representatives from ATGPAC and ATGLANT have come together to produce a standardized set of course materials for a TDT instructor training module to be incorporated into a shipboard instructor training course given to all SBTTs. Moreover, representatives from the other military services, the nuclear power industry, and the Federal Aviation Administration have expressed interest in examining TDT in other team environments (e.g., air traffic control). As the use of TDT becomes more widespread, we look forward to having opportunities to validate empirically its effectiveness and applicability to a number of team environments.

Lessons Learned and Future Directions

This chapter described a training strategy that uses guided team self-correction to improve teamwork-related knowledge and skills in conjunction with exercise-based team training. Guidelines for the optimal use of this strategy, TDT, were offered; the guidelines were based on our experiences during seven shipboard demonstrations and numerous shore-based simulations. Finally, results from our efforts to validate the effectiveness of TDT were summarized. Following is a set of lessons learned and directions for future research.

Lesson 1: Facilitator-Led Guided Team Self-Correction Is a Promising Strategy for Improving the Quality of Team Debriefing Sessions

Facilitator guidance, structured around a topical outline, appears to be useful in helping teams avoid the pitfalls that prevent team self-correction sessions from being constructive. We have found that, when used correctly, guided team self-correction can help to (a) keep team discussions focused, (b) facilitate a climate in which team members feel comfortable communicating questions and concerns, (c) increase the level and representativeness of team member participation, and (d) improve the constructiveness of team member input.

Lesson 2: Guided Team Self-Correction Can be Used to Develop Shared Mental Models of the Teamwork Processes that Are Linked to Performance Outcomes

Our data indicated that instructors' mental models of teamwork became significantly closer to the data-driven TDT model after a 1-day workshop. These findings were consistent across two groups of instructors from different team environments (i.e., combat systems and seamanship). Moreover, the hands-on practice portion of the training proved to be critical. Instructors who did not have the opportunity to practice extracting and categorizing concrete examples from a live exercise displayed only half as much change as did their counterparts.

Although the preliminary data described here are encouraging, further research is needed to investigate the maintenance of instructors' knowledge over time and the impact of TDT on team member mental models of teamwork. In addition, future research should investigate the use of guided team self-correction to develop other types of team knowledge (e.g., task expectations, teammate-specific characteristics).

Lesson 3: Guided Team Self-Correction that Focuses on Teamwork Processes Appears to Aid Teams in Diagnosing Team Problems, Focusing Their Practice on Specific Goals, and Generalizing Lessons Learned

As illustrated by the case study described in this chapter, we have found that the deficiencies noted by team members in TDT debriefs are highly consistent with those identified by instructor observers. Furthermore, improvement on specific teamwork behaviors tends to coincide directly with team goals set in previous TDT debriefs. Finally, data from our case studies as well as reports from shipboard instructors indicate that teams are better able to generalize lessons learned through TDT to novel situations in subsequent exercises.

Whereas these findings are encouraging, further research is needed to investigate empirically the impact of TDT on team performance and to examine the relationship between teamwork mental models and the development of team skills. In addition, guided team self-correction, and TDT in particular, should be examined as a strategy for facilitating continuous learning on the job. It is expected that once team members develop shared, accurate mental models of teamwork, they will be able to facilitate their own guided team self-correction sessions in the work environment.

Lesson 4: TDT Is a Viable Strategy for Objective-Based Team Training

TDT is a concrete example of how an EBAT approach can be applied to objective-based training for teamwork skills. The TDT model of teamwork serves as the training objective; exercise events are designed specifically to allow practice and assessment of these objectives; and component behaviors within the four dimensions are used as performance criteria that ultimately form the basis for developmental feedback. Over the course of 3 years, TDT has been refined to the point that it is considered both effective and practical to use for shipboard training. Moreover, TDT is expected to complement the objective-based technical training currently being implemented by afloat training groups. To streamline further the process of objective-based training, ongoing research and development efforts are investigating instructor handheld computers and other advanced technologies for data collection and storage (chapter 15, this volume).

Lesson 5: Active Practice and Feedback Are Critical for Imparting the Procedural Knowledge and Feedback Skills Necessary to Facilitate a Successful TDT Evolution

We have found that instructors' ability to use TDT improves dramatically after three or four experiences with it. A number of common mistakes initially made by instructors were listed earlier in our description of the components of TDT. Although each of these mistakes is routinely discussed during the classroom portion of the TDT instructor workshop, each one is typically made by at least one instructor in the subsequent hands-on portion of the training. It appears that effective facilitation of TDT requires some amount of trial-and-error learning. The impact of these errors (e.g., failing to record enough detail in their notes, offering too many of their own examples) on the quality of a TDT debrief (e.g., reduced participation, accusations, frustration) becomes clear once instructors make such an error themselves or witness a fellow instructor making one.

Moreover, the results from our instructor validation study clearly indicate that hands-on practice is critical for developing the procedural knowledge necessary to apply the TDT model. This point becomes obvious as one observes the dramatic reduction in the time it takes a group of instructors to classify examples of teamwork across multiple postexercise consensus meetings.

TDT may depart significantly from the type of training many instructors are familiar with. Given this, and the fact that we expect instructors to make certain mistakes early on, we feel it is important that instructors be given the opportunity initially to practice using TDT with guidance and feedback from an experienced TDT instructor in a relatively nonthreatening environment. This can be accomplished during the hands-on "lab" portion of the instructor workshop either in a simulated team environment or on board ship.

Lesson 6: TDT Is Best Used When a Team Is Given (a) the Objective and Forum to Discuss Teamwork-Related Objectives, (b) Sufficient Time (30 Minutes) to Conduct a Thorough Team Debrief, and (c) the Opportunity to Have All Key Team Members Participate

TDT is intended to supplement, *not replace*, other types of training (e.g., individual technical training). This is one of the most common misconceptions about the technique. TDT works best when individual team members have a sufficient baseline level of technical expertise and are prepared to focus on building the skills that enable them to function effectively as a collective.

When the team's objective is specifically to improve their teamwork processes, the members need a sanctioned forum to focus on teamwork (postexercise debrief) and sufficient time to engage in guided team self-correction with all key team members in order to reap the greatest reward from TDT. Although TDT debriefs tend to be longer than typical debriefs, it is argued that the accelerated learning they provide will result in the need for fewer exercises, ultimately reducing time in training. Moreover, we have found that a 30-minute TDT exercise is sufficient to generate more than enough concrete teamwork examples. In sum, a shorter exercise allows for more time to debrief, and it is during the debrief that most of the learning is expected to occur during a TDT evolution.

Lesson 7: Frequent Communication and Forming Partnerships With the Ultimate Beneficiaries of Our Research Are Essential for Translating Sound Theories into Usable Tools and Strategies

TDT is based on several sound theories of adult learning in general and team learning specifically (e.g., feedback theories, shared mental model theory). However, the mechanics of how best to implement TDT were derived from shipboard demonstrations and ongoing discussions with the more than 400 Navy instructors who participated in TDT workshops. As our relationship with the fleet strengthened and our understanding of their unique implementation issues grew, so did their interest in TDT. This interest provided us with rare opportunities to move the method out of the lab and on board ship. The specific terms used to describe components of teamwork, the format of TDT materials, and the guidelines for optimal use of TDT all went through numerous iterations (too many to count) over the course of 3 years. Had this not been the case, TDT would have remained a great idea in theory that never reached the fleet. The success of TDT is a direct result of our partnership with members of the fleet training community and their enthusiasm and commitment to improving Navy training.

Conclusion

As the use of work teams continues to rise and organizational structures flatten, tools and strategies that support effective team self-correction will become more and more critical. This chapter described one such strategy, TDT, which was developed under TADMUS. Preliminary data suggest that TDT is a promising strategy for enhancing teams' ability to diagnose and correct problems stemming from faulty teamwork processes. Such processes have been shown to play a significant role in effective team decision making, particularly in stressful environments. We hope this chapter will stimulate continued research in this exciting area and encourage studies that examine the generalizability of our results to other team environments.

References

Baldwin, T. T. (1992). The effects of alternative modeling strategies on outcomes of interpersonal-skills training. *Journal of Applied Psychology, 77*, 147–154.

Blickensderfer, E., Cannon-Bowers, J. A., & Salas, E. (1997a). Fostering shared mental models through team self-correction: Theoretical bases and propositions. In M. Beyerlein, D. Johnson, & S. Beyerlein (Eds.), *Advances in interdisciplinary studies in work teams series* (Vol. 4). Greenwich, CT: JAI Press.

Blickensderfer, E., Cannon-Bowers, J. A., & Salas, E. (1997b). *Training teams to self-correct: An empirical investigation.* Paper presented at the 12th annual meeting of the Society for Industrial and Organizational Psychology, St. Louis, MO.

COMAFLOATRAGRUPAC. (1996, March 18). Navy Message 181600Z. San Diego, CA: United States Navy.

Dwyer, D. J., Fowlkes, J. E., Oser, R. L., Salas, E., & Lane, N. E. (1997). Team performance measurement in distributed environments: The TARGETS methodology. In M. T. Brannick, E. Salas, & C. Prince (Eds.), *Team performance assessment and measurement: Theory, methods, and applications* (pp. 137–153). Mahwah, NJ: Erlbaum.

Dwyer, D. J., Oser, R. L., Salas, E., & Fowlkes, J. E. (in press). Performance measurement in distributed environments: Initial results and implications for training. *Military Psychology.*

Gentner, D., & Stevens, A. L. (1983). *Mental models.* Hillsdale, NJ: Erlbaum.

Kozlowski, S. W. J., Gully, S. M., Smith, E. M., Nason, E. R., & Brown, K. G. (1995). *Learning orientation and learning objectives: The effects of sequenced mastery goals and advanced organizers on performance, knowledge, metacognitive structure, self-efficacy and skill generalization of complex tasks.* Paper presented at the 10th annual conference of the Society for Industrial and Organizational Psychology, Orlando, FL.

Rentsch, J. R., Heffner, T. S., & Duffy, L. T. (1990). What you know is what you get from experience: Team experience related to teamwork schemas. *Group and Organization Management, 19*, 450–474.

Rouse, W. B., & Morris, N. M. (1986). On looking into the black box: Prospects and limits in the search for mental models. *Psychological Bulletin, 100*, 349–363.

Rumelhart, D. D., & Ortony, A. (1977). The representation of knowledge in memory. In R. C. Anderson & R. J. Spiro (Eds.), *Schooling and the acquisition of knowledge* (pp. 99–135). Hillsdale, NJ: Erlbaum.

Smith-Jentsch, K. A., Campbell, G., Ricci, K., & Harrison, J. R. (1997). *Applying guided team self-correction to enhance teamwork mental models and performance.* Unpublished manuscript.

Smith-Jentsch, K. A., Payne, S. C., & Johnston, J. H. (1996). Guided team self-corrections: A methodology for enhancing experiential team training. In K. A. Smith-Jentsch (Chair), *When, how, and why does practice make perfect?* Symposium conducted at the 11th annual conference of the Society for Industrial and Organizational Psychology, San Diego, CA.

Smith-Jentsch, K. A., Salas, E., & Baker, D. (1996). Training team performance-related assertiveness. *Personnel Psychology, 49*, 909–936.

Stout, R. J., Salas, E., & Carson, R. (1994). Individual task proficiency and team process: What's important for team functioning. *Military Psychology, 6*, 177–192.

12

Cross-Training and Team Performance

*Elizabeth Blickensderfer, Janis A. Cannon-Bowers,
and Eduardo Salas*

At times, experienced teammates seem to read each others' minds. They are aware of what their teammates are doing and when their teammates need assistance. A technical term for this is *implicit coordination* and it refers to teammates coordinating without overt communication (Kleinman & Serfaty, 1989). Although this team characteristic is exhibited frequently in many team environments (e.g., sports teams), it has received little research attention. With its emphasis on the fast-paced world of the combat information center (CIC), the Tactical Decision Making Under Stress (TADMUS) program investigated how teams anticipate each other's needs and coordinate their actions without overt communication. This included considering training strategies that would foster the shared understanding teams need to coordinate implicitly (i.e., "to read each other's minds"; (Cannon-Bowers, Salas, & Converse, 1993). The purpose of this chapter is to discuss one such strategy: cross-training. We first define cross-training, we then review two empirical tests of cross-training interventions, and we discuss the lessons we learned from these tests.

Theoretical Drivers

CIC teams and other teams in fast-paced environments perform highly complex tasks, and teammates depend on each other to perform effectively. Not only must team members be concerned with their individual tasks, they also must be concerned with how the team performs as a whole. This team-level performance includes communicating needed information among teammates, integrating information both from electronic displays and from other team members, backing teammates up in times of high

The views expressed herein are those of the authors and do not represent the official position of the organization with which they are affiliated. We thank Cathy Volpe and Kim Travillian for their efforts in conducting the cross-training research as well as their insight regarding the ideas presented in this chapter. We also thank Paul Radtke for his helpful comments on the manuscript.

workload, and making effective decisions for the team. Complicating the problem, teams perform these activities under severe time pressure. To perform effectively in these environments, researchers argue that an essential skill for teams is *adaptability*. Effective teams are able to use information from the task environment to adjust team strategies by backing each other up, making mutual adjustments, and reallocating team resources (Cannon-Bowers et al., 1993). Furthermore, researchers argue that *implicit coordination* is the mechanism that helps teams adapt in these complex environments. Implicit coordination is present when team members work together effectively without overt strategizing. Team members know implicitly what to do, how to compensate for their teammates' limitations, and what information or materials they must provide to their teammates. This knowledge enables team members to carry out their duties with minimal communication.

A shared understanding of the task goals, team members' roles, and how and why the team operates as it does may be the basis for implicit coordination (Cannon-Bowers et al., 1993). As individuals gain experience on a team, they build shared understanding of how and why the team operates as it does. This knowledge allows them to build expectations of upcoming actions and events. These expectations may range from highly specific predictions about an individual teammate's actions under certain conditions to more global expectations about the team's performance. In other words, individuals develop a *mental model of the team* analogous to mental models of mechanical systems (described by Wickens, 1992). To the extent that individual teammate's mental models overlap, they have a shared understanding or a shared mental model (Cannon-Bowers et al., 1993).

Researchers argue that shared mental models enable team members to predict their teammates' actions and informational requirements. These predictions, in turn, enable the team to adapt and coordinate implicitly. The construct of shared mental models has been adopted by a number of team researchers as the mechanism that allows successful teams to coordinate and to have smooth, implicit interactions (Duncan et al., 1996; Kraiger & Wenzel, 1997; Orasanu, 1990; Rentsch & Hall, 1994). A growing body of research on the shared mental model theory exists, and interested readers are recommended to read Cannon-Bowers and Salas (chap. 2, this volume) for a more detailed review of the literature.

One aspect necessary for shared mental models is interpositional knowledge (IPK) (Baker, Salas, Cannon-Bowers, & Spector, 1992). Interpositional knowledge is knowledge about the roles, responsibilities, and requirements of other positions in the team. For example, the task goals, methods, and tools for the different positions are aspects of interpositional knowledge. The opposite of interpositional knowledge is interpositional uncertainty (Baker et al., 1992). That is, if team member A has little or no knowledge regarding the duties of team member B or how team member B goes about accomplishing those duties, team member A has a high degree of interpositional uncertainty regarding team member B. To develop shared mental models of the task and team, teammates need a certain degree of IPK. This includes a clear understanding of how the team functions as a unit and how the individual team member's tasks and responsibilities interrelate with those of the other team members. One way to foster team IPK is through *cross-training*.

Volpe, Cannon-Bowers, Salas, and Spector (1996) define *cross-training* as an instructional strategy in which each team member is trained in the duties of his or her teammates. This training includes fostering the basic knowledge necessary to perform the tasks, duties, or roles of the other team members and the overall team framework. Greenbaum (1979) added that participation in cross-training can provide team members with a sense of a shared "common bond." Thus, the shared experiences and feelings resulting from "walking in another's shoes" may be critical in the establishment of morale, cohesion, and confidence (Travillian, Volpe, Cannon-Bowers, & Salas, 1993). Finally, Lawler (1982) noted that cross-training is one of the mechanisms that contribute to communication, coordination, and self-control by encouraging people to learn and understand the operations of those around them. Overall, cross-training is an approach to increasing the level of IPK among team members.

Types of Cross-Training

Cross-training can be divided into three types on the basis of the depth of information provided. The three types of cross-training are positional clarification, positional modeling, and positional rotation.

Positional Clarification

Positional clarification is a form of awareness training aimed at providing members with general knowledge of each member's general position and associated responsibilities. The end result is knowledge about the overall team structure or architecture and knowledge about the general responsibilities of each member's respective role and the general requirements of the team. Training methods for positional clarification include discussion, lecture, and demonstration.

Positional Modeling

Positional modeling is a training procedure in which the duties of each team member are discussed and observed. Because the emphasis is on direct observations, positional modeling provides details beyond what is learned in positional clarification. This technique provides information involving the general dynamics of the team, knowledge about each member's duties, and an understanding of how those duties relate to and affect those of the other team members. Individuals are assigned to observe tasks of another teammate in a simulated situation. This method, behavior observation, has been used successfully to teach concrete behaviors such as how to operate equipment, assemble a machine, and perform a surgical procedure (for review see Tannenbaum & Yukl, 1992).

Positional Rotation

Through direct, hands-on practice, positional rotation provides members with a working knowledge of each member's specific tasks and how those tasks interact. Although expertise in all positions may not be necessary or feasible, team members are trained to a basic competency level for specific tasks of each of their teammates. This method is similar to job rotation in that team members gain first-hand knowledge and experience in the specific tasks of others. The goal is for members to gain improved understanding of the interactions between team members and to develop different perspectives of the task. Ideally, team members would be trained in those tasks that demand cooperation and high interdependency among teammates. This explicit demonstration of team interdependencies should give teammates an increased understanding of the interconnectedness of the team.

Classification Scheme for Cross-Training Strategies

The methods of cross-training differ in utility depending on the degree of interdependency and the technical expertise required of a particular team. As shown in Table 1, team tasks vary in interdependency levels. We suggest that the level of interdependency required in a team task (low, medium, and high) should drive the selection of cross-training method. Low-interdependence teams operate with minimal amounts of communication and coordination and with little or no mutual monitoring or back-up behaviors. Feedback, if provided, is directed to individuals rather than the team as a whole. A team orientation is not required to perform the task, and the team can function without a great deal of leadership. Examples of low-interdependence teams include quality control circles, advisory groups, and review panels. The level of IPK required on a low-interdependence team is general knowledge of the overall structure of the team. For this type of team, we suggest that positional clarification is the appropriate method of cross-training. That is, engaging the team members in a discussion of team members' general responsibilities and roles or using lectures to present general team structure would foster the general understanding needed for low-interdependence teams.

Medium-interdependence teams have some necessity for timed inputs and interactions. Teams of this type have a fairly constant internal exchange of information and resources. Positions are differentiated and are somewhat functionally dependent on each other for successful performance. Some communication and coordination is required, and monitoring and feedback procedures among members are important. Leadership, team orientation, and back-up behaviors impact team performance. Examples of medium-interdependent teams include mining teams, flight attendant crews, research teams, task forces, and data-processing crews. Along with an understanding of the overall team structure and general dynamics of the team, these team members need working knowledge of each member's responsibilities and the interaction process among team members. We suggest the positional modeling method to foster this level of interpositional knowledge.

Finally, high-interdependence teams are teams with a critical need for cooperation and coordination. These teams usually require direct verbal communication or physical interaction. Positions are functionally unique and the team's task accomplishment is impossible without interactions. Monitoring, feedback, and back-up behaviors are critical for effective performance. Effective team leadership is crucial for successful performance, and orientation has great impact on processes and performance. Examples of high-interdependence teams include surgical teams, cockpit crews, sports teams, fire-fighting teams, and Special Weapons and Tactics (SWAT) teams. Because of the high level of interdependence among team members, to perform effectively these teams need knowledge of the specific job activities of their teammates and an understanding of how each job interacts and impacts other team members. This type of knowledge allows team members to develop commonly shared expectations regarding interactions among the team and other aspects of the team's performance. We suggest positional rotation as the method of cross-training that will allow teammates to build this level of interpositional knowledge and enable the team to develop shared expectations.

As a final note, we acknowledge that as interdependencies increase, tasks also increase in general complexity. This makes it neither impossible nor desirable to provide full cross-training. For example, we list surgical teams as highly interdependent teams and we recommend positional rotation as a cross-training strategy. However, a detailed positional rotation training for a surgical team is inconceivable. The implication is that researchers and trainers must identify particular duties to highlight during cross-training. We suggest highlighting duties that demand the greatest degree of cooperation and high interdependence among teammates.

To review, cross-training is a type of training intervention that can foster team IPK. Different methods of cross-training exist, and we have suggested that the method selected for a particular team depends on the level of interdependency required of that team. The next portion of this chapter describes the efficacy of cross-training interventions. We begin with a brief review of two early studies and follow with a description of two empirical tests of cross-training interventions that were conducted as part of the TADMUS program.

Does Cross-Training Work?

Two related studies of the effectiveness of cross-training appeared in early team research. Hemphill and Rush (1952) hinted at the usefulness of cross-training to build IPK among team members. They argued that ef-

Table 1. Classification Scheme for Cross-Training Strategies

Level of interdependence	Types of teams	Team characteristics	Type of IPK required	Potential cross-training methods
Low interdependence: Few demands for synchronized or coordinated interactions, internal exchange of information and resources minimal; positions exist as independent units with some limited functional grouping.	Quality control circles Advisory groups Review panels	Minimal amount of interactive communication and coordination with little or no monitoring or back-up behaviors Feedback, if provided, is at individual rather than team level. Orientation is not important. The team can function without leadership.	General knowledge of the overall team structure or architecture	Positional clarification: Discussion and clarification among members concerning the general responsibilities of their respective roles Lectures and audio/visual aids can be used to present the general requirements of the team.
Medium interdependence: Necessity for timed inputs and interactions; internal exchange of information and resources is fairly constant. Positions are differentiated and somewhat functionally dependent on each other for optimal performance.	Mining teams Flight attendant crew Research teams Task forces Data-processing crews	Some interactive communication and coordination is required. Monitoring and feedback procedures among members become more important. Leadership and team orientation impact performance. Back-up behavior facilitates performance.	Working knowledge of overall team structure and general dynamics Working knowledge of each member's responsibilities Basic understanding of the interaction process	Positional modeling: A multistep procedure in which selected duties of each member are discussed and observed The emphasis is observational learning.

High interdependence: Critical need for cooperation and coordination; direct verbal communication or physical interaction required; positions are functionally unique and goal attainment impossible without interdependent interactions.	Surgical teams Cockpit crews Sports teams Fire-fighting crews SWAT teams Combat information center teams	Interactive communication and coordination are critical for goal attainment. Monitoring, feedback, and back-up behaviors become critical. Team leadership is necessary for optimal performance. Team orientation has a critical impact on performance.	Knowledge of team members' specific job activities and how each job interacts with and affects other members Shared expectations regarding interactions and performance	Positional rotation: Members are trained in the specific tasks of their own positions as well as some of each team member's. First-hand knowledge and experience is provided through active participation in a "learn-by-doing" framework.

Note. IPK = interpositional knowledge; SWAT teams = Special Weapons and Tactics teams.

fective coordination among aircrew personnel depended on the amount of knowledge each member had of the other crew members' duties. They asked B-29 aircrew members questions about the tasks performed by all seven crew positions. Team members answered questions both about their own tasks and about their teammates' duties. Teams with higher degrees of overlap in understanding of the tasks were rated by instructors as performing better than the teams with less overlap in knowledge. Team overlap scores correlated positively with supervisor ratings of competence of the unit, effectiveness of working with others, conformity with standard operating procedures, performance under stress, attitude and motivation, and overall effectiveness. This supports the proposition that IPK is related to team coordination and overall performance.

Cream and Lambertson (1975) inadvertently provided some evidence of the IPK–performance relationship (Volpe et al., 1996). They examined aircrews' expectations regarding the appropriate functional requirements of the other crew members. This study indicated that each team member needed accurate functional expectations for the team to perform effectively. Furthermore, to develop accurate functional expectations, team members needed to understand the operational demands and interactional demands of the other team members. Thus, this study also supports the proposition that to perform most effectively, team members need some degree of knowledge about their teammates' tasks and responsibilities.

TADMUS Study 1

Until addressed by the TADMUS program, no direct test of the effect of cross-training on team performance had been attempted. The TADMUS program conducted two empirical studies of cross-training. The first experiment (Volpe et al., 1996) examined the interaction of positional rotation cross-training (presence or absence) and workload (high or low). The experiment required two-member crews of undergraduates to fly a PC-based F-16 aircraft simulation and "shoot down" enemy aircraft ($n = 40$ teams). The functions were divided, and each team member was responsible for specific tasks. The cross-trained teams were informed about and given practice time on all tasks pertaining to their own and their teammate's functional responsibilities. Participants not cross-trained received training only on the responsibilities of their particular position. In addition to the tasks normally required by their primary flight duties, workload was manipulated by requiring both team members to locate information on their displays and relay it orally to the "base." The dependent variables were teamwork behavior (i.e., technical coordination, team spirit, interpersonal cooperation, and cross-monitoring), communication (i.e., volunteering information, acknowledging teammates' comments, agreeing with a teammate, giving information, and providing task-relevant remarks), and task performance (i.e., time to shoot down the first enemy, number of times the enemy "locked on" with radar, the number of times the team's aircraft had the enemy within range and locked on, and the total number of enemies destroyed).

The results indicated that cross-training positively affected overall team performance; teams that were cross-trained performed more effectively than those not cross-trained. Importantly, the study demonstrated that a fairly simple cross-training intervention improved teamwork behaviors. The cross-training intervention also caused differences in team communication. In particular, teams that were cross-trained volunteered more information in advance of requests. The cross-trained teams were also more successful in task performance (i.e., shooting down enemy aircraft, etc.). The results did not show a relationship between cross-training and workload, but Volpe et al. (1996) argued that the lack of relationship may have been an artifact of the study. Overall, the study provided evidence that cross-training has a positive impact on certain aspects of team performance.

TADMUS Study 2

The second cross-training study (Cannon-Bowers, Salas, Blickensderfer, & Bowers, 1998) was a replication and extension of the Volpe et al. (1996) study. First, we reexamined the relationship between cross-training (presence or absence) and workload (high or low). Second, to provide a better test of the positional rotation cross-training method, we selected a task that was more highly interdependent than the task used by Volpe et al. The experimental task was a low-fidelity simulated radar control task that was designed to simulate specific functions and interdependencies found in interactive teams such as those in CIC environments (Weaver, Bowers, Salas, & Cannon-Bowers, 1995). The task required the team members to share information to identify and determine the intent of multiple targets on a radar screen. This demanded a high degree of IPK for teammates to develop expectations that would enable them to communicate efficiently. In addition, we used three-member teams (as opposed to dyads) of Navy recruits ($n = 40$). Participants also completed an IPK test to determine the level of knowledge gained from the cross-training intervention.

Team members were trained in their respective stations. The teams in the cross-training condition then proceeded to rotate to each of their teammate's stations and were trained to a level of proficiency on these tasks as well. Once training was completed, the team performed the task. The dependent variables included both task performance and team processes. The team process measures were similar but not precisely the same as those used by Volpe et al. (1996).

Consistent with the earlier study, the results supported the effectiveness of cross-training. The cross-trained teams scored significantly higher on the IPK test than did those not cross-trained. Differences between the training conditions also appeared on both task performance and team processes. Specifically, cross-trained teams engaged a greater number of targets correctly and spent less time accessing information than non-cross-trained teams (i.e., they spent less time looking at information before passing it on or making a decision). In terms of team communication, similar to the Volpe et al. (1996) study, cross-trained teams volunteered significantly more information than the teams without cross-training.

Interestingly, the Cannon-Bowers et al. (1998) study revealed an interaction between cross-training and workload on two dependent variables: the team score and the overall team process quality. For both team score and overall teamwork processes, those teams that were cross-trained were able to maintain their performance under high workload, whereas the performance of non-cross-trained teams suffered under high workload. No differences appeared between the teams under low workload.

What Have We Learned About Cross-Training Teams?

The two TADMUS studies described herein provide empirical support for the effectiveness of cross-training as a training intervention for improving team performance. We now discuss a number of lessons learned concerning cross-training.

Lesson 1: Cross-Trained Teams Are Better Able to Anticipate Each Other's Needs (Volpe et al., 1996)

Team members who become familiar with each other's knowledge, skills, and task requirements improve their ability to anticipate teammates' task, informational, and interpersonal needs. Thus, cross-training seems to help build team-shared mental models (i.e., Cannon-Bowers et al., 1993). Experienced teams coordinate without the need to communicate overtly (Orasanu, 1990; Rouse, Cannon-Bowers, & Salas, 1992), and cross-training appears to accelerate the development of this ability (Cannon-Bowers et al., 1998; Volpe et al., 1996). Team members with shared understanding of the task and team role can anticipate each other's behavior. These shared expectations and intentions enable them to coordinate implicitly (Kleinman & Serfaty, 1989).

Lesson 2: Cross-Training Fosters Interpositional Knowledge (Cannon-Bowers et al., 1998)

Research has provided evidence that the positional rotation method of cross-training builds overlap in understanding of individual duties and responsibilities of team members. This interpositional knowledge is considered to be a building block for shared mental models (Cannon-Bowers et al., 1993).

Lesson 3: Cross-Training Should Be Used in Combination With Team Process Training to Provide Maximum Benefit

Knowing that a task should be performed does not guarantee that it will be done correctly. For example, team members not only need to know when the task requires communications among teammates, but also they need to know how to communicate effectively.

Lesson 4: Length of Cross-Training Intervention Is Not Necessarily Related to Value of the Intervention

Interventions as short as 30 minutes have produced positive results.

Lesson 5: Cross-Training Interventions Should Be Designed on the Basis of the Interdependency Requirements of the Task

Although the studies reported here examined only the effectiveness of positional rotation, we suggest that teams with high interdependencies be given positional rotation, whereas teams with few interdependency requirements may need only basic knowledge of team structure. This recommendation is conditional on the need to examine the effectiveness of other cross-training methods on other types of teams.

Lesson 6: A Number of Guidelines Regarding the Training Objectives and Content Can Be Based on the Cross-Training Research

Cross-training should do the following (adapted from Salas, Cannon-Bowers, & Blickensderfer, 1997):

1. provide team members with an understanding of how other team members operate, why they operate as they do, and the manner in which they are dependent on teammates for information and input;
2. provide team members with exposure to the roles, responsibilities, tasks, information needs, and contingencies of their teammates' tasks;
3. provide team members with practice on the roles and tasks of teammates, highlighting the interdependencies of the positions as requested; and
4. provide feedback during cross-training exercises that allows teams members to formulate accurate explanations for their teammates' behavior and reasonable expectations for their teammates' resource needs.
5. To determine the specific content of cross-training, a team task analysis should be conducted. This will help to identify interdependencies in the task and to identify what interpositional knowledge is necessary to help teammates coordinate.

Conclusion

The research described in this chapter, together with the growing research on shared mental models, gives strong support for cross-training. Cross-training is a relatively straightforward strategy that can be used to build interpositional knowledge. This interpositional knowledge, in turn, provides a foundation on which team members can build a team-shared understanding. Team-shared understanding appears to enable teams to communicate and coordinate smoothly and efficiently in fast-paced, complex environments (Cannon-Bowers et al., 1993; Orasanu, 1990). Although the research thus far gives a solid empirical foundation for the usefulness of cross-training, a number of research directions still merit attention.

First, cross-training needs to be tested in the field with complex tasks. Along with establishing the generalizability of cross-training, tests of that nature can help identify the information that is most necessary to include in cross-training as well as information that may be omitted. In addition, the impact of cross-training on individual performance needs to be examined. We do not want to create individual performance decrements as we attempt to improve team processes. Finally, tests combining team-training interventions that affect various aspects of team performance will give insight into developing a comprehensive team-training strategy.

Overall, cross-training appears to be a sound, theoretically driven, and empirically tested training strategy for teams. Although it has not been tested in the field, we suspect strongly that the results will generalize to other interdependent team tasks. Indeed, the cross-training methodology exemplifies the nature of products from the TADMUS program, which merit transitioning to shipboard as well as commercial settings.

References

Baker, C. V., Salas, E., Cannon-Bowers, J. A., & Spector, P. (1992, April). *The effects of interpositional uncertainty and workload on team and task performance.* Paper presented at the annual meeting of the Society for Industrial and Organizational Psychology, Montreal, Canada.

Cannon-Bowers, J. A., Salas, E., Blickensderfer, E. L., & Bowers, C. A. (1998). The impact of cross-training and workload on team functioning: A replication and extension of the initial findings. *Human Factors, 40,* 92–101.

Cannon-Bowers, J. A., Salas, E., & Converse, S. A. (1993). Shared mental models in expert team decision making. In N. J. Castellan, Jr. (Ed.), *Individual and group decision making: Current issues* (pp. 221–246). Hillsdale, NJ: Erlbaum.

Cream, B. W., & Lambertson, D. C. (1975). *A functional integrated systems trainer: Technical design and operation* (AFHRL-TR-75-6[II]). Brooks Air Force Base, TX: HQ Airforce Human Resources Laboratory.

Duncan, P. C., Rouse, W. B., Johnston, J. H., Cannon-Bowers, J. A., Salas, E., & Burns, J. J. (1996). Training teams working in complex systems: A mental model-based approach. In W. B. Rouse (Ed.), *Human/technology interaction in complex systems* (Vol. 8, pp. 173–231). Greenwich, CT: JAI Press.

Greenbaum, C. W. (1979). The small group under the gun: Uses of small groups in battle conditions. *The Journal of Applied Behavioral Science, 15,* 392–405.

Hemphill, J. K., & Rush, C. H. (1952). *Studies in aircrew composition: Measurement of cross training in B-29 aircrews* (AD No. B958347). Columbus, OH: Ohio State University.

Kleinman, D. L., & Serfaty, D. (1989). Team performance assessment in distributed decision making. *Proceedings of the Symposium on Interactive Networked Simulation for Training* (pp. 22–27). Orlando, FL: University of Central Florida and Florida High Technology and Industry Council.

Kraiger, K., & Wenzel, L. H. (1997). Conceptual development and empirical evaluation of measures of shared mental models as indicators of team effectiveness. In M. T. Brannick, E. Salas, & C. Prince (Eds.), *Assessment and measurement of team performance: Theory, research, and application* (pp. 63–84). Hillsdale, NJ: Erlbaum.

Lawler, E. E., III. (1982). Increasing worker involvement to enhance organizational effectiveness. In P. S. Goodman (Ed.), *Change in organizations* (pp. 280–315). San Francisco, CA: Jossey-Bass.

Orasanu, J. (1990, October). *Shared mental models and crew performance.* Paper presented at the 34th annual meeting of the Human Factors Society, Orlando, FL.

Rentsch, J. R., & Hall, R. J. (1994). Members of great teams think alike: A model of team effectiveness and schema similarity among team members. In M. M. Beyerlein & D. A. Johnson (Eds.), *Advances in interdisciplinary studies of work teams: Theories of self-managing work teams* (Vol. 1, pp. 223–262). Greenwich, CT: JAI Press.

Rouse, W. B., Cannon-Bowers, J. A., & Salas, E. (1992). The role of mental models in team performance in complex systems. *IEEE Transactions on Systems, Man, and Cybernetics, 22,* 1296–1308.

Salas, E., Cannon-Bowers, J. A., & Blickensderfer, E. L. (1997). Enhancing reciprocity between training theory and practice: Principles, guidelines, and specifications. In J. K. Ford & Associates (Eds.), *Improving training effectiveness in work organizations* (pp. 291–322). Hillsdale, NJ: Erlbaum.

Tannenbaum, S. I., & Yukl, G. (1992). Training and development in work organizations. *Annual Review of Psychology, 43,* 399–441.

Travillian, K. K., Volpe, C. E., Cannon-Bowers, J. A., & Salas, E. (1993). Cross-training highly interdependent teams: Effects on team process and team performance. *Proceedings of the 37th Annual Human Factors and Ergonomics Society Conference* (pp. 1243–1247). Santa Monica, CA: Human Factors and Ergonomics Society.

Volpe, C. E., Cannon-Bowers, J. A., Salas, E., & Spector, P. (1996). The impact of cross training on team functioning. *Human Factors, 38,* 87–100.

Weaver, J. L., Bowers, C. A., Salas, E., & Cannon-Bowers, J. A. (1995). Networked simulations: New paradigms for team performance research. *Behavior Research Methods, Instruments, and Computers, 27,* 12–24.

Wickens, C. D. (Ed.). (1992). *Engineering psychology and human performance.* New York: Harper Collins.

Part IV

Future Applications and Advanced Technology Implications From TADMUS

13

Cognitive Task Analysis and Modeling of Decision Making in Complex Environments

Wayne W. Zachary, Joan M. Ryder, and James H. Hicinbothom

Decision theory has been characterized, for much of its history, by a debate about whether human decision processes are inherently flawed. The remarkable part of this debate is that for virtually its entire duration, it has been conducted without reference to detailed data on how people actually make decisions in everyday settings. In recent years, this issue has come to the forefront in the work of Cohen (1981); Barwise and Perry (1983); Klein, Orasanu, Calderwood, and Zsambok (1993); and others, who have pointed out the fundamental differences between decision making as it has been studied using traditional decision theory and as it occurs in socially situated naturalistic settings. The resulting *naturalistic decision theory* has emphasized highly detailed, almost ethnographic, studies of decision processes in specific domains. This had resulted in dense data but primarily prose representations and analyses.

In parallel to the rise of naturalistic decision theory, cognitive science and human–computer interaction (HCI) researchers were developing increasingly powerful analysis methods that collectively were called *cognitive task analysis* techniques. The purpose of these techniques was to analyze and model the cognitive processes that gave rise to human task performance in specific domains, as the basis for design and evaluation of computer-based systems and their user interfaces. The Tactical Decision Making Under Stress (TADMUS) project provided a unique opportunity for these two avenues of inquiry to come together. This chapter describes research that combined the highly formal methods and tools of the HCI community with the theoretical orientation of naturalistic decision theory. The aim of the research reported here was to create a detailed and domain-

The authors acknowledge the contributions made by Janine Purcell to the research reported here as well as the cooperation and effort of the many individuals who acted as participants in the data collection effort. The efforts of Don MacConkey and John Pollen in supporting the data collection and analysis effort on the Navy side are also gratefully acknowledged.

specific model of decision making in the anti-air warfare (AAW) domain that was also sufficiently formal to be used to drive the design and development of systems to support and train decision making in that domain.

Cognitive task analysis and modeling were undertaken in the TADMUS program for three reasons (see also Cannon-Bowers and Salas, chap. 2, this volume). The first reason is theoretical. Theories of situated cognition and naturalistic decision making form a major theoretical underpinning of the larger TADMUS program. These theories posit that the form and content of decision processes are highly determined by the organization of the domain in which the decision maker is operating. From this theoretical position, one cannot begin to develop decision-making interventions, such as decision-training tools or decision support systems, without a detailed understanding and formal representation of the relationship between the decision-making expertise and knowledge that is unique to experts in that domain.

The second reason is historical. It is a virtual truism, based on decades of collective experience in human factors engineering, that the design of new or modified systems that include human operators should begin with a detailed mapping of what the human beings are (or should be) doing. This notion of task analysis is so strong as to be perhaps the single most unifying principle of human factors. More important is the repeated observation that systems that have been built or redesigned with a sound task analysis at the onset are much more usable, lead to higher human performance, and require less training. Thus, there is every reason to suspect that a task analysis, albeit one that includes cognitive as well as observable acts, would be necessary in the TADMUS case as well.

The third reason is technological. The larger TADMUS program sought ultimately to develop actual systems, specifically decision support systems and team decision-training tools, that would improve empirical decision making. Many advanced technologies that can be incorporated into such systems require detailed models and analyses of the decision strategies of the human operators. For example, embedded user models can be used to create intelligent or adaptive user interfaces to the decision support system (e.g., Rouse, Geddes, & Curry, 1987), and the support system itself can incorporate or be designed from models of user strategies (see Kieras, 1997). The development of a detailed and accurate cognitive analysis would thus be an enabling condition for application of a broad range of potentially useful technologies for improving decision performance.

This chapter discusses both the method for and the results of detailed cognitive analysis and modeling in complex, tactical domains of the kind considered by the TADMUS program. The first part of the chapter is theoretical. It begins with a set of logical requirements for the cognitive analysis process and then describes a framework used to meet those requirements. This framework, called COGNET (Cognition as a Network of Tasks), is discussed in terms of its theoretical underpinnings, its description, and data collection and analysis methods. The remainder of the chapter presents the cognitive analysis of tactical decision processes that was conducted in the TADMUS research, in the form of the COGNET model that resulted from the analysis. The applications of the model are discussed in the conclusion.

Logical Requirements for Cognitive Analysis and Modeling

Cognitive analysis and modeling is a relatively new subject, and it means different things to different people. Many techniques for analyzing human cognitive processes and decision making have been developed (see Essens, Fallesen, McCann, Cannon-Bowers, & Dorfel, 1995; Meyer & Kieras, 1996), and there is no clear standard method that is appropriate for all situations and domains. Rather, the method used must be selected (or developed) to meet the specific needs of the analysis. (The major needs for a cognitive analysis in TADMUS are cited above.) From these needs, the requirements of the analytic method were defined. Specifically, the cognitive analysis had to be able to represent four major aspects of tactical decision making:

- *Real time*: In tactical domains, data arrive and must be processed in real time, so decisions have temporal constraints. Making the right decision too late is as bad (or worse!) than making the wrong decision in a timely manner. The cognitive analysis had to make clear how decisions were temporally organized and related to the flow of external events in the problem environment.
- *Opportunistic and uncertain decision processes*: Although the tactical decision maker will have clear goals, the external events to be faced will typically be unpredictable. This means that it will be unclear exactly what decisions may be required until the situation unfolds. Moreover, the results of actions taken by the person are uncertain (i.e., they may or may not have the desired result). The decision maker thus must adapt both to the unfolding situation and to the results of actions taken. The cognitive analysis had to be able to capture this opportunistic aspect of the decision process.
- *Multitasking*: The pace of events and the uncertain nature of the process require the decision maker to be prepared to interrupt any cognitive activity to address a more critical decision at any time. This will typically result in a weakly concurrent multitasking, in which the decision maker may have several decision processes underway at a time (with one processing and the others suspended). Managing competing demands for (limited) attention is a critical part of the decision process requiring the cognitive analysis to address the ways in which decision makers manage and share attention.
- *Decision processes situated in computer-based and verbal interactions*: The majority of information available to the tactical decision maker comes not from direct sensation of the problem environment, but rather through information displayed at computer-based

workstations and verbal messages from teammates. Similarly, decisions are implemented not through direct action, but as interactions with the computer workstation or verbal messages to other persons. The cognitive analysis had to be able to capture decision processes that are based on these types of input–output stimuli.

In addition to these requirements and constraints imposed by the tactical decision process itself, the cognitive analysis also needed two other properties: to be able to *integrate behavior and cognitive processes*. The cognitive analysis must feed both into training and decision support interventions, both of which function by relating decision maker *actions* to decision maker cognitive states. Thus the analysis to be undertaken had to decompose and describe not just cognitive processes but also the way in which those cognitive processes were linked to observable behavior. The analysis also had to be *generic (predictive) rather than situation specific*. Many forms of conventional and cognitive task analysis attempt only to describe and decompose human behavior in the context of a specific exemplar situation, called a scenario. However, the analysis required here clearly needed to be at a more general level. Because it needed to be used in training and decision support interventions that could apply to *any* scenario, the analysis and resulting cognitive model also had to be able to predict the decision processes and their associated observable actions in any scenario. Finally, of course, the analysis had to be undertaken with a technique that is able to deal with the complexity of a difficult naturalistic setting such as naval command and control.

The COGNET Framework

To address these requirements, the cognitive analysis was undertaken with an adaptation of a cognitive analysis and modeling method developed by the authors and colleagues in prior research. This framework, called COGNET, is a theoretically based set of tools and techniques for performing cognitive task analyses and building models of human–computer interaction in real time, multitasking environments (Zachary, Ryder, Ross, & Weiland, 1992). COGNET had been developed, applied, and refined in a series of earlier studies. The original development and application was to a vehicle tracking domain (Zachary, Ryder, Ross, & Weiland, 1992; Zubritzky, Zachary, & Ryder, 1989). COGNET's ability to represent and predict attention-switching performance was empirically demonstrated in a validation study based on the vehicle-tracking model (Ryder & Zachary, 1991). The COGNET vehicle-tracking model was also successfully translated into an embedded user model for an intelligent, adaptive human–computer interface (Zachary & Ross, 1991). The framework was subsequently applied to several other complex domains, including enroute air traffic control (Seamster, Redding, Cannon, Ryder, & Purcell, 1993) and telephone operator services (Ryder, Weiland, Szczepkowski, & Zachary, in press). The COGNET analysis of air traffic control was used as the basis for redesigning a training curriculum (Australian Civil Aviation Authority, 1994), whereas the analysis of operator services was used to design new interfaces and decision support tools.

The COGNET framework is summarized below in terms of its theoretical base; its description language (in which the knowledge is actually represented); and its data collection, knowledge elicitation, analysis, and knowledge representation methods.

Theoretical Underpinnings

The theoretical underpinnings of COGNET research lie in cognitive science research, particularly the symbolic computation branch, which views cognitive processes as the operation of a specific computational mechanism on a set of symbols that are themselves a representation of sensation, experience, and its abstraction (see, e.g., Newell, 1980, and Pylyshyn, 1984, for theoretical discussions of this viewpoint). Thus COGNET presumes

- an underlying mechanism of a specific structure with clear principles of operation (henceforth referred to as the *cognitive architecture*) and
- a set of underlying symbols on which it operates (henceforth termed the *internal knowledge*) that are organized in a specific representational scheme (henceforth termed the *knowledge representation*).

Both are largely developed from the work of Newell (see Card, Moran, & Newell, 1983; Newell, 1990; Newell & Simon, 1972), which in its simplest form breaks human information processing into three parallel macro-level mechanisms—perception, cognition, and motor activity—shown as the ovals in Figure 1. Perception (which, in COGNET, includes the phys-

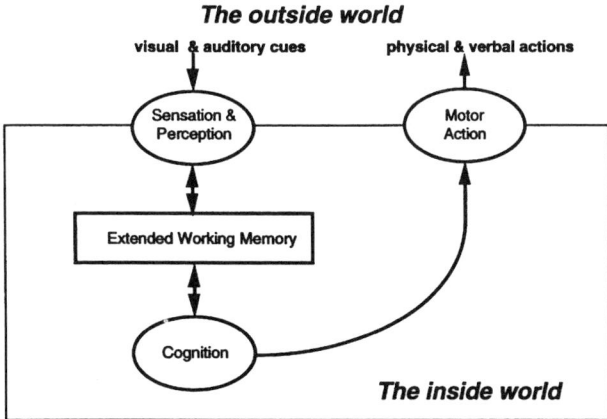

Figure 1. Conceptual view of COGNET (Cognition as a Network of Tasks) cognitive architecture.

ical process of sensation) receives information from the external world and internalizes it into the symbolic or semantic information store that is accessed by both the perceptual and cognitive mechanisms through an information store that is shared by both. As used in COGNET, this symbol store corresponds to what has come to be called extended working memory (see Ericsson & Kintsch, 1995). This shared store is depicted in Figure 1 because it is shared by both mechanisms, but it is not the only information store in the COGNET architecture. Both the cognitive and sensory–perceptual mechanisms incorporate other information stores that are accessed by each mechanism (i.e., long-term memory accessed by the cognitive mechanism and acoustic/visual information stores accessed by the perceptual mechanism).

A completely parallel cognitive process manipulates this internal symbolic representation of the external world, using previously acquired procedural knowledge. The cognitive process thus operates on an internal mental model of the world, not on direct sensations of the world. The cognitive process also modifies the mental model, as a result of cognitive reasoning processes (induction, deduction, and abduction). The problem representation thus is affected both by the perceptual processes and the cognitive processes. The cognitive process, in addition to being able to modify the problem representation, can also invoke actions through commands or instructions to the motor system. This system operates outside the scope of the problem representation (i.e., does not have access to and does not depend on the contents of the current extended working memory). The motor system provides manipulation of physical instrumentalities that in turn manipulate the environment (Card et al., 1983, provide a detailed empirically based argument for this underlying structure).

Each person is presumed to possess and use the same mechanisms described above (subject to individual variations in parameters of the mechanism). Thus, mechanism by itself does not help differentiate the novice from the expert or the ability to make decisions in one domain from another. Given this observation, it must be the other component of the system—the internal knowledge—that differentiates performance among domains and expertise levels. Thus, the goal of cognitive task analysis must be to understand and represent the internal knowledge of experts in the domain of interest. The COGNET framework has been developed to provide a practical tool that can be used to pursue this goal in complex, real-world domains.

COGNET presumes a certain organization and representation of internal knowledge, based on the architecture in Figure 1 and the emerging cognitive theory of expertise, as discussed generally by Chi, Glaser, and Farr (1988); Ericsson and Smith (1991); Hoffman (1992); Ryder, Zachary, Zaklad, and Purcell (1994); VanLehn (1996); and Zachary and Ryder (1997). In real-time, multi-tasking, HCI-based decision domains, the person interacts with the external problem environment through the medium of the machine system (and, specifically, through the person–machine interface). The person is implicitly assumed to be in a work setting and therefore to be pursuing some high-level mission or goal with regard to the external environment. Within this overall goal, the activities of the expert human operator of the person–machine system appear as a set of tasks with complex interrelationships. These tasks represent chunks of knowledge that the expert has compiled from lower level procedures and rules for use in a broad range of situations or cases. They are analogous to the various case strategies that are the basis for the case-based reasoning theory of highly expert decision making and planning (e.g., Kolodner, 1988). Some of these tasks may compete to be performed simultaneously, whereas others may be complementary; still others may need to occur essentially in sequence. Each task represents a specific local goal that the operator may pursue to achieve or maintain some aspect of the overall mission or goal.

The way in which a task is accomplished depends heavily on the evolution of the current problem instance to that point and the current problem situation at each instantiation of the task. The knowledge in the task contains the expert's understanding of how the task's goal can be achieved in different contexts. The knowledge that makes up the task also includes the knowledge needed to recognize the situations or contexts in which the task goal is relevant. In this regard, the COGNET tasks are activated by a recognitional process, analogous to that used in Klein's (1989) recognition-primed decision making. In a real-time domain, however, multiple cognitive tasks may be recognized as needing to be done without there being enough time to actually carry out each task. Thus, the tasks must compete with each other for attention, with each task that has been recognized as relevant shrieking for the focus of the person's attention. This shrieking process is analogous to the pandemonium process for attention originally postulated by Selfridge (1959). Even when a task gains the focus of the attention and begins to be executed by the cognitive processor, the process of tasks competing for attention continues unabated. The result is that tasks may (and often do) interrupt one another, and a given task may be interrupted and resumed several times as the operator copes with the evolving sequence of events.

What unites these separate chunks of procedural knowledge or tasks into a more global problem-solving strategy is a common declarative representation of the overall situation and its evolution (Hayes-Roth, 1985; Hayes-Roth & Hayes-Roth, 1979). This common problem representation is highly interactive with the individual tasks. As a given task is performed, the person gains knowledge about the situation and incorporates it into the current problem representation; similarly, as the problem representation evolves, it can change the relative priority among tasks and lead one task to come to the front and require immediate attention. At the same time, much of the information in the current problem representation is obtained from perceptual processes, for example, by scanning and noting information from displays, external scenes, or auditory cues, encoding it symbolically, and adding it onto the declarative problem knowledge. The procedural knowledge in each task includes knowledge about when and how to activate specific actions at the workstation or in the environment.

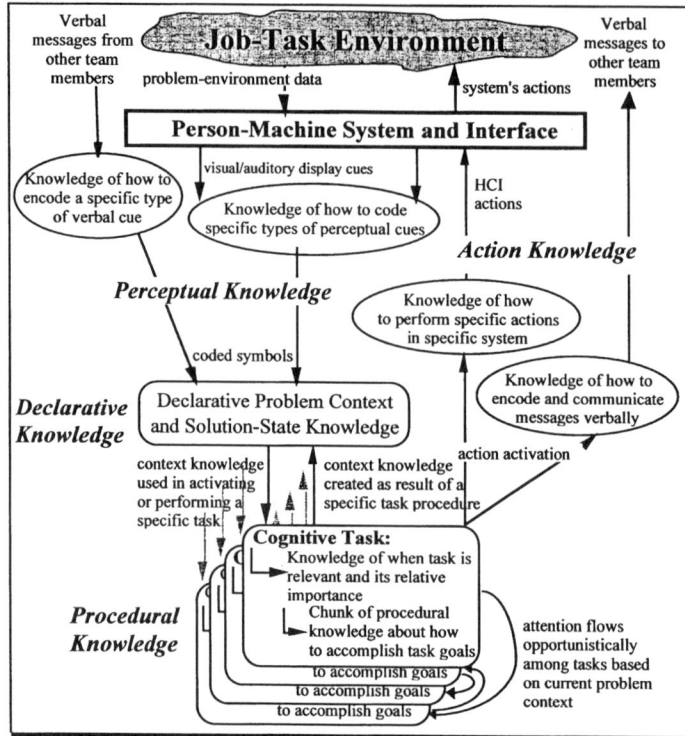

Figure 2. COGNET (Cognition as a Network of Tasks) knowledge framework.

These action activations are passed to the motor system where they are translated into specific motor activity (e.g., button presses).

This conceptual view of the types and organization of knowledge is pictured in Figure 2. It gives COGNET the structures necessary to link sensation/perception, reasoning and decision making, and action into a common framework. The conceptual structure shown in Figure 2 was created to deal with individual-level decision making. In anti-air warfare (AAW)—the specific domain of interest in TADMUS—decision making is highly distributed throughout a team. Thus, this preexisting COGNET framework had be enhanced to address the team nature of command and control decision making. The concept of perceptual knowledge was broadened to include verbal communication among team members. In addition, the concept of declarative knowledge was broadened to include a representation of other team members and their roles within the team.

COGNET Description Language for Cognitive Task Analysis

One main reason for developing a theoretical framework for human information processing and decision making is that the framework can provide a means of decomposing empirical phenomena in a way that permits their more formal description. The process of constructing a formal description for a specific set of human activities in a specific domain constitutes a form of that mainstay of human factors, the task analysis. In particular, it is a cognitive form of task analysis because it relates cognitive constructs and mechanisms to the observed behaviors. To conduct a cognitive task analysis (with COGNET or any other framework), it is necessary to have a set of constructs that is to be identified and described and notation with which to describe the constructs. The knowledge framework shown in Figure 2 identifies the set of constructs that is necessary for real-time, multitasking performance. To construct a cognitive task analysis using this framework, specific notations are used to describe each of the four major types of knowledge included in Figure 2: perceptual knowledge, declarative knowledge, procedural knowledge, and action knowledge. These notations (summarized below) have been derived from existing notations within cognitive science and knowledge engineering wherever possible. The attention process, which is really an epiphenomenon of the knowledge description process, is also summarized.

Declarative knowledge, in COGNET, refers specifically to a person's internal representation of the current problem, including its history or evolution to the current point, all declarative information related to the solution strategy for that current problem, and all long-term knowledge about things in the environment (e.g., system characteristics). Solution strategy information includes, for example, the features of a plan that is being developed and expectations about future events. Representing this kind of knowledge has been widely studied under the rubric of blackboard systems (e.g., Carver & Lesser, 1994; Englemore & Morgan, 1988; Nii, 1986a, 1986b), and the COGNET framework uses blackboard notation to describe declarative knowledge. A declarative knowledge blackboard is a collection of individual hierarchical structures, each of which is called a panel. Each panel consists of a hierarchy of knowledge elements, called levels, that are conceptually related. Often, but not always, the levels represent different degrees of abstraction in the conceptual space defined by the panel, although they may also represent a simple partitioning of the conceptual space into different aspects. Each level represents a dynamic collection of different individual concepts that provide specific instances of information at that level of the panel's overall conceptual space. For example, a panel may represent a tactical situation, and different levels may partition that conceptual space into air tracks, surface tracks, and sub-surface tracks. The air track level, in this example, would then contain individual concepts that correspond to the individual air tracks of which the person is aware in the current situation. Each individual concept will have a number of attributes that are common to concepts to the panel or level where it is located (also termed where it is *posted*). The attributes of air tracks, for example, might be different from the attributes of surface tracks (e.g., the latter not having an altitude attribute). The attribute values differentiate the individual concepts from one another at a given panel or level. In addition, concepts on the blackboard can be semantically associated with other concepts. For example, an air track can have the relationship "taken off from" a surface track. However, every concept at a given level may not have a relationship of each kind (e.g., some air tracks may not have taken off from any surface track, but may have taken off from a land location or from another air track).

Two additional pieces of terminology are applied to the declarative knowledge blackboard in COGNET. First, all the messages on the blackboard at any one time constitute the *context* for cognitive processes at that time. This momentary context drives the way in which procedural knowledge is activated and applied. Second, the structure in which the declarative knowledge in the blackboard is organized is sometimes termed the person's *mental model* of the domain. This usage is primarily metaphorical, helping explain the construct to domain experts from whom knowledge must be acquired and who must help validate the model. It is not intended as a strong theory of mental models in the cognitive science sense.

In COGNET, the information processing activity is presumed to occur through the activation and execution of chunks of *procedural knowledge*, each of which represents an integration or compilation of multiple lower level information processing operations around a high-level goal specific to the work-domain being modeled. This combination of the high-level goal and the procedural knowledge needed to fulfill it is referred to as a *cognitive task*. All the knowledge compiled into each task is activated whenever the high-level goal defining that task is activated. Each task-level goal includes metacognitive knowledge that defines the contexts in which that task is relevant. This metacognitve knowledge is simply a description of the contexts (as defined above) under which the goal should be activated. Thus, the high-level goals are activated according to the current problem context, as defined by this metacognitive "trigger." In addition to this trigger, another piece of metacognitive knowledge defines the relative priority of that goal in the current context so that attention can be allocated to the goal with the highest priority given the current context. This second piece of metacognitive knowledge is called the priority expression. These common features provide the structure for describing a cognitive task in COGNET. Each task has two main parts, the task definition, and the task body. The task definition identifies the high-level goal involved and a specification of when that goal is activated and the priority it will have when activated. A cognitive task is defined in the following form:

TASK <task-goal-name> . . . *activation condition / Priority (formula)*
<task body>.

The task body is a hierarchy of lower level information processing operators, based strongly on the GOMS (goals-operators-methods-selection rules) notation of Card et al. (1983), but with customizations to allow for: (a) manipulation of concepts on the blackboard, (b) evaluating GOAL conditions on the basis of the blackboard context, and (c) interrupting and suspending the current task.

As in GOMS, the GOALs can be either sequential or subordinated to one another, forming a hierarchical structure that defines branch points in the procedural logic. The lowest level GOALs have only nongoal operators as "children". These operators, when executed, accomplish that goal. Other COGNET operators fall into three groups: action, cognitive, and metacognitive.

- Action operators involve interactions with the workstation, both generic (e.g., Point, Enter, etc.) and workstation specific (Perform <FUNCTION>). Verbal actions are denoted with the "communicate" operator. Action operators comprise *action knowledge*, in that they define how actions are to be performed in a specific job-task environment.
- Cognitive operators create (i.e., POST), delete (i.e., UNPOST), or manipulate (i.e., TRANSFORM) declarative knowledge on the blackboard or encapsulate lower level cognitive processes that are included in a task only by reference (e.g., determine ["Is_track_flying_commercial_air_route?"]).
- Metacognitive operators actually affect the execution of the procedure through conditional suspensions (i.e., suspend).

As in GOMS, frequently used goal–operator subhierarchies can be subsumed into separate units, called *methods*, that can be invoked directly by name. It should be noted that not all of the notation is necessarily needed to describe any specific domain. Domains with a great deal of explicit human–computer interaction may require more use of the action operators, whereas those in which the cognitive processes are not closely coupled with machine controls may make little or no use of these action-level operators.

The final element of COGNET is a notation for describing the person's *perceptual knowledge*. In Figure 2, perceptual processes are assumed to operate in parallel with the cognitive processes, and the information registered by perceptual processes is passed to the cognitive subsystem by being entered directly onto the problem representation blackboard. Thus, information enters the purview of the cognitive processes through spontaneous events or activities of the perceptual systems that POST objects onto the blackboard. These events are modeled in COGNET simply as production rules, of the form

IF <environmental event> Then POST <panel:level:attributes>.

Each of these rules is termed a *perceptual demon* or sometimes a *perceptual monitor*. A demon is spontaneously activated and executed whenever the corresponding sensory event, typically a verbal or visual cue, is sensed.

In COGNET, the person's *attention* resides, at any given moment, in some specific cognitive task that is being performed. Attention can shift in one of two ways. First, attention can remain at one task until it is *captured* by another task. This results when a change in the problem representation causes some second cognitive task to be activated (i.e., causes its trigger condition to be satisfied) and results in it having higher priority (as defined by its priority expression) than the currently executing task. When this occurs, the second task will capture the focus of attention from the first task, which will remain suspended until it, once again, has the highest priority or until its activation TRIGGER condition is no longer satisfied.[1] Second, attention can be suspended. A given procedure within a task can involve events or actions that involve expected delays (e.g., giving a navigational direction and then waiting for it to be carried out). Attention in the current task can then be deliberately suspended. This then allows whatever other task has highest current priority to gain the focus of attention. When the expected event occurs, the suspended tasks must then again compete to recapture attention from whatever task that has the focus of attention.

Data Collection and Analysis Methodology

The main purpose of the COGNET framework is to facilitate the cognitive task analysis and description of specific work domains.[2] The COGNET analysis notation described above is supported by a methodology for collecting, analyzing, and reducing empirical data on behavioral and cognitive processes so that they can be represented in the COGNET notation. The general approach is naturalistic, in that the behavior of an expert decision maker in a realistic problem-solving context constitutes the data used in the analysis. In more concrete terms, experts in the domain are typically asked to make a series of decisions, normally representative problems embedded in the form of scenarios in the domain of interest. Although this can be accomplished in either the actual environment or a simulated equivalent, the latter is usually chosen because it affords more experimental control. The selection of both the scenarios and the operational personnel must reflect the range of problem solving challenges posed by the actual operational environment and the range of strategies to meet those challenges. This scope ensures that the diversity and complexity of the environment will be captured by the COGNET model.

For each real or simulated scenario, the activities of each subject matter expert (SME) are observed and recorded for subsequent analysis in conjunction with verbal introspective data collected using knowledge elicitation methods. The verbal data, in the form of thinking-aloud protocols and question-answering protocols (obtained while reviewing the recorded behavior), are taken immediately after the problem or simulation has been completed. This is done because in high workload, time-constrained domains of the kinds studied here, taking the verbal protocol *during* the problem-solving process is too intrusive to be practical. Experience has shown, however, that high-quality protocols can be obtained in response

---

[1]This is the main attention process in COGNET and is based on the Pandemonium model originally developed by Selfridge (1959).

[2]This perhaps distinguishes it from other frameworks such as SOAR (Laird, Newell, & Rosenbloom, 1987; Newell, 1990) and EPIC (Kieras & Meyer, 1995), which have been principally developed as vehicles to develop and test psychological theory.

to recordings of actual behavior, particularly if they are taken shortly afterward when the problem is still fresh in the individual's mind. In these verbal protocols, individuals are asked to introspect and recount their internal decision process. Specific verbal probes are often made to clarify these accounts (at which time the replay of the problem is temporarily halted). These primary verbal data are supported by unstructured debriefs by participants and interviews and critiques by SMEs from the domain (especially instructors), particularly during the data analysis process.

This method places COGNET somewhere along a continuum between purely experimental methods and purely naturalistic ones. Experimental paradigms often rely on data-gathering techniques that use subjects who are unfamiliar with the problem domain and experimental problems that are deliberately artificial. This affords great experimental control as well as convenient statistical analysis, but excludes most or all features of domain-specific context. Naturalistic paradigms, on the other hand, typically rely on observation of behavior in its usual (or "natural") context, or verbal accounts of behavior in that context. This affords a great ability to understand the role of context in decision making and other cognitive processes, but usually excludes very detailed data needed for quantitative, computational, or statistical analysis.

Once the problem solving and verbal data have been collected, the analysis of this data proceeds. Identification of available motor actions (e.g., buttons to press, messages to speak, and displays to observe) and required perceptual demons (e.g., information content of visual displays, messages overheard, and other perceived events) proceeds as early as possible. These model elements are usually identified from system documentation, job descriptions, and information environment audits drawn from the recorded data and interviews. The initial stages of the analysis decompose the decision processes in the problem domain into a set of cognitive tasks that organize the decision maker's procedural knowledge and an initial problem representation structure (i.e., blackboard panels and levels). This is done by reviewing the sequences of problem-solving behavior (either through video or audio recordings or use of computer-generated problem replays) in conjunction with verbal protocols from the SMEs. For the task decomposition, the principal focus is on recreating expert-level, context-sensitive attention shifts among competing cognitive tasks. With regard to the blackboard structure, the focus is on identification of the primary principles for structuring domain knowledge to support decision making.

The detailed modeling of each task is undertaken, in general, as a traditional GOMS analysis. Unlike GOMS analysis, however, a COGNET analysis involves extensive definition of specific aspects of the task representations on the basis of the declarative knowledge incorporated into the blackboard structure. Thus, the blackboard structure is defined iteratively as its elements (i.e., concepts, their attributes, and semantic links relating them to each other) are needed to specify the trigger condition, priority expression, and GOAL conditions in the various cognitive tasks. As this is done, the necessary cognitive operators to post, unpost, and transform the concepts on the blackboard must be inserted into the tasks and perceptual demons. The final phase of the analysis involves reviewing, validating, and revising the model to increase its completeness and quality. A major means of doing this is through a series of walk-throughs using domain experts and (ideally) a set of problem scenarios not used in the initial data collection. The model is walked through against the scenario step-by-step, assessing with the domain expert whether the actions predicted by the model are valid and complete. The model is then revised as specific problems are identified, with the walk-through process continuing until an acceptable level of model validity is reached.

In the remainder of this chapter, we describe the COGNET analysis undertaken for the AAW domain that was the focus of the TADMUS program.[3]

Applying COGNET to the AAW Domain

The abstract methodology described above needed to be operationalized before it was possible to conduct a COGNET analysis of the AAW domain studied in the TADMUS program. In particular, it was necessary to do the following:

- identify a specific decision maker or set of decision makers to analyze,
- define a setting in which baseline decision performance could be observed and recorded, and
- collect and analyze an appropriate body of data on AAW decision performance using the COGNET method and description language.

This customization of the general methodology to the AAW problem is discussed below.

Selecting an AAW Role to Analyze and Model

AAW decisions are made at multiple organizations by multiple individuals in a naval battle group. Even with the TADMUS focus on ship self-defense, many individual roles are involved in the AAW process. The commanding officer (CO) and tactical action officer (TAO) have overall tactical responsibility for all ship activities and so make AAW decisions. Within the combat information center (CIC) several teams focus on different tactical areas, one of which is AAW. The AAW team consists of many roles that deal with specific parts of the AAW problem, such as identifying tracks, observing and interpreting electronic emissions from tracks, controlling combat air patrol (CAP), which is a patrol of fighter aircraft assigned to an AAW role, and so on. The activities of all the individual roles are integrated and organized by the AAW coordinator (AAWC).

---

[3]This analysis is reported in greater detail by Zachary, Zaklad, et al. (1992).

The COGNET analysis was focused on the AAWC for several reasons. The main function of the individual standing watch as AAWC is the coordination of ownship AAW resources to defend ownship (and to defend the battle group as well, if so directed). The AAWC works under the command of the ownship CO/TAO, and under control of a battle group AAW commander in battle group AAW operations. The AAWC is the highest level individual on the AEGIS ship who is concerned with just AAW. The AAWC performs this job mainly through interactions with the computer-based workstation, hooking (i.e., selecting) and interrogating (i.e., requesting data on hooked) track symbols on the display, moving attention among different parts of the tactical scene, trying (on the one hand) to actively push the process through the detection-to-engage cycle for individual tracks, and (on the other) to coordinate the use of resources for AAW, usually through voice traffic. At any given time, the AAWC must choose which of many tracks or tasks to attend to, internalize implications of decisions and actions of others on the AAW team, and infer the intents and implications of the enemy and the overall tactical situation as shown on screen.

Defining a Data Collection Setting

A major problem for naturalistic methodologies such as COGNET is the difficulty in gathering data on realistic, *in situ* behavior. Purely introspective approaches such as the critical decision method used by Klein and colleagues (e.g., Klein et al., 1989) avoid this problem by relying only on recollections of past decision situations. COGNET analysis, however, relies on the long-standing protocol analysis approach of integrating verbal introspective accounts with physical task performance. Thus, the analysis needs to work with, as primary data, expert level individuals solving realistic problems in the setting in which those problems are usually encountered. In domains such as AAW, this can create some problems in data acquisition.

Formerly, there were several ways in which data on experienced AAW teams working realistic AAW problems could be gathered. One was onboard Navy ships. The logistics involved with this and the intrusiveness of the data-recording process to shipboard life ruled this option out almost immediately. Other options all involved simulated problem solving. One simulation possibility was the Decision-Making Evaluation Facility for Tactical Teams (DEFTT) simulation system that was developed for the TADMUS program (see Johnston, Poirier, & Smith-Jentsch, this volume, chap. 3). However, DEFTT had drawbacks, most notably the fact that it was still under development at the time the data for this analysis had to be collected. Even if it had been available, though, two problems still remained. First, it was believed that DEFTT posed too low a fidelity in mission and (particularly) workstation simulation. A model build with DEFTT might be complete, but would be difficult to apply back to the more complex shipboard systems for which the analysis was ultimately directed. Second, DEFTT contained no communication networks (internal to the

CIC or external to the ship), thus removing one major source of perceptual cues for the operators.

Another simulation source was the embedded training simulations used for onboard training of ship crews. This option, however, raised the same logistical problems as the original onboard data collection option. The one remaining option—shore-based, team-training simulators—fortunately provided a workable solution. These simulators use high-fidelity mission simulations and the same workstations as used on shipboard, contain both internal and external communication networks, run realistic scenarios of varying complexity, and have powerful built-in facilities for recording data on operator performance. There were still difficulties, of course. Team training occurred at irregular and often widely spaced intervals, and access to the facility required complex scheduling arrangements. When a team was in the facility, its main purpose was training, and other needs (such as COGNET data collection!) had to be piggy-backed onto the schedule of the team and the data collection plans of the training facility staff. In general, however, the team-training simulator provided ample opportunities to observe and record data on experienced teams solving realistic AAW problems in their "natural" setting.

Data Collection and Analysis

The COGNET analysis of the AAWC role is based on the following specific data sources and types:

- observations and recordings of AAWC actions during parts of four, 3-week, team-training classes at a high-fidelity team-training simulator;[4]
- debriefings by participants in observed team-training sessions;
- verbal protocols collected in the context of replays of the recordings of observed AAWC behavior; and
- reviews and walk-throughs of the evolving analysis by a variety of experts in AAW, Navy tactical experts, members of several ships' crews, and civilian AAW trainers.

The observed behavior was recorded on videotape, which captured

- the AAWC video displays' contents, including the spatial display of tracks from the radar display and most of the textual data from the character read-out display;
- voice communications, including the ship's internal command network communications and the cross-ship AAW network plus much of the AAWC's direct voice communications with the other AAW team members seated nearby;[5] and

---

[4]Specifically, the simulator at the Combat Systems Engineering Development Site in Moorestown, New Jersey, was used.

[5]The AAWC typically listens to *both* of these networks simultaneously using the headset, which can receive a different communication channel on each earpiece!

- physical interaction with console and environment, including a wide-angle view of AAWC seated at the console.

The analysis was conducted using the methodology described above, with one modification. The verbal introspective data were collected not from the original individuals who performed the tasks, but rather from a second group of experts. This was done primarily because the original participants were generally unavailable for the verbal data collection process. The observed behavior that was recorded was collected as part of operational team training at a working Navy training facility. The team members whose actions were observed simply had to return to other duties shortly after the training simulations were completed and did not have time to participate in reviews and verbal protocol procedures. Although this was questionable from a theoretical standpoint, the logistical problems of doing it any other way (e.g., collecting recordings of observed behavior with experienced teams using realistic equipment and scenarios onboard ship) were insurmountable. In lieu of the original participants, a group of mastery-level experts were obtained from another source,[6] and used to collect the verbal data. As experienced trainers, these individuals were cognitively familiar with the task of observing, inferring, and commenting on the decision processes of individuals standing watch in the AAWC (and other AAW) role. The only modification of the methodology discussed above was that each expert saw each recording twice: the first time without comment to familiarize himself with the problem and what the participant did and the second time immediately afterward to provide the verbal protocol.

The verbal data were then analyzed to identify cognitive tasks and the procedural bodies of these tasks, a blackboard structure, and a set of perceptual demons that linked the cognitive processes to the external audio and visual cues. The results of this analysis are given in the next section.

Analysis Results: The AAWC COGNET Model

The result of a COGNET cognitive task analysis, as discussed earlier, is a COGNET model that represents the underlying knowledge that an expert maintains and uses to make and implement decisions in a specific domain. Following the structure of the conceptual framework and description language, a COGNET model has three key parts:

- the blackboard;
- the set of cognitive tasks, each of which includes a definition of its associated triggers, priority expression, and procedural bodies; and
- the set of perceptual demons that filters relevant information from the environments and posts it on the blackboard.

---

[6]The experts were from the Navy's Surface Warfare Development Group.

THREAT RESPONSE MANAGEMENT PANELS

Threat Status

| Killed Tracks |
| Engaged Tracks |
| Engageable Tracks |
| Action Tracks |
| Interest Tracks |
| Unknown Tracks |

Plans

| Posture and Strategy |
| Rules of Engagement |
| Preplanned Responses |
| Expected Indicators |
| Coordinations |
| Resource Assignments |
| Resource Actions |

SITUATION PANELS

Geo-Political Picture

| Political Situation |
| Areas and Boundaries of Interest |
| Geography and Physical Environment |

Tactical Picture

| Missions and Objectives |
| Threat and Movement Axes |
| Groups and Patterns |
| Tracks |
| Track Data Elements |

Resource Status

| Ownship Weapons and Countermeasures |
| Ownship Sensors |
| Other Controlled Assets |
| Computer Systems |
| Communication Nets |

Team Relationships

| Battle Group |
| Ownship |
| AAW Team |

Figure 3. Anti-air warfare coordinator (AAWC) COGNET (Cognition as a Network of Tasks) model blackboard.

Descriptions of each of these three components are provided next. Following these is an example of a realistic AAW situation describing how that situation would be modeled dynamically.

Blackboard. Figure 3 shows the panel/level organization of the blackboard portion of the AAWC COGNET model. The structure of this blackboard implies a conceptual framework used by the AAWC for organizing domain knowledge and implies a strategy for applying the knowledge in job conduct. The contents of the blackboard at any point in time represent the AAWC's dynamic understanding of the tactical situation and his or her plan for handling it. This blackboard is composed of six panels where different categories of concepts get posted. The blackboard organization corresponds to the two primary aspects of the AAWC's job of monitoring and assessing the tactical situation and evaluating and responding to threats. Threat response management includes evaluating air targets for threat status, determining those to engage, monitoring the progress of engagements, as well as plans for engaging threatening targets and determining that they have been destroyed. The two panels dealing with threat response management are

- *Threat status*: information on tracks that are potentially or actually threatening; must be monitored or acted on or used in an engagement

- *Plans*: strategies and plans for responding to anticipated or actual threats.

Situation assessment includes maintaining an understanding of the tactical picture; its geo-political context; the status of resources needed to obtain information or conduct an engagement; and ownership and force command structure, coordination agreements, and communication links. Four panels represent the AAWC's understanding of the tactical situation, as follows:

- *Geo-political picture*: information on the geo-political context in which the battle/watch takes place,
- *Tactical picture*: snapshot of the current status of an evolving battle/watch
- *Resource status*: status of all resources needed or under control
- *Team relationships*: the AAWC's relationships with other players with whom he or she must coordinate, including methods of communication and coordination

Each of the six panels are composed of a number of levels on which specific concepts are posted. The individual concepts posted in each level are structured by specific attributes. The detailed definition of one blackboard panel is given below to provide a flavor of the blackboard contents. Each description takes the form of a specification, for each level on that panel, of the structure of concepts posted at that level, as

> [main attribute, modifying attribute1, modifying attribute2, . . .
>   modifying attribute n]
> optional parameters are given as

The threat status panel contains information on tracks that are potentially or actually threatening and that must be monitored or acted on because of their relationship to a threat (e.g., a friendly track whose position must be considered in threat planning). A separate panel, the tactical picture panel, contains all tracks in the area of operation. The threat status panel contains only a subset of the tracks contained on the tactical picture panel. Thus, it is a mental construct for reducing the amount of information that must be attended to by tagging those tracks that are of interest, in that they must be monitored or some action must be taken regarding them. The ordering of the levels within a given panel often can provide further organization of the knowledge situated on that panel. For example, the ordering of the levels often represents a *constructive* process whereby a solution to some significant decision problem or aspect of the overall situation is constructed by building the solution knowledge. In this panel, the level hierarchy represents a progression, from bottom to top, in understanding what, if any, threat the track poses and progress in eliminating the track as a threat.

The six levels of the threat status panel are shown in Exhibit 1. New

Exhibit 1.   Threat Status Panel

Killed tracks: tracks that have been destroyed
  [track #, location, weapon]
Engaged tracks: tracks that have been engaged
  [track #, location, characteristic(s) of interest, weapon engaged with]
Engageable tracks: tracks that have met the rules of engagement
  [track #, location, characteristic(s) of interest, weapons capable of engaging]
Action tracks: tracks requiring some action to be taken (e.g., warnings, reports)
    [track #, location, category (air, surface, subsurface), classification (hostile,
    neutral, friendly, etc.), characteristic(s) of interest, action needed]
Interest tracks: tracks that must be monitored because they are potential threats
    or are friendly tracks that the AAWC must be aware of for coordination in
    an engagement
    [track #, location, category, classification, characteristic(s) of interest]
Unknown tracks: unevaluated tracks
    [track #, category, location, status (new, unknown)]

Note.   AAWC = anti-air warfare coordinator.

or unevaluated tracks are posted on the bottom level by a perceptual demon. The tasks of "manage battlespace" and "evaluate track" include cognitive operations to evaluate tracks that are or could be threats. Those
that the AAWC determines must be monitored are promoted to the interest
track level by unposting them on the unknown tracks level and posting
them on the interest track level with the appropriate attributes added. If
at any time a response is required to a track behavior or threatening intention, the track is promoted to the action track level. If unknown tracks
are identified as commercial air or if a potentially hostile track turns away,
these tracks are deleted from the threat status panel, although they would
remain as part of the tactical picture panel. When a track meets the rules
of engagement and is classified as hostile, it is promoted to the engageable
tracks level, which triggers the task "take track." Once the engagement
has begun, the track is promoted to the engaged tracks level, and if it is
killed, it is promoted to the killed tracks level. Messages for killed tracks
only remain on the blackboard until the evaluation is complete and the
track is deleted from the system.

Cognitive Tasks

The overall role of the AAWC is to monitor and evaluate air targets for
threat value and engage and destroy all air threats (under direction of the
CO or TAO and battle group AAW commander). In a COGNET analysis,
tasks are the primary units of cognitive activity and are defined as a single
unit of goal-directed activity that would be performed to completion if un-
interrupted. Thus, each cognitive task encapsulates a logically self-
contained procedure, which is formalized as a set of subgoals that are
sequentially pursued to attain the overall task goal. Ten tasks resulted
from the COGNET analysis of AAWC. They are as follows:

1. Manage battle space: Scan tracks in larger context of evolving scenario.
2. Evaluate track: Identify/classify an individual track in terms of its tactical significance.
3. Plan specific threat response: Plan a response to a specific track that is a potential or actual threat. The plan may include specific actions to take if the track becomes hostile at any of various points along its projected path.
4. Plan posture for expected threat: Plan posture and strategies for handling expected classes of threats, includes determining assets needed for expected threats, establishing preplanned responses in accordance with battle orders, and understanding of the geopolitical situation.
5. Neutralize/control potential threat: Get a potentially hostile track to conform to your needs/wishes.
6. Cover track: Map a specific tactical response/targeting solution to a track of interest.
7. Take (engage) track: Engage in a hostile track.
8. Position AAW assets to maintain desired posture: Position assets in accordance with plan.
9. Manage combat air patrol status: Monitor and maintain combat air patrol in readiness for expected or actual threats, includes maintaining adequate fuel and weapon load.
10. Manage resources: Monitor and maintain AAW resources, including sensor, weapon/countermeasure, computer, and communication systems.

A task description includes the task name, which defines the goal associated with the task, and the trigger, which defines the conditions under which the task is activated for performance. The body of the task is described in the COGNET description language reviewed above. The format for representing references to the blackboard in the triggers, goal/subgoal conditions, and cognitive operators (POST/UNPOST/TRANSFORM) is

[<object> posted on PANEL: *Level*]

The full blackboard panel names (from Exhibit 1) are abbreviated in these PANEL:*level* references as follows: threat status abbreviated as THREAT and tactical picture abbreviated as TAC PIC.

The analysis only decomposes the tasks into subgoals and, in some cases, sub-subgoals, thus giving the primary skeletal structure of the model.[7] Where appropriate, cognitive and communicative operators are included to illustrate how changes are made to the contents of the black-

---

[7]This level of detail is sufficient to indicate the content of the cognitive processing within each activity without providing an overwhelming level of detail. It also allows the cognitive organization of the task to be examined without reference to the (sometimes sensitive) details of the particular combat system involved.

board and how task subrogation occurs. Part of 1 of the 10 individual task models, evaluate track, is as follows:

> **TASK: Evaluate Track** . . . *IF <new track> posted on THREAT: Interest Track OR query/command regarding specific track or locational fix OR (<new track data> posted on TAC PIC: Track Data Elements AND track on THREAT: any level)* **OR lost track**
> GOAL: Locate track
> GOAL: Review track data
> Operator: Determine how much time available to evaluate track (distance from ownship)
> GOAL: Assess intentions and threat capability of track . . . *if time available*
> > Operator: TRANSFORM <track data> on TAC PIC: *Track Data Elements to* <tracks> on TAC PIC: *Tracks*
> GOAL: Determine if track is part of group or pattern . . . *if time available*
> > Operator: TRANSFORM <tracks> on TAC PIC: *Tracks to* <track groups> on TAC PIC: *Groups and Patterns>*
> GOAL: Determine composition of group . . . *if part of group or pattern*
>
> • • •

This task has a number of conditions under which it is appropriate, one of which is that a new track has been posted on the "interest track" level of the threat status panel of the blackboard. The body of the task is composed of goals, including "locate the track [on the display]" and "review the track data [shown on the display]." Some cognitive operators are included that indicate decision processes (e.g., "TRANSFORM <track data> on TAC PIC: *Track Data Elements* to <tracks> on TAC PIC: *Tracks*" indicates that the AAWC would at that point in the task evaluate data about a track from multiple potential sources and form an integrated picture of the track, posting the result on a higher level.

Perceptual Demons

Perceptual processing involves translating sensed information to symbolic terms and making it accessible to the cognitive system by posting it on the problem representation blackboard. Each perceptual demon describes *how* a specific class of cue is processed by the perceptual system, indicating what information is posted or transformed on the blackboard as a result of the processing of a specific cue. Once the information is on the blackboard, it may affect the flow of attention because the task triggers are based on patterns of information on the blackboard. This provides the mechanism for situational changes to affect selection and sequencing of tasks (put differently, this provides the mechanism for data-driven cognitive processes, whereas the task models provide for goal-driven processes).

The COGNET analysis of the perceptual processes of the AAWC identified 19 key perceptual demons, which respond to either visual cues (dis-

Exhibit 2. Perceptual Demons in the Anti-Air Warfare Coordinator (AAWC) COGNET (Cognition as a Network of Tasks) Model

• Radar acquires new track	• Radar loses track
• New track acquired through datalink	• Datalinked track lost by reporting unit
• Combat air patrol reports of track behavior	• AAW readiness reports
• State reports	• Doctrine setup
• Link reports	• Electronic warfare reports (from electronic warfare supervisor)
• Identification reports (from identification supervisor)	• Reports from air intercept controller, air control supervisor
• Missile system supervisor reports	• Battle group AAW commander orders
• Tactical action officer orders	• Commercial air track coming out of commercial air corridor
• Course change report (from tactical action officer)	• Query about track, track group, or engagement
• Change in weapon/warning status	

Note. AAW = anti-air warfare.

play events at the workstation) or auditory cues (voice communication from team members), as listed in Exhibit 2. In the COGNET notation, each demon is modeled as a production rule in which the sensing of a specific cue is the antecedent condition and a (possibly conditional) blackboard operation forms the consequent perceptual process. For example when the event is a radar's acquisition of a new air track, the resulting demon would be formalized as

> IF air track
> POST: "New air track [track number] at time [mission-time] held by Radar with Bearing [bearing], Course [course], Range [range], Speed [speed], Altitude [altitude], and Track Quality [track quality]" on TAC PIC:*Tracks* and THREAT:Unknown *Tracks*

Model Dynamics

The structure and content of blackboard, cognitive tasks, and perceptual demons components define the knowledge that the person needs to perform the AAWC job. However, a COGNET model is intended to be used dynamically as well and can be used to analyze how this knowledge would be applied dynamically in the context of a specific problem situation.[8] The analyst can trace arbitrarily long threads of information processing, up to and including an entire problem, through the model. For example, a set

---

[8]At the time of this research, the underlying mechanism had not yet been reduced to a fully executable architecture. Subsequently, however, an executable architecture was created (Zachary, Le Mentec, & Ryder, 1996) and is being used to create fully executable versions of the model described here, as discussed in the conclusion to this chapter.

of display events would result in the corresponding perceptual demons being fired, which would result in certain information being internalized and posted on the blackboard. As soon as there is information on the blackboard, the analyst can check (each time the blackboard changes) to see if any cognitive task has been triggered. If it has, the procedural knowledge in that task can be stepped through, one goal, method, or operator at a time. These procedural traces will proceed through the task, executing the behavioral operators (e.g. PERFORM function) and thus indicating which observable actions should be taken as a consequence of the internal decision process. Each time the blackboard changes, either through the firing of a new demon or through execution of a post, unpost, or transform operator, all of the cognitive task triggers will have to be reexamined to determine whether another task has been activated for execution. If and when this happens, the priority formulae of the newly triggered tasks will have to be evaluated and compared with the priority formula of the currently active task to determine whether one of the newly activated tasks would have more priority in that context and therefore capture attention away from the currently active one.

A simplified example of how this model works might start with a perceptual demon posting a message (in the "unknown tracks" level of the "threat status" panel) that a new unknown track has been spotted on the AAWC's display. This message might then trigger the "evaluate track" task, if it had a high enough priority at that point. This task would use other messages from other panels and levels to decide that this track is interesting and post a message to the interest tracks level of the threat status panel. Then, the task might determine to take some action on the track (e.g., investigate with CAP) and post a new message to the "action tracks" level (removing the relevant interest tracks message) and post a CAP investigation plan message to the resource assignments level of the "plans" panel. The task would then likely release (or lose) AAWC's attention, waiting for the plan to be fleshed out and implemented. Once a visual identification from the CAP is received and perceived (triggering another perceptual demon to post a message about it), the "evaluate track" task might be triggered again (assuming it has high enough priority). The fact that the track is a commercial airliner would cause removal of the message about this track from the threat status panel and posting of an updated message (including the fact that it is an airliner) to the tracks level of the "tactical picture" panel as a track that no longer has to be monitored.

Lessons Learned

1. Although the work done by the AAWC and other personnel in the ship's CIC certainly focused on the computer-based watchstation, we learned that that work is even more strongly tied to the dynamics of the larger decision-making team. Although there is always a specific individual in command of the team, all decision making is not performed by that individual. The complex nature of the information, assets, and tasks inherent in AAW require that multiple individuals be involved in the

decision-making and command-and-control process.[9] These interdependencies also require that information must flow to many places and that many individuals must coordinate their activities to avoid chaotic or ineffective operations. The Navy doctrine of "command by negation" also plays a key role in distributing decision making across a team. Under the command-by-negation philosophy, subordinates are responsible for making tentative decisions and announcing them as intentions that are passed upward on the command-and-control hierarchy; the superior individual may then accept the decision by saying nothing or countervail it by explicit negation (which is often followed by an alternative decision or intention).

Analyzing this type of team-situated decision making requires advancing the COGNET theoretical base beyond a traditional human–computer interaction framework to one that deals with team-based interactions as well. Two specific conceptual modifications were made to the COGNET framework to accommodate team-situated decision making. The first was a broadening of the concept of perceptual knowledge beyond simple workstation-based cues (displays and auditory alerts) to include communications with other individuals on the decision-making team. This expansion required the cognitive task analysis notation to deal with loosely structured linguistic information as well as the highly structured display and alert information. The second broadening of the theoretical framework was to expand the notion of the mental model from a representation of just the problem to a representation of the problem and of the team. In team settings, people maintain declarative knowledge about the team and its structure, organization, and roles, along with knowledge about the problem being solved, and we found that this knowledge is just as important in organizing the decision process as is knowledge of the tracks, rules of engagement, and so forth.

2. Although there is a strong tradition of using behavioral task analysis in the design of systems and (manual and automated) operator aids, we found that cognitive task analyses provided different kinds of data that have to be applied in different ways. We learned that cognitive analysis results were able to affect design at a much deeper level than traditional behavioral analyses, which are most easily applied to surface features of the system design, particulary person–machine interface layout and functionaltiy. During the course of this research, various means were explored for using the cognitive task analysis data to direct and inform the design of the decision-support system and human–computer interface (HCI) for AAW decision makers.

The ways of using cognitive models in Decision Support System (DSS) and HCI design were codified as a set of design principles (see Zaklad & Zachary, 1992). These principles, which concerned both functionality of the DSS and its structure, were articulated at an abstract level for general use but also were tailored to TADMUS application at a more specific

---

[9]This is further complicated by the intertwining of AAW and other aspects of ship and battle group operations, such as anti-surface warfare, anti-submarine warfare, and air resources coordination.

level by more detailed principles derived from the content of the COGNET AAW model. For example, one general principle was that the DSS should support coordination among team members. At this level, the principle could apply to any team-oriented DSS, but additional, more detailed principles relating it to the TADMUS—AAW case were derived. Detailed specifics of this principle suggest that the coordination support specifically focused on:

- the sharing of mental models of team activities (derived from the blackboard representation),
- the transmission and acknowledgment of intentions across the AAW team because of the prominent role that such communications play in the timely triggering of cognitive tasks, and
- helping each operator in the team know when it would be appropriate to notify or ask for permission from higher authorities.

3. The depth of understanding of the decision-making process that was provided by the AAW COGNET model was also useful in support of training. Because the COGNET model provided a model of an expert-level decision strategy, it can be used analytically to derive observable characteristics of the decision process that can in turn be codified and incorporated into performance assessment and measurement instruments. A team outcome instrument, the Anti-Air Team Performance Index (ATPI), identified preliminary outcomes from the COGNET cognitive task goals and subgoals (Dwyer, 1992). Preliminary individual outcomes contributing to the team outcomes were also determined and codified into an individual outcome measurement instrument, the Sequenced Actions and Latencies Index (SALI), drawing on the same COGNET components for the AAWC portion (Johnston, Poirer, & Smith-Jentsch, chap. 3, this volume).

Because COGNET models of expertise can be used as a benchmark of expertise at a particular position, they can be used to derive measurement instruments for performance assessment. Specifically, the cognitive tasks indicate actions that should be performed and the contexts in which they should be performed. Thus, performance predictions can be derived for a given training scenario.

4. In the course of applying the paper-and-pencil cognitive task analysis in the ways described, we learned that there were many potential applications of a cognitive model that could not be accomplished with a paper analysis alone, but that instead required an executable version of the model. An *executable* model refers to a software program that can simulate the cognitive, perceptual, and motor activities of a person. Such an executable model can be embedded inside the software of a training system or decision support system to provide many new capabilities, including the following:

- generating expectations of desired trainee performance, against which actual performance is measured;

- identifying the possible cognitive bases for trainee performance that does not achieve the expected levels;
- determining possible content for training interventions to improve training performance in the future; and
- acting as surrogate operators in team training settings when all roles cannot be staffed by human operators.

During the TADMUS program, such a capability to execute COGNET models did not exist. However, through the efforts to apply it to team training and DSS design, we learned the importance of having such a capability for future use.

Executable cognitive models embedded in training or decision support systems promise to enhance the capabilities of such systems. Thus, development of executable cognitive architectures capable of generating executable models may enhance future systems. The research described here stimulated the development of an executable software version of the underlying cognitive architecture of COGNET as shown in Figure 2 (see Zachary et al., 1996, for details). The executable architecture provides a software environment in which a COGNET model can be executed, given appropriate sensory stimuli and a (digital) means to implement its actions. When an executable model of the AAWC is provided with the visual and auditory stimuli available to a person at the AAWC watchstation, and is given the ability to implement its actions at the workstation digitally, the model is able to emulate (expert-level) human decision-making performance in that problem.

Conclusion

As a final point in this chapter, two observations are offered about the development of executable cognitive models. The first is simply that the required data collection, model-building, and model testing processes become much more complex when the result of a cognitive task analysis becomes a piece of executable software. The original COGNET analysis of the AAWC yielded a paper-and-pencil model that was intended to be read, understood, and applied by a human system or training designer. The model builders could therefore rely on the abilities of the human reader of the model to deal with minor inconsistencies, points that were implied but not explicit, and varying levels of detail. This is not so when the reader of a model is a piece of software that emulates an underlying cognitive architecture. In that case, the model-building process becomes as painstaking as any other software development effort. In particular, each goal within each task has to be fully decomposed to the point that every keystroke needed to work an AAW problem at the AAWC watchstation is included. All ambiguity must be removed, everything must be made explicit, and all details must be specified to the same level or the model will not be able to execute as intended. The result is that the cognitive task analysis effort expands greatly (perhaps by as much as a factor of two!) but also blends into the software development sphere much more so than when the result was (only) a paper-and-pencil analysis.

This leads to the second observation. With the advent of executable cognitive architectures, which include not only COGNET but others such as SOAR (Laird, Newell, & Rosenbloom, 1987), EPIC (Meyer & Kieras, 1996), and ACT-R (Anderson, 1993), cognitive task analysis may be poised to assume a much more prominent role in the development of future systems for training and decision support. The executable architecture allows the (more extensive) cognitive task analysis to be rapidly and directly transitioned into working software that can be incorporated into the application itself, in addition to its existing use as an analytical support for design and evaluation. Such a transition does not eliminate the current uses of cognitive analyses and cognitive models in system design, but rather enhances these roles by integrating cognitive analysis and modeling into most phases of the system life cycle, including the three largest: implementation, maintenance, and support. The changes that this enhanced role can bring are potentially enormous, although the full implications are perhaps difficult to imagine.

References

Anderson, J. R. (1993). *Rules of the mind*. Hillsdale, NJ: Erlbaum.

Australian Civil Aviation Authority. (1994). *Air Traffic Control Abinitio Training Program*. Canberra, Australia: Civil Aviation Authority, Air Traffic Services, ATS Human Resources.

Barwise, J., & Perry, J. (1983). *Situations and attitudes*. Cambridge, MA: MIT Press.

Card, S., Moran, T., & Newell, A. (1983). *The psychology of computer–human interaction*. Hillsdale, NJ: Erlbaum.

Carver, N., & Lesser, V. (1994). Evolution of blackboard control architectures. *Expert Systems with Applications, 7*, 1–30.

Chi, M. T. H., Glaser, R., & Farr, M. J. (1988). *The nature of expertise*. Hillsdale, NJ: Erlbaum.

Cohen, L. J. (1981). Can human irrationality be experimentally demonstrated? *The Behavioral and Brain Sciences, 4*, 317–370.

Dwyer, D. J. (1992). An index for measuring naval team performance. *Proceedings of the Human Factors Society 36th annual meeting* (pp. 1356–1360). Santa Monica, CA: The Human Factors Society.

Englemore, J., & Morgan, T. (1988). *Blackboard systems*. Reading, MA: Addison-Wesley.

Ericsson, K. A., & Kintsch, W. (1995). Long-term working memory. *Psychological Review, 102*, 211–245.

Ericsson, K. A., & Smith, J. (Ed.). (1991). *Toward a general theory of expertise: Prospects and limits*. Cambridge, England: Cambridge University Press.

Essens, P., Fallesen, J., McCann, C., Cannon-Bowers, J., & Dorfel, G. (1995). *COADE—A framework for cognitive analysis, design, and evaluation* (Technical Report No. AC/243). Brussels, Belgium: NATO Defence Research Group.

Hayes-Roth, B. (1985). A blackboard architecture for control. *Artificial Intelligence, 26*, 251–321.

Hayes-Roth, B., & Hayes-Roth, F. (1979). A cognitive model of planning. *Cognitive Science, 3*, 275–310.

Hoffman R. R. (1992). *The psychology of expertise: Cognitive research and empirical AI*. New York: Springer-Verlag.

Kieras, D. E. (1997). Task analysis and the design of functionality. In A. Tucker (Ed.), *The computer science and engineering handbook* (pp. 1401–1423). Boca Raton, FL: CRC.

Kieras, D., & Meyer, D. E. (1995). *An overview of the EPIC architecture for cognition and performance with application to human–computer interaction* (EPIC Technical Rep. No. 5 [TR-95/ONR-EPIC-5]). Ann Arbor, MI: University of Michigan, Electrical Engineering and Computer Science Department.

Klein, G. (1989). Recognition-primed decisions. In W. Rouse (Ed.), *Advances in man–machine systems research* (Vol. 5, pp. 47–92). Greenwich, CT: JAI Press.

Klein, G., Calderwood, R., & MacGregor, D. (1989). Critical decision method for eliciting knowledge. *IEEE Transactions on Systems, Man, and Cybernetics, 19,* 462–472.

Klein, G., Orasanu, J., Calderwood, R., & Zsambok, C. (Eds.). (1993). *Decision making in action: Models and methods.* Norwood, NJ: Ablex.

Kolodner, J. (1988). *Proceedings of the case-based reasoning workshop.* San Mateo, CA: Morgan Koffman.

Laird, J. E., Newell, A., & Rosenbloom, P. S. (1987). SOAR: An architecture for general intelligence. *Artificial Intelligence, 33,* 1–63.

Meyer, D. E., & Kieras, D. E. (1996). *A computational theory of executive cognitive processes and human multiple-task performance: Part 1. Basic mechanisms* (EPIC Report No. 6 [TR-96/ONR-EPIC-06]). Ann Arbor, MI: University of Michigan Press.

Newell, A. (1980). Physical symbol systems. *Cognitive Science, 4,* 135–183.

Newell, A. (1990). *Unified theories of cognition.* Cambridge, MA: Harvard University Press.

Newell, A., & Simon, H. (1972). *Human problem solving.* Englewood Cliffs, NJ: Prentice Hall.

Nii, P. H. (1986a). Blackboard systems: The blackboard model of problem solving and the evolution of blackboard architectures. Part One. *AI Magazine, 7,* 38–53.

Nii, P. H. (1986b). Blackboard systems: The blackboard model of problem solving and the evolution of blackboard architectures. Part Two. *AI Magazine, 7,* 82–106.

Pylyshyn, Z. (1984). *Computation and cognition.* Cambridge, MA: The MIT Press.

Rouse, W., Geddes, N. D., & Curry, R. E. (1987). An architecture for intelligent interfaces: Outline of an approach to supporting operators of complex systems. *Human–Computer Interaction, 3,* 87–122.

Ryder, J. M., Weiland, M. Z., Szczepkowski, M. A., & Zachary, W. W. (in press). Cognitive engineering of a new telephone operator workstation using COGNET. *International Journal of Industrial Ergonomics.*

Ryder, J. M., & Zachary, W. (1991). Experimental validation of the attention switching component of the COGNET framework. In *Proceedings of the Human Factors Society 35th annual meeting* (pp. 72–76). Santa Monica, CA: Human Factors Society.

Ryder, J. M., Zachary, W. W., Zaklad, A. L., & Purcell, J. A. (1994). *A cognitive model for Integrated Decision Aiding / Training Embedded Systems* (IDATES; Technical Report No. NTSC-92-010). Orlando, FL: Naval Training Systems Center.

Seamster, T. L., Redding, R. E., Cannon, J. R., Ryder, J. M., & Purcell, J. A. (1993). Cognitive task analysis of expertise in air traffic control. *International Journal of Aviation Psychology, 3,* 257–283.

Selfridge, O. G. (1959). Pandemonium: A paradigm for learning. In *Proceedings of the Symposium on the Mechanization of Thought Processes* (pp. 511–527).

VanLehn, K. (1996). Cognitive skill acquisition. *Annual Review of Psychology, 47,* 513–539.

Zachary, W., Le Mentec, J.-C., & Ryder, J. (1996). Interface agents in complex systems. In C. Ntuen & E. H. Park (Eds.), *Human interaction with complex systems: Conceptual principles and design practice* (pp. 35–52) Norwell, MA: Kluwer Academic Publishers.

Zachary, W., & Ross, L. (1991). Enhancing human–computer interaction through use of embedded COGNET models. In *Proceedings of the 35th annual meeting of the Human Factors Society* (pp. 425–429). Santa Monica, CA: Human Factors Society.

Zachary, W., & Ryder, J. (1997). Decision support systems: Integrating decision aiding and decision training. In M. Helander, T. Landauer, & P. Prabhu (Eds.), *Handbook of human–computer interaction* (2nd ed., pp. 1235–1258). Amsterdam, The Netherlands: Elsevier Science.

Zachary, W., Ryder, J., Ross, L., & Weiland, M. (1992). Intelligent human–computer interaction in real time, multi-tasking process control and monitoring systems. In M. Helander & M. Nagamachi (Eds.), *Human factors in design for manufacturability* (pp. 377–402). New York: Taylor & Francis.

Zachary, W., Zaklad, A., Hicinbothom, J., Ryder, J., Purcell, J., & Wherry, R., Jr. (1992). *COGNET representation of tactical decision-making in ship-based anti-air warfare* (CHI Systems Technical Report No. 920211.9009). Springhouse, PA: CHI Systems, Inc.

Zaklad, A., & Zachary, W. (1992). *Decision support design principles for tactical decision-making in ship-based anti-air warfare* (CHI Systems Technical Report No. 920930.9000). Springhouse, PA: CHI Systems, Inc.

Zubritzky, M., Zachary, W., & Ryder, J. (1989). Constructing and applying cognitive models to mission management problems in air anti-submarine warfare. In *Proceedings of the Human Factors Society 33rd annual meeting* (pp. 129–134). Santa Monica, CA: Human Factors Society.

14

Improving Decision-Making Skills Through On-the-Job Training: A Roadmap for Training Shipboard Trainers

J. Thomas Roth

As the size of the active Navy changes in response to the evolving world situation, and as force structure is realigned in response to new visions of the Navy's missions, readiness becomes increasingly important. Readiness for the Navy's missions depends largely on effective training.

In the future, more training will be accomplished afloat (i.e., shipboard) than has been the case in recent decades. The capability of shipboard personnel to conduct training will become increasingly critical.

Much afloat training is conducted using a team approach. Organized shipboard training teams, such as damage control training teams (DCTT), engineering casualty control training teams (ECCTT), and combat systems training teams (CSTT), are responsible for preparing ships' personnel to accomplish their roles under all conditions (see Smith-Jentsch, Zeisig, Acton, & McPherson, chap. 11, this volume). Highly qualified watchstanders or supervisors are normally assigned as members of training teams.

Afloat training teams are responsible not only for ongoing training of current team members but also for integrating new team members into shipboard teams. Newly assigned personnel—who may or may not have previously received team training—must be prepared to fill their roles on teams.

Whereas training team members are generally highly qualified technical experts in their occupational specialties, they may not have all of the skills needed to conduct training for a team, as a member of a training team. Many skills acquired in other instructional settings may transfer to the afloat team training situation, but added skills are often needed for personnel to function effectively as afloat trainers.

The main goal of this chapter is to summarize research that identifies critical skills for shipboard trainers. A secondary goal is to describe desirable characteristics of training to develop afloat trainers in their roles as trainers. Finally, the characteristics of a candidate training program to accomplish such training, and a proposed related research program, are presented.

Afloat Tactical Decision Making: Overview and Trainers' Roles

Afloat, tactical decision-making responsibility is distributed among members of combat systems teams (CSTs), with final shoot—no shoot decisions typically the responsibility of the commanding officer (CO) or the tactical action officer (TAO). These principal decision makers are supported in the combat information center (CIC) by several subteams, each of which is responsible for a specific warfare area (e.g., surface, air defense, submarine, strike), that in turn gain enormous amounts of information from numerous sensor systems (radar, active and passive sonar, electronic warfare, data link, etc.). Individual members of combat systems subteam members make critical decisions in their more bounded areas of expertise. In addition to competence in individual positional skills, it is crucial that CST members function with highly effective teamwork.

U.S. Navy-sponsored research (e.g., Burgess, Riddle, Hall, & Salas, 1992; Driskell & Salas, 1991; Glickman et al., 1987; Guerette, Miller, Glickman, Morgan, & Salas, 1987; Hogan, Peterson, Salas, & Willis, 1991; Orasanu & Salas, 1992) has identified a number of factors critical to effective teamwork. These include efficient and timely communication among team members, close coordination of team members' activities, understanding of one's own and critical others' roles in team functioning, and an acute understanding of the team's task and the contributions of one's own tasks to it. Much of the knowledge essential to teamwork may be represented to some extent as mental models constructed by team members (Cannon-Bowers & Salas, 1990; Converse & Cannon-Bowers, 1991; Rouse, 1991). Recent evidence (Duncan & Rouse, 1993) suggests that team members' shared mental models should contain compatible elements that support teamwork functions.

Afloat instructors will be required to convey teamwork skills, as well as position- and specialty-specific individual skills, to team members. Initiatives developed under TADMUS and other Navy-sponsored research provide tools, techniques, and guidelines that contribute to teamwork training (see Blickensderfer, Cannon-Bowers, & Salas, chap. 12, this volume; Kozlowski, chap. 6, this volume; Serfaty, Entin, & Johnston, chap. 9, this volume; Smith-Jentsch, Zeisig, Acton, & McPherson, chap. 11, this volume). Afloat instructors will be charged with conveying teamwork-critical instruction and principles to CST members. Some form of training that enables the afloat instructor to accomplish this task should be provided.

Related research has established the centrality of situation awareness and assessment to decision making in many contexts (Klein, 1992; Zsambok et al., 1992). Prototype methods for enhancing information displays and providing situation assessment and decision-making aids have been developed and tested with some success (Hutchins, 1996; Morrison, Kelly, Moore, & Hutchins, chap. 16, this volume). CST members may in the future use such enhanced displays and decision aids embodying the same or similar principles to perform some aspects of their tasks. To effectively use these tools, CST members will require training (some likely provided afloat) beyond that now provided. Afloat instructors may themselves require specialized training to help CST members acquire the requisite knowledge and skills.

Wider Implications

Training objectives and programs such as those described in this chapter may be applicable in other on-the-job training (OJT) contexts. Although no attempt is made here to compare the training needs of Navy shipboard trainers with those of trainers in other OJT situations, such an analysis would likely reveal many similarities. Many training objectives for OJT trainers discussed here represent objectives that can be appropriate for training OJT trainers in other contexts. That is, trainers in other OJT situations may require much of the same general skills and knowledge (with, of course, differences in detail to suit particular situations) required by Navy shipboard trainers. Additional discussion of this prospect appears at the end of this chapter.

The Roadmap Effort

The research discussed here had the objective of creating a foundation for development of a comprehensive deployable (i.e., exportable, stand alone) training program for afloat trainers. Training for afloat training-team members is currently carried out principally by shore-based organizations. As noted elsewhere (Collyer & Malecki, chap. 1, this volume), the focus of operational and tactical training will, in the future, continue to shift to the shipboard domain; at least some instructor training is likely to do so as well. Afloat training teams may to be required to self-train to some extent, including the task of building afloat-trainer skills for new training-team members. Currently, some support is beginning to be provided for afloat training team self-training, but additional support for the self-training function is needed. This work provides an initial framework and suggests a development path for research leading to the creation of comprehensive, deployable training support for afloat trainers.

Research Elements and Results

Several sequential research elements were performed to attain the goal of this effort: creating a foundation for the future development of deployable training programs for afloat training teams. These elements and their associated results are discussed below.

Knowledge Base Synthesis

One assumption in this work was that the TADMUS program and allied Navy research would produce results that could alter how shipboard personnel perform their duties. This would mean that alterations in training goals and methods might also be necessary. Therefore, likely products of Navy-sponsored research work at large were reviewed and synthesized, and their potential impacts on performance and training were assessed. The summaries and synthesis were documented in a working paper (Evans & Roth, 1993).

From the synthesis, we concluded that afloat trainers may be required to perform several novel kinds of training as a result of possible adoption of research-based interventions. These include the following:

1. training to enable members of decision-making teams to develop appropriate mental models of (a) situations, (b) unique and reciprocal roles and responsibilities of team members, (c) global and situation-relevant experience, and (d) other critical declarative and strategic knowledge elements on which performance may depend (Afloat instructors need methods to enable them to identify when defects in individuals' mental models are responsible for less than adequate performance, and to administer training to re-form faulty mental models into more nearly correct ones.);

2. training to understand and use innovative and advanced information coding and display techniques derived from decision-aiding and interface-design research (see Morrison, Kelly, Moore, & Hutchins, chap. 16, this volume); (Afloat instructors may be responsible for providing training to enable team members to effectively use improved displays and decision-aiding features of shipboard systems.); and

3. training to develop essential elements of teamwork (Formal, organized, teamwork-building training per se is only now beginning to be provided to Navy shipboard teams [see Smith-Jentsch, Zeisig, Acton, & McPherson, chap. 11, this volume]; afloat training teams may become the principal providers of teamwork training.)

Preliminary Assessment of Instructor Responsibilities

Concurrent with the synthesis of the knowledge base, a preliminary assessment of afloat-instructor responsibilities was performed. This assessment was developed from two sources:

1. a logical analysis of the duties, responsibilities, and tasks of instructor personnel in a closed, hands-on training system. (By *closed* we mean an instructional system that is largely self-contained, with little or no routine recourse to outside assistance for training resources or guidance. The afloat-training situation is

a conceptually closed one, although external resources are sometimes available to afloat trainers. We assumed that afloat trainers are essentially on their own, particularly when deployed.)

2. observation of organized precommissioning team training for Combat Systems Team (CST) members from an AEGIS cruiser. This included direct observation of all stages of training conducted during this shore-based event. These included
   - pre-exercise briefing development and presentation
   - setup and preparation for training exercises
   - exercise conduct (with particular attention to the roles and activities of instructors, not ship's company, who were training CST members)
   - on-station postexercise debriefing of individual watchstanders and some subteams of the CST (e.g., the Anti-Air Warfare [but note that this is now referred to as Air Defense Warfare] subteam)
   - diagnosis of weaknesses in performance and selection of future (training) experiences to improve performance
   - team debriefing (with no special orientation to teamwork processes, rather just a debrief of the entire team at once).

In addition to direct observation of training, numerous ad hoc interviews and discussions were conducted with members of the ship's CST; personnel acting as trainers of the ship's CST; and other available, knowledgeable personnel.

From these analyses and observations, a model of an idealized afloat-training (OJT) instructional cycle was developed. This model depicts activities that ideally are performed by OJT instructional teams. The model is presented in Figure 1.

Initial Identification of Implications for Afloat-Trainer Training

From the training cycle model, implications and opportunities to enhance training for afloat trainers were identified. An analysis was performed to identify candidate ways to exploit the opportunities. Approaches were categorized as either requiring further research to explore their ramifications or as easily dealt with (by current knowledge) in designing a training system for afloat trainers.

Several concepts for ways to address the challenges of providing exportable training for afloat training teams were identified. These were developed from review of the model and analysis of comments from interviews during observation of training. These were considered concurrently with the identified opportunities and implications and contrasted with instructor duties and responsibilities (from the model). This enabled identifying what (OJT) instructors should do at various stages of the training cycle, how to train and support instructor functions, and initiatives required to attain the needed training and support.

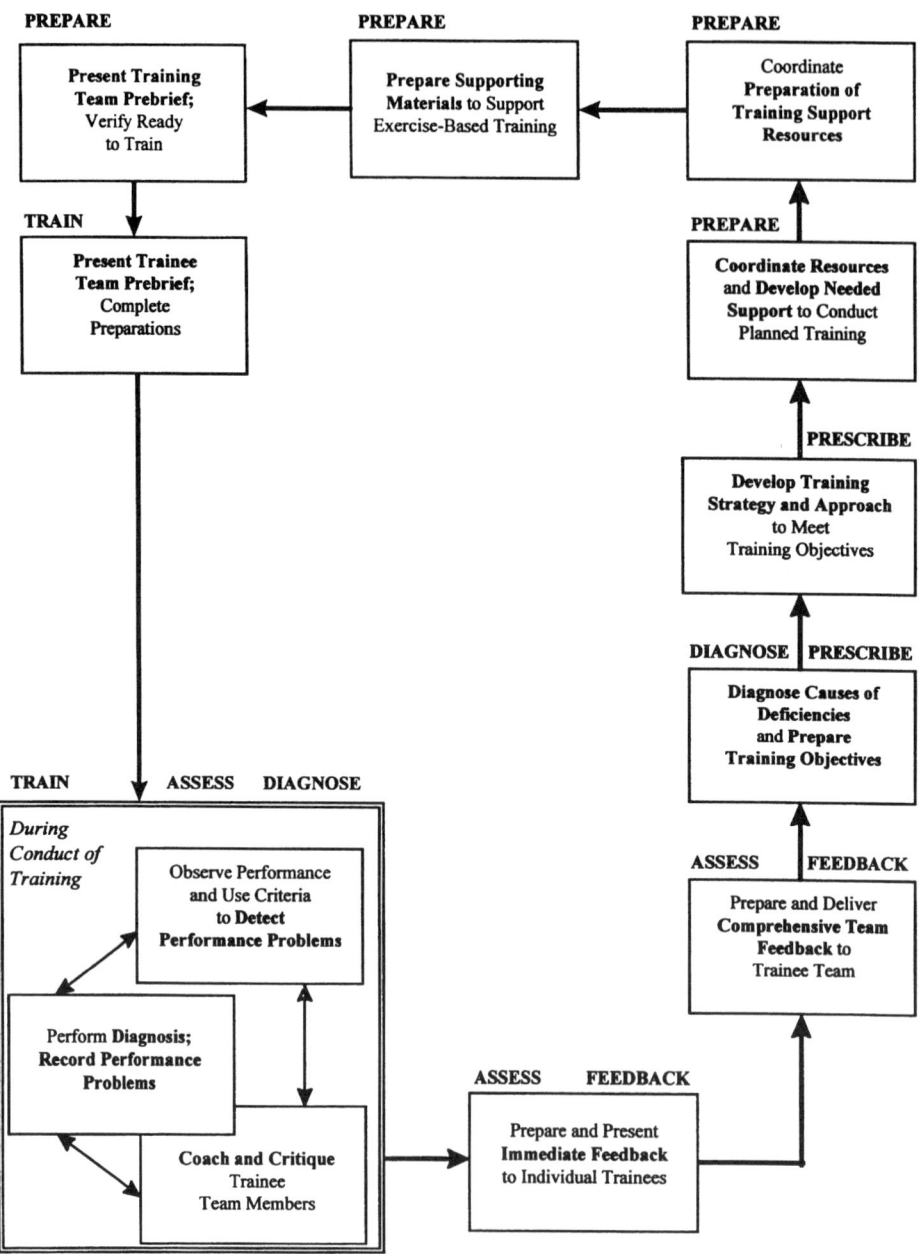

Figure 1. A cyclical process model for shipboard training.

Verification of Afloat-Instructor Duties and Responsibilities

Structured interviews were conducted with representatives of several Navy ashore training organizations to verify the duties and responsibilities inventory. Personnel interviewed were directly involved in training and assessment of CSTTs for several classes of combatant ships or had recent (within 6 months) experience as CSTT members.

Interviews were held in a group format and were tape-recorded for later content analysis. The interviews were conducted in three stages. First, interviewees were given the current version of the instructor duties and responsibilities inventory and asked to critique it for completeness and comprehensiveness. Suggestions for rephrasing items, and recommendations for additions to the inventory, resulted from each interview.

Next, interviewees were asked whether, to their knowledge, explicit training or other preparation was (at that time) provided for each of the instructor duties and responsibilities. If interviewees indicated that training was currently provided, they were asked to identify the organizations providing the training and describe the methods used to provide the training. This provided insights into how afloat instructors were (then) trained to conduct their responsibilities as instructors and identified areas where additional training or other preparation might be necessary.

The final part of each interview was an open-ended forum in which we solicited interviewees' ideas on how best to provide deployed training to afloat instructors and instructional teams. After the interviews were complete, the results were used to revise the instructor duties and responsibilities inventory and annotate the revised inventory with comments and remarks about ways to improve afloat-instructor training.

The interviews yielded 85 discrete topical statements, which were coded by topic and content analyzed. Fourteen categories (listed in Table 1) were identified in the content analysis. The topic categories in Table 1 were derived from the comments and statements made by interviewees, rather than an a priori categorization scheme. The comments in each topic category were reviewed to identify recurrent themes. Then, each theme

Table 1. Content Analysis Categories Derived From Interviews

Category	Topic
1	CSTT membership (6)
2	Training provided for CSTT members (7)
3	Evaluation of CSTT performance (4)
4	Command emphasis and priorities (2)
5	How CSTTs now do Combat Systems Training (13)
6	How CSTTs should do Combat Systems Training (3)
7	Support resources for Combat Systems Training (6)
8	Critical elements for training emphasis (6)
9	Methods for training Combat Systems Teams (13)
10	Feedback methods and techniques (4)
11	CSTT roles and responsibilities (5)
12	Support and training needs for CSTTs (11)
13	Training detractors encountered (4)
14	Other (1)

Note. *n*s following the topic titles in parentheses are the number of statements made in each category. CSTT = Combat Systems Training Teams.

was rewritten as a single sentence or short paragraph that captured the sense and context of comments made on one particular topic. These theme statements were annotated with suggestions as to how to improve the aspect of afloat training that was the theme's topic.

After all themes were extracted from interview comments, the discrete themes in each category were examined together to identify any additional implications for deployable afloat-instructor training. Some of these implications were noted as topics for possible additional data gathering or for future research.

Identifying Characteristics of a Candidate Afloat-Instructor Training System

Using the results of all analyses and information gathered to this point, the characteristics of an idealized training system for afloat-instructor teams were synthesized. The list of characteristics attempts to take into account the spectrum of constraints found in the afloat environment, as well as desirable training system features. Concurrently, a model curriculum outline for afloat-instructor training was created.

These products provide the basis from which detailed exploration of the feasibility and potential of a deployable instructor training system can be considered. The two products are presented in Tables 2 and 3.

Evaluating Development Potentials for the Proposed Instructional System

We next prioritized the instructional elements included in the model curriculum. The goal was to identify relatively low-cost, high-payoff elements that would be straightforward to develop and implement. The elements that were evaluated and prioritized are those listed in the left-hand column of Table 3. These correspond to conceptual units of instruction and can each be implemented in different appropriate media or using different instructional methods.

To prioritize the elements, we first determined how each might be implemented, in terms of training media and approach. To do this, the content topics in each of the 12 instructional elements were reviewed by a panel of three instructional designers. The panel made one blanket recommendation for implementation approach (see Table 4) for each of the 12 elements. The recommendation for an element could include multiple instructional media, but if multiple media were recommended, they should implement a unified instructional strategy.

The media identified as candidates in Table 4 are not, of course, the only acceptable media for training these objectives. These media were selected as potentially effective for conveying the objectives in a research-oriented, OJT–instructor training program. Other media might be equally effective.

Table 2. Desirable Characteristics of a Training and Support System for OJT Instructors

Characteristic	Description
1	Supports acquisition and sustainment of skills needed for technically qualified personnel to be effective members of an OJT training team
2	Enables effective diagnosis of instructor-skills training needs and prescription of training to develop skills mastery
3	Provides effective, efficient, self- and team instructional material that supports the acquisition and sustainment of instructor skills
4	Uses media and training methods appropriate for training particular skills
5	Preferentially uses training aids that promote incidental learning and reinforcement of skills, supplementing dedicated training material
6	Dedicated instructor-training material is highly modularized to enable training in the time periods available
7	Is closely tied to enabling instructor roles as coach, evaluator, mentor
8	Incorporates instructor-qualification element (if required)
9	Is primarily oriented toward preparing instructors to train teams, but also enables instructors to develop the skills required to support individual training
10	Requires minimum administration and record keeping
11	Requires minimum space and volume (especially if shipboard)
12	Amenable to easy update of perishable instructional contents

Note. OJT = on-the-job training.

Six variables were selected for prioritizing the instructional elements. They are

- potential *breadth of application* of the instructional element—ranging from the CSTTs of one ship class to many kinds of training teams on many ship classes;
- *development effort required* to create a research prototype of the instruction—ranging from under one half professional staff year to over 1.6 staff years;
- *estimated cost of developing instructional media* for a research prototype of the instructional element—ranging from under $50,000.00 to over $150,000.00;
- *likelihood of early implementation* of training on the basis of a research prototype for the instructional element—an estimated probability figure;
- *resource requirements* for widespread implementation of training that is based on a research prototype for the instructional element—ranging from existing deployed training support resources to new, risky high-technology support resources; and

Table 3. Lessons and Topics for Model OJT-Instructor Curriculum

Lesson purpose and contents (instructional elements)	Topics
Learning the duties, responsibilities, and roles of training team members	Instructor roles (coach, evaluator, controller, mentor)
Acquiring training exercise prebriefing skills	Essential elements Prebriefing formats Procedures
Developing training objectives and evaluation criteria, and conducting error diagnosis	Using evaluation criteria in observation and performance assessment Identifying causes of performance errors Coordinating performance error diagnosis with other training team members Relating performance error observations to training objectives
Effective coaching and feedback for individual instruction	Principles for coaching and feedback Techniques of coaching and feedback Problem solving with the trainee
Fostering accurate mental models and mental model evaluation	Mental models concept Shared mental models as a foundation for team performance When to suspect faulty mental models Developing correct mental models through training Assessing mental models for appropriateness Techniques for correcting faulty mental models
Providing effective feedback on teamwork	Diagnosing teamwork problems and combining separate observations Team feedback principles Use of debriefing or replay capabilities Discovery methods for problem identification
Diagnosing performance problems and planning remediation	Collective diagnosis: combining observations to diagnose teamwork problems Problem attribution: to individual watchstanders, subteams, and teams Developing training objectives Using training objectives as the foundation for training plans
Training and resource planning	Selecting appropriate training strategies Identifying resources required for training Sequencing and scheduling training Training under constraints Preparing trainees for training
Preparing to train	Preparing prebriefs for training team and trainees Verifying of readiness to train

Table 3 continues

Table 3. *(Continued)*

Lesson purpose and contents (instructional elements)	Topics
Using embedded training capabilities	Capabilities and limitations of embedded training components
	Operation, maintenance, and update procedures
	Available baseline scenarios
	Modifying baseline scenarios
	Creating new baseline scenarios
	Modifying evaluator tools and aids
	Developing and validating casualty drills
	Integrating casualty drills into scenarios
	Using replay and data products for debriefing
Using adjunct training equipment	Identity, capabilities, and operation
	Constraints on equipment use
	Liaison with training-support personnel
	Available baseline scenarios
	Planning scenario modifications
	How to accomplish scenario modifications
	Integrating casualty drills with adjunct training
Training team self-evaluation and self-training	Performance criteria and assessment
	Diagnosis and developing prescriptions
	Developing self-training objectives and plans

Note. OJT = on-the-job training.

- whether empirical information or other bases of content for developing a research prototype of the instructional element are (or may be) available—ranging from current availability to availability after roughly 2 years of additional research.

Table 4. Candidate Media for Instructional Elements

Instructional elements	Proposed media
Training team duties and responsibilities	Paper-based training aid and checklists
Exercise prebriefing	Videotape and training aid and checklists
Training objectives, criteria, and errors	Workbook and videotape
Individual coaching and feedback	Videotape and training aid
Mental model evaluation and assessment	Interactive video and training aid
Preparing and delivering team feedback	Interactive video and training aid
Diagnosing and planning remediation	Interactive video and training aid
Training and training resource planning	Self-study workbook and checklist
Preparing to train	Training aid and hands-on instruction
Using embedded training capabilities	Workbook and hands-on instruction
Using strap-on training capabilities	Workbook and hands-on instruction
Training team evaluation and training	Diagnostic guides and training aids

Each instructional element in Tables 3 and 4 was prioritized using a three-step process to arrive at a figure of merit (see Table 5) for each element.

1. Each instructional element was first classified high, moderate, or low, with regard to where it falls in the qualitative range implied by each of the 6 prioritization variables.
2. The three possible classifications for each of the six prioritization factors were matched to either a positively ordered or a negatively ordered set of cost–benefit coefficients. The positively ordered assignments (1 = low, 2 = moderate, 3 = high) applied to the three factors that were judged to be net contributors to high payoff relative to cost: Breadth of Application, Likelihood of Implementation, and Content Availability. The negatively ordered assignments (3 = low, 2 = moderate, 1 = high) applied to the three factors that were judged to be net contributors to high cost relative to payoff: Development Effort, Cost of Development, and Resource Requirements. The coefficients were assigned once only, independently of the variables' application to the instructional elements.
3. The coefficients assigned for the classifications on all six prioritization factors were then added to produce a figure of merit (see Table 5) for each instructional element.

Table 5 summarizes the priority evaluation of the 12 instructional elements. Shown here is the assessment of each instructional element on each of the six prioritization dimensions, in the form of both range judgments and score assignments. It was not possible to make definitive estimates about 2 of the instructional elements on the Breadth of Applicability dimension or for 2 others on the Data or Content Availability dimension. The priority estimates for these instructional elements are therefore incomplete. This is indicated in the column titled priority figure of merit by an asterisk. Had estimates been available for these dimensions, the priority scores for these instructional elements would have been higher.

Personnel who participated in the validation interviews stressed some instructor-training topics as having particularly high priority and value from their perspective. This is indicated in the priority figure of merit column of Table 5 by the notation (C).

Creating a Roadmap for Research and Development

The final element of the work was to synthesize the results of previous steps to develop a research roadmap for exploring the development of deployable afloat-instructor training capabilities. To develop the roadmap, we jointly considered what had been learned about the following factors:

1. current practice in the training and evaluation of CSTTs;
2. afloat-instructor duties and responsibilities;

Table 5. Prioritization Evaluation of Instructional Elements

Instructional element	Breadth of applicability	Development effort	Cost of development	Likelihood of implementation	Resource requirements	Content availability	Priority figure of merit
Training team duties and responsibilities	H 3	L 3	L 3	H 3	L 3	H 3	18 (C)
Exercise prebriefing	H 3	M 2	M 2	M 2	M 2	H 3	14 (C)
Training objectives, criteria, and errors	Unknown ?	M 2	M 2	M 2	L 3	M 2	11*
Individual coaching and feedback	H 3	M 2	M 2	H 3	L 3	H 3	16 (C)
Mental models: Evaluation, assessment, and correction	M 2	H 1	H 1	M 2	H 1	L 1	8
Preparing and delivering team feedback	M 2	H 1	H 1	M 2	H 1	M 2	9 (C)
Diagnosis and planning remediation	M 2	H 1	H 1	M 2	H 1	L 1	8
Training and training resource planning	Unknown ?	M 2	M 2	H 3	L 3	M 2	12*
Preparing to train	L 1	L 3	L 3	H 3	H 3	Unknown ?	13*
Using embedded training capabilities	L 1	H 1	H 1	M 2	L 3	L 1	9
Using strap-on training capabilities	L 1	H 1	H 1	M 2	L 3	L 1	9
Training team evaluation and training	L 1	L 3	L 3	H 3	L 3	Unknown ?	13*

Note. H = high priority on assessment dimension; L = low priority on assessment dimension; M = medium priority on assessment dimension.
? indicates our inability to assign priority on assessment. Asterisks indicate incomplete priority estimates.

3. commonalities and differences in training support capabilities (particularly embedded training and strap-on training support equipment) between ship classes;
4. instructor training requirements considered likely to arise as a result of implementing research-based interventions;
5. recommendations for training media and methods to be considered in implementing deployable afloat-instructor training;
6. the general constraints of the shipboard environment to support additional equipment, training materials, and so forth;
7. constraints on time available for instructors to train (shipboard) for their roles as instructors.

Following are the principal conclusions of this synthesis.

1. Additional afloat-instructor self-training capability would be welcomed and used if it is thoughtfully implemented. One implication is that implementing such training should impose minimum administrative requirements.
2. Little is known for certain about shipboard capability to accommodate and support introduced advanced technical media for training delivery, to be dedicated to training purposes. Therefore, introducing afloat-instructor training that depends on such media should be done with caution, particularly if the training delivered is to extend to a broad audience (i.e., aboard different ship classes). Acceptance and use of training that depends on such technical capabilities could vary, depending on the perceived burden of accommodating and supporting the equipment.
3. There is wide variability in the capabilities and employment of embedded training components across ship classes. Embedded training capability can be one of the more valuable assets available for combat systems team instruction. Including instruction on the use of embedded training components in a deployable afloat-instructor training system may be problematic because of the many embedded training configurations installed aboard various ship classes and the different ways in which these components can be—and are—used to provide training. However, including such training is clearly desirable so that the most value can be realized from these sophisticated and capable tools for training.

A similar finding, with parallel implications and cautions, applies to currently available strap-on training support equipment, such as pierside training equipment. And, similar issues may arise when introducing new training support capabilities (e.g., Battle Force Tactical Training), particularly if equipment and operating implementations vary across ship classes.

On the basis of these conclusions, we developed an incremental, experimental approach to introducing elements of deployable afloat-instructor training. In this approach, widely applicable instructional elements that (a) can be largely self-administered by afloat instructors, (b) require only simple instructional support, and (c) do not require advanced technical media for instructional delivery would be developed and implemented on an experimental basis. The training effectiveness and acceptability of these elements would be assessed, and any issues that arise would be used as guidelines for later developments.

If the initial, simple instructional elements proved to be acceptable and effective, further research would be conducted with elements that require more advanced approaches and media. A similar process of experimental implementation, accompanied by measurement of effectiveness and acceptability and assessment of problems and issues, would take place successively with several instructional elements. The collective experience and lessons learned from the experimental implementations would become broad-based guidelines for deployable afloat-instructor training.

The Roadmap

The roadmap for research assumes that developing exportable instructor training must be based on empirically established principles and guidelines. Such are not currently well established. The goal is to perform experiments and investigations to develop needed guidelines and, ultimately, develop a comprehensive exportable training program for afloat instructors. The roadmap consists of three successive stages or phases. These are described in the following paragraphs, and an outline of the roadmap is shown in Figure 2.

Stage 1 is concerned with exploring parameters of acceptability and instructional effectiveness of exportable shipboard training. In this stage, several instructional packages would be designed, developed, and experimentally implemented aboard ship for use by afloat instructors. These packages would have the following characteristics:

1. Each would be a stand-alone element of training, designed to be used to train a specific subset of afloat-instructor knowledge and skills. Although packages would stand alone for experimental purposes, they would be designed as elements of a complete training system. Therefore, they can be easily integrated into a more complete deployable afloat-instructor training system.
2. The packages would use different instructional approaches, suited to the training objectives. The objectives trained in the packages may range from those requiring simple recall to more advanced cognitive skills requiring application of concepts, rules, and principles to develop plans and implementation strategies.
3. Packages would be implemented in different types of instructional media appropriate to the training approaches chosen and learning objectives. Media might sample from the full range of instructional support approaches.

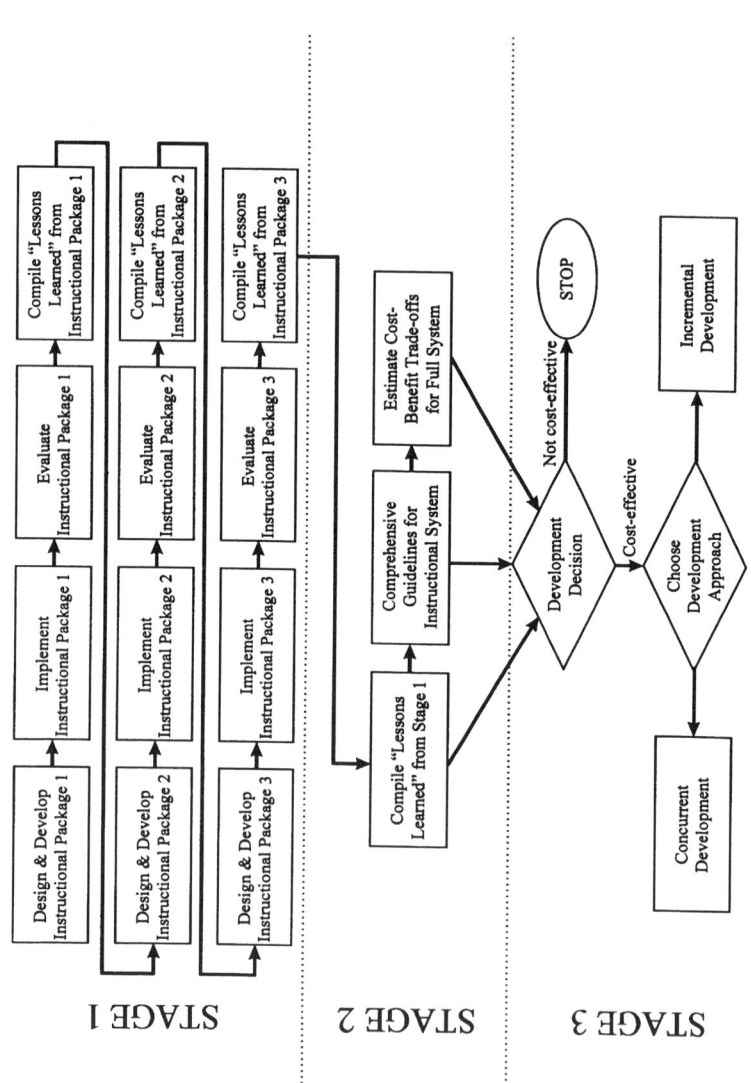

Figure 2. Schematic of a roadmap for developing and evaluating a deployable training system for shipboard instructors.

4. Packages would be designed to be used with as large an audience as possible, other constraining factors considered. For example, a package presenting an introduction to instructor duties and responsibilities could be developed to be applicable for essentially all instructors, with some minor tailoring. Other packages (e.g., dealing with ship- or class-specific training equipment) might have less broad application.

This can help to assure that the experimental afloat-training products have practical, residual value beyond that of the contemplated research. Instructional packages designed for broad applicability can be implemented in other situations while a decision to proceed with development of a comprehensive training system is pending. This supports an incremental investment strategy, in parallel with continuing experimentation.

Stage 1 training packages would be developed and implemented sequentially. This enables applying interim lessons learned from packages implemented in the design of subsequent packages. The concept is that three or more packages be developed and evaluated in Stage 1.

Stage 2 involves first compiling the lessons learned from, and ways of addressing issues that arise during, Stage 1. These would be used to develop guidelines to support developing a full-scale deployable training system. During Stage 2, the costs and anticipated benefits of developing such a system would be estimated, and this information used to decide whether and how to develop such a system. A decision to go forward with development would lead to Stage 3.

For Stage 3, there are two alternatives. The first is to proceed with concurrent development of the remaining elements of the instructor-training system. This includes refining and updating elements developed in Stage 1. The second alternative is to continue an incremental approach, producing and deploying one or a few instructional elements at a time. No particular choice between these two approaches is favored. The concurrent approach has the advantage of deploying a complete training system all at once. This avoids problems that can arise from multiple implementations. However, investment costs will be incurred all at once, and the evaluation of benefits can be more difficult than with an incremental approach.

The incremental approach spreads the investment costs of the training system over a longer time. This might be a preferred approach given declining budgets and reduced resources. The incremental approach also has the advantage of stepwise demonstration of instructional effectiveness and acceptability and of continuing the collecting of lessons learned.

Lessons Learned: Implications for OJT

OJT is becoming an ever more common approach to training. Often, OJT supplements more traditional (e.g., classroom or laboratory-oriented) training by providing structured on-the-job experiences to facilitate learning of job-critical skills and knowledge. In some cases, OJT may be the predominant (or even the only) form of training provided. OJT can be a less costly alternative to traditional training approaches. However, developing and implementing OJT should receive the same thoughtful analysis and insightful crafting as for any other training.

In some cases, trainers who present or mediate OJT experiences may require little training beyond that required to master their principal jobs (actual job activities, as opposed to training). In other cases, particularly those involving highly technical job specialties or advanced training-support technology (e.g., embedded training), significant additional instructor training may be needed. In practically all cases, OJT instructors can benefit from training in providing effective feedback, coaching methods, and other common instructor skills (as with common sense, such skills are not truly common, but must be learned).

The results of the work discussed here provide a high-level structure of skills and knowledge elements that may apply to the design of training programs for OJT instructors in other contexts. Obviously, additional analyses and decisions are required to adapt this high-level structure to any particular case. And, not all of the elements of skill and knowledge identified here will be appropriate for training OJT instructors in all cases. Consideration should be given to using these results as a point of departure for thinking about appropriate content for training programs for OJT instructors and designing such training. Some specific "lessons learned" from this work that apply to many OJT contexts follow.

Lesson 1: Job Performance Does Not Impute Instructional Competence

Instructors who conduct OJT are likely to be technical experts in their job specialties. However, they are also not likely to have had much if any training in how to be an effective instructor. In other words, OJT instructors may lack skills and knowledge that would make them more effective instructors. At a minimum, the skills and knowledge that should be trained for OJT instructors should include methods and techniques for objectively assessing performance against relevant measures and criteria and methods and techniques for providing objective, nonpersonal, and trainee-involving feedback on performance. Although this lesson has become a truism in the training community, it bears repeating for the benefit of those who may not realize the consequences of ignoring it.

Lesson 2: Performance-Quality Feedback Methods and Approaches Are Critical

OJT instructors should particularly have skills to involve trainees in problem diagnosis and performance problem solving through discovery or Socratic methods of presenting objectively based feedback. In this regard, some sort of objective record of performance can be a tremendous asset to both the instructor and the trainee, because it promotes objectivity and can help to avert the common "Who struck John?" arguments about performance in context of delivering feedback. Technically sophisticated performance records (e.g., those obtainable through many training simulators and simulations; see Burns, Pruitt, Smith-Jentsch, & Duncan, 1996; Cannon-Bowers, Burns, Salas, & Pruitt, chap. 15, this volume) will not always be available for this purpose. However, even simple checklists summarizing desirable or essential characteristics of performance, when objectively and conscientiously completed, can serve this purpose to some extent. Serious attention should always be given to effectively supporting objective performance assessment and feedback in any OJT context. If this is not done (or if instructors and trainees alike do not understand that objectivity in assessment is strongly promoted and encouraged), less objective criteria are likely to come into use and less constructive methods of providing feedback may come to dominate OJT-based training sessions. Too often, there is a tendency for untrained OJT instructors to migrate toward a personalized and subjective mode of feedback delivery—one that is unlikely to result in effective and desirable behavior change.

Lesson 3: Assessing Teamwork May Require Different Methods Than Assessing Individuals

Depending on the context and the domain, OJT performance assessment for teams of job performers may require multiple instructors (perhaps, as in the analysis reported here, instructor teams) assigned to specific roles for assessing both individual task performance and performance on teamwork aspects of jobs or tasks. So that all essential aspects of OJT performance be addressed, methods are being devised (see among others Burns, Pruitt, Smith-Jentsch, & Duncan, 1996; Pruitt, Burns, Rosenfeld, Bilazarian, & Levas, 1996; Smith-Jentsch, Zeisig, Acton, & McPherson, chap. 11, this volume) to effectively combine the observations of multiple assessors and provide comprehensive feedback on both individual and teamwork aspects of performance. One aspect of these methods is a structured debrief or feedback methodology that has multiple phases. The first phase of feedback after a training session is typically a brief one-on-one at-the-workstation session between one instructor and one trainee team member, emphasizing both effective and ineffective aspects of the trainee's individual performance. Following this, instructors involved in observing teamwork aspects of performance caucus to combine their observations and prepare a comprehensive debriefing on teamwork aspects of performance in the training session. This debriefing is ideally presented to all members of the trainee team in a didactic group problem-solving format.

References

Burgess, K., Riddle, D., Hall, J., & Salas, E. (1992, March). *Principles of team leadership under stress.* Paper presented at the annual convention of the Southeastern Psychological Association. Atlanta, GA.

Burns, J. J., Pruitt, J. S., Smith-Jentsch, K. A., & Duncan, P. C. (1996). The role of personal digital assistants in advanced training systems. *Proceedings of the Human Factors and Ergonomics Society 40th annual meeting*. Santa Monica, CA: Human Factors and Ergonomics Society.

Cannon-Bowers, J., & Salas, E. (1990, June). *Cognitive psychology and team training: Shared mental models in complex systems*. Symposium address presented at the annual meeting of the Society for Industrial and Organizational Psychology, San Antonio, TX.

Converse, S., & Cannon-Bowers, J. (1991). Team member shared mental models: A theory and some methodological issues. *Proceedings of the Human Factors Society 35th annual meeting*. Santa Monica, CA: Human Factors Society.

Driskell, J., & Salas, E. (1991). Overcoming the effects of stress on military performance: Human factors, training, and selection strategies. In R. Gale & A. Mangelsdorff (Eds.), *Handbook of military psychology* (pp. 183–193). New York: Wiley & Sons.

Duncan, P. C., & Rouse, W. B. (1993, April). *Shared mental models training*. Presentation to TADMUS Technical Advisory Board, Dahlgren, VA.

Evans, D. C., & Roth, J. T. (1993). *Tactical decision making under stress (TADMUS): Relevant findings and integration* (Working paper developed under Contract N61339-92-C-0104, Task 0155-02P09). Butler, PA: Applied Science Associates.

Glickman, A., Zimmer, S., Montero, R., Guerette, P., Campbell, W., Morgan, B., & Salas, E. (1987). *The evolution of teamwork skills: An empirical assessment with implications for training* (Tech. Rep. No. 87-016). Orlando, FL: Naval Training Systems Center.

Guerette, P., Miller, D., Glickman, A., Morgan, B., & Salas, E. (1987). *Instructional processes and strategies in team training* (Tech. Rep. No. 87-017). Orlando, FL: Naval Training Systems Center.

Hogan, J., Peterson, A., Salas, E., & Willis, R. (1991). *Team performance, training needs and teamwork: Some field observations* (Tech. Rep. No. 91-0007). Orlando, FL: Naval Training System Center.

Hutchins, S. (1996). *Principles for intelligent decision aiding* (Tech. Rep. No. 1718). San Diego, CA: Naval Command, Control, and Ocean Surveillance Center.

Klein, G. (1992). *Decision making in complex military environments* (Tech. rep. prepared for Naval Command, Control and Ocean Surveillance Center). Fairborn, OH: Klein Associates.

Orasanu, J., & Salas, E. (1992). Team decision making in complex environments. In G. Klein, J. Orasanu, & R. Calderwood (Eds.), *Decision making in action: Models and methods*. Norwood, NJ: Ablex.

Pruitt, J. S., Burns, J. J., Rosenfeld, J. P., Bilazarian, P., & Levas, R. (1996). Enhancing a paper-based team-performance measure through technology. *Proceedings of the Human Factors and Ergonomics Society 40th annual meeting*. Santa Monica, CA: Human Factors and Ergonomics Society.

Rouse, W. (1991). *Mental models in training for complex tactical decision making tasks*. (Report for Contract No. DAAL03-86-D-0001 prepared for Naval Training Systems Center). Norcross, GA: Search Technology.

Zsambok, C., Klein, G., Beach, L., Kaempf, G., Klinger, D., Thordsen, M., & Wolf, S. (1992). *Decision-making strategies in the AEGIS combat information center* (Tech. rep. prepared for Naval Command, Control, and Ocean Surveillance Center). Fairborn, OH: Klein Associates.

15

Advanced Technology in Scenario-Based Training

Janis A. Cannon-Bowers, John J. Burns,
Eduardo Salas, and John S. Pruitt

Much of the research that followed the initial work under the Tactical Decision Making Under Stress (TADMUS) project has focused on scenario-based training. By *scenario-based training* we mean training that relies on controlled exercises or vignettes, in which the trainee is presented with cues that are similar to those found in the actual task environment and then given feedback regarding his or her responses. We selected this focus as a basis for advanced research and development (R&D) for several reasons. First, from a theoretical perspective, the notion of scenario-based training is consistent with the need to provide novice decision makers with a variety of task episodes as a means to augment their experience and accelerate their development (Cannon-Bowers & Bell, 1997). Second, from a practical standpoint, scenario-based training is consistent with the Navy's goal of moving more training to the more cost-effective shipboard environment, where traditional training resources are not readily available. The purpose of this chapter is to describe briefly the R&D being conducted to support the delivery of scenario-based training for building expertise. To do this, we first describe the components of scenario-based training. Next, we briefly document applied R&D efforts in this area.

Components of Scenario-Based Training

A major difference between scenario-based training and more traditional training is that in scenario-based training situations there is no formal curriculum; instead, *the scenario itself is the curriculum*. This means that the scenario must be crafted, and training executed around it, in a manner that accomplishes desired training objectives. Figure 1 graphically depicts the cycle of scenario-based training. The cycle begins with specification of training objectives and associated competencies (see circle 1 in Figure 1). This process must rest on sound analytical approaches designed to provide

The views expressed herein are those of the authors and do not reflect the official position of the organizations with which they are affiliated.

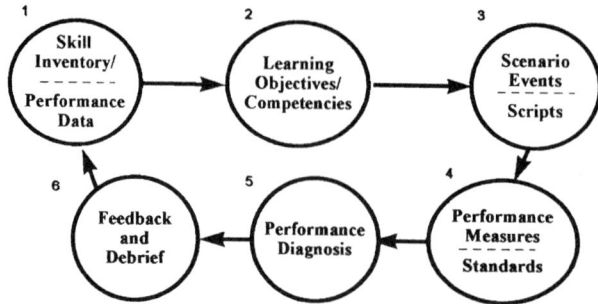

Figure 1.   Components of scenario-based training.

information regarding the knowledge, skills, and attitudes (KSAs) that underlie effective performance. Techniques such as traditional task analyses, cognitive task analyses, and team task analyses are useful here (see Cannon-Bowers & Salas, 1997, for more detail).

Once training objectives and competencies are established and selected for an exercise, these must drive the development of the training scenario events and scripts (circle 2 in Figure 1). That is, the scenario must be crafted carefully to ensure that training objectives are actually addressed (Stretton & Johnston, 1997). This means that events must be scripted into the scenario that will allow team members to perform targeted skills so that their mastery can be assessed. If the scenario is not developed explicitly to exercise training objectives, valuable training time will be lost on nonessential skills.

The next step in the cycle (see circle 3 in Figure 1) is development of performance measures that will indicate whether trainees are mastering targeted materials. The importance of accurate performance measures cannot be overstated (see Smith-Jentsch, Johnston, & Payne, chap. 4, this volume; Cannon-Bowers & Salas, 1997). This is because effective performance measures not only allow assessment of whether the trainees learn materials, but also they help determine why performance occurred as observed. This process leads to, and supports, diagnostic mechanisms that allow an assessment of the causes of observed performance (see circle 4 in Figure 1)—that is, to determine which deficiencies might exist in the trainees' knowledge or skill that could explain their behavior. Diagnostic mechanisms must be developed on the basis of a detailed understanding of performance in the domain so that observed performance can be mapped onto underlying competencies. In this manner it is possible to draw conclusions about knowledge and skill levels (even in more cognitive tasks) by assessing observed performance.

Once diagnoses are formulated, feedback must be constructed and delivered so that trainees have a basis on which to improve in subsequent episodes (represented by circle 5 in Figure 1). Obviously, the specific form and content of the feedback must be based on the initial training objectives and on the diagnosed performance deficiencies. For example, if the diagnostic process suggests that trainees are unfamiliar with system operations, feedback will be directed at this deficiency. Further, decisions about when to provide feedback (e.g., during performance episodes, during post-performance debriefs, etc.) can be based on the nature of knowledge and skill being taught and the diagnosis of trainee performance (see Smith-Jentsch, Zeisig, Acton, & McPherson, chap. 11, this volume, for a description of the use of post-exercise feedback).

To close the loop, trainee performance information must be incorporated into future training sessions to ensure that trainees are exposed to training objectives that build on, but do not duplicate, what they have already learned (see circle 6 in Figure 1). In many current training situations, effective use of prior training data is not made. Instead, we advocate a strategy in which the results of training episodes drive the selection of training objectives in subsequent instruction. Such a strategy is efficient because training resources are directed at bona fide training requirements.

Advanced R&D in the area of scenario-based training has focused on developing technologies and methodologies that can support the scenario-based training cycle documented in Figure 1. These are described further in the next section (also see Oser, Cannon-Bowers, Dwyer, & Salas, 1997; Fowlkes, Dwyer, Oser, & Salas, 1998; Stretton & Johnston, 1997).

Advanced Scenario-Based Training

TADMUS results provided a basis on which to commence the development and demonstration of an intelligent scenario-based training system to support individual and team training in a shipboard environment (Zachary, Bilazarian, Burns, & Cannon-Bowers, 1997). Briefly, TADMUS taught us that performance measurement is paramount to training (Cannon-Bowers & Salas, 1997); that effective measurement not only describes and evaluates behavior, but also diagnoses the causes of effective performance (Cannon-Bowers & Salas, 1997; Smith-Jentsch, Johnston, & Payne, chap. 4, this volume); and that sound training techniques can be developed on the basis of sound assessment (see Smith-Jentsch, Zeisig, Acton, & McPherson, chap. 11, this volume). Furthermore, our theoretical foundation in TADMUS (Cannon-Bowers & Salas, chap. 2, this volume), and subsequent empirical results, provides a basis to expect that team-level training directed at improving the development of shared mental models (e.g., see Blickensderfer, Cannon-Bowers, & Salas, chap. 12, this volume; Serfaty, Entin, & Johnston, chap. 9, this volume) can result in better decision making. Taken together, these results led us to conceive of a system that would streamline and improve the measurement of performance and provide a vehicle for team-level training interventions. The system we envisioned consists of a number of technologies and methodologies, including the following:

1. sophisticated training management software that enables instructors to implement event-based training scenarios and to track student progress,
2. automated performance-monitoring technologies (e.g., eye tracking and speech recognition) to aid in performance measurement,
3. advanced computer-modeling techniques that allow for dynamic representation of human performance as a basis for assessment and feedback,
4. software-aided, automated and semiautomated performance assessment and diagnosis, and
5. handheld computing devices to aid instructors.

The following sections describe these technologies and method in more detail.

Intelligent Scenario Management

As we have indicated, the scenario is integral to team training. Currently, the scenario development and event selection or creation process is manual and left to the user. This places a nearly impossible demand on instructors to ensure that scenarios and supporting products are related to the mission, training objectives, and past performance. Indeed, although this task is within the capabilities of expert instructors, it is very time consuming and not often done.

R&D in this area seeks to provide the capability to tailor scenarios to individual and team-performance deficiencies and allows shipboard instructors to easily create scenarios that contain events that are linked to training objectives, performance measures, and feedback. In addition, it is desired that the human interface for tasks such as scenario generation be streamlined to minimize demands on the user (Stretton & Johnston, 1997). As such, this capability directly supports the efficient and effective application of training resources.

Performance-Monitoring Technologies

According to Cannon-Bowers and Salas (1997), dynamic assessment involves the real-time collection and interpretation of performance data to support the training process. One of the key features of the shipboard environment in which we are working is the tremendous volume of raw performance data that is available to support training. For example, the existing operational system traps operator keystrokes (made at the rate of over a thousand per hour), although these data are not processed or interpreted to support the training process. Current research is aimed at developing and testing automated keystroke data-reduction techniques (see Zachary et al., 1997). To date, we have demonstrated the ability to record operator keystrokes and then interpret these at the functional level. For example, a pattern of keystrokes is automatically interpreted as the trainees' attempt to interrogate a radar contact. The appropriateness of this action can then be interpreted in light of scenario events.

A second source of data we are tracking is trainee communication behavior. Here we take advantage of the fact that tactical communications are based on a restricted vocabulary and that operators in this environment are trained to be precise and efficient in all communication behavior. These characteristics, coupled with increasingly sophisticated speech recognition hardware and software (that allow the technology to work in noisier environments than was previously possible) make voice recognition a viable technology for inclusion in our advanced scenario-based training system.

A final source of raw performance data that we are working to harness is perceptual information. We believe that information about where and when operators are looking at (or not looking at) key information on their consoles may provide a vital piece of information necessary to understand performance in complex environments. Eye-tracking technology necessary to support our efforts has developed to the point where it can be inserted into the operational environment without dramatically impacting the operator's ability to perform the task.

Taken together, these sources of data are being tracked and integrated to shed light on the trainees' progress. These data can then be further interpreted by a human instructor, or fed into an automated performance-assessment process (described next). In either case, the important point is that raw performance data are recorded and used to support the assessment and diagnostic processes.

Expert Performance Models

Perhaps the most ambitious component of our vision for an advanced scenario-based training system is the development of the embedded expert operator models that will function in the automated performance-recording and diagnosis system. This effort is detailed elsewhere in this volume (see Zachary, Ryder, & Hicinbothom, chap. 13, this volume). Briefly then, as the key to the automated performance recording and diagnosis system, expert models for each member of the team are developed. These models are capable of generating expert-level solutions to the simulated missions that are being worked by the team.

The difficulties and challenges associated with such an approach are too involved to discuss here. Suffice it to say that we are attempting to use human performance models as a means to implement intelligent scenario-based training. Our position is that our cognitive modeling capability is sufficiently advanced such that we will be able to significantly automate the diagnosis of at least some aspects of team performance. However, given the complexity of the environment and team task, we believe that it will be necessary to funnel some portion of performance data to the human instructor to support the task of high-level team assessment and diagnosis.

Automated Performance Diagnosis

Another software-based tool currently being researched takes advantage of the volume of data available in our scenario-based system by using individual- or team-level performance records that it then uses to compute deviations from expected actions. Many general and task-specific measures of performance can be computed directly from these deviations. Moreover, some of the performance measures will be diagnostic of behavioral deficiencies, whereas others of these can be diagnostic of cognitive deficiencies as well.

Essentially, this process takes advantage of the performance-data-collection capabilities and expert models (described earlier) to provide the instructor with information that allows him or her to pinpoint performance breakdowns by focusing on the preceding actions by an individual or another team member that may have influenced performance. Using this information, cognitive diagnosis by the instructor can also be supported as the system can help to determine which conditions or constraints were violated and which informational cues that specify those conditions or constraints were missed or misinterpreted.

Automated Instructor Aids

As the preceding discussion suggests, team training using complex scenario-based systems places enormous demands on instructors to make relevant and memorable interventions to the team in real time, or shortly after performing. Some of these demands include monitoring multiple individuals, diagnosing errors, delivering feedback, and specifying remediation. Even if the instructor's task were limited to teaching one individual, the difficult tasks of assessing understanding, interpreting misconceptions, and tailoring training to the individual's specific needs would remain. When we consider these facts, it is clear that instructors are overloaded in team training situations where they must simultaneously measure process and performance, detect problems, deliver feedback, and tailor remediation. However, as we have already argued, if instructors are not able to do these things, scenario-based training will not be optimized.

The Shipboard Mobile Aid for Training and Evaluation (ShipMATE) is a prototype software application designed to aid shipboard instructors in the task of preparing for, conducting, and debriefing objective-oriented, scenario-based training (Pruitt, Burns, Wetteland, & Dumestre, 1997). ShipMATE runs on an off-the-shelf, pen-based, hand-held computer; a computer that can easily serve multiple purposes in addition to hosting ShipMATE. Importantly, this extremely lightweight and portable system has infrared and radio frequency communications capability that allows the instructor to move freely about the ship, untethered to any physical equipment.

ShipMATE currently supports the team dimensional training (TDT) process, which has been empirically validated and is currently being institutionalized in the Navy's training curriculum (see Smith-Jentsch, Zeisig, Acton, & McPherson, chap. 11, this volume). TDT incorporates a training methodology consistent with the scenario-based training cycle shown in Figure 1 and, using ShipMATE, enables an instructor to do the following:

1. make written and spoken observations of trainee performance,
2. capture team communications and graphic displays of the scenario related to those observations,
3. track and preview significant scenario events, and
4. make specific, event-based observations with cuing from the system.

In addition, TDT and ShipMATE provide the instructor with guidance in organizing and delivering pre-exercise and post-exercise briefs. Such capabilities support the instructor in providing accurate and appropriate feedback to the trainees, thus facilitating the learning of specific training objectives.

A Scenario-Based Training System

According to Figure 2, current R&D efforts will culminate in development of a user-friendly system for easily creating scenarios that will allow instructors to tailor their training to assessed deficiencies or new mission requirements. It will also automate the capture and interpretation of performance data (to the extent possible), while leaving the human instructor in the loop to assess those competencies that the automated system cannot. The continuous performance record created by this process will then be interpreted using expert performance models, and a diagnosis of performance will be formulated. Again, having the instructor included as a viable part of the system will ease the diagnostic burden from the automated processes, particularly for higher order and teamwork skills. The system will then provide immediate feedback in the form of on-line cues or hints where appropriate. The system will also provide instructor aids to organize, streamline, and improve both on-line coaching and post-exercise debriefs. These enhancements to the training cycle are predicted to improve performance on subsequent scenarios. Empirical data will be collected to determine whether this prediction of improved performance will be a reality.

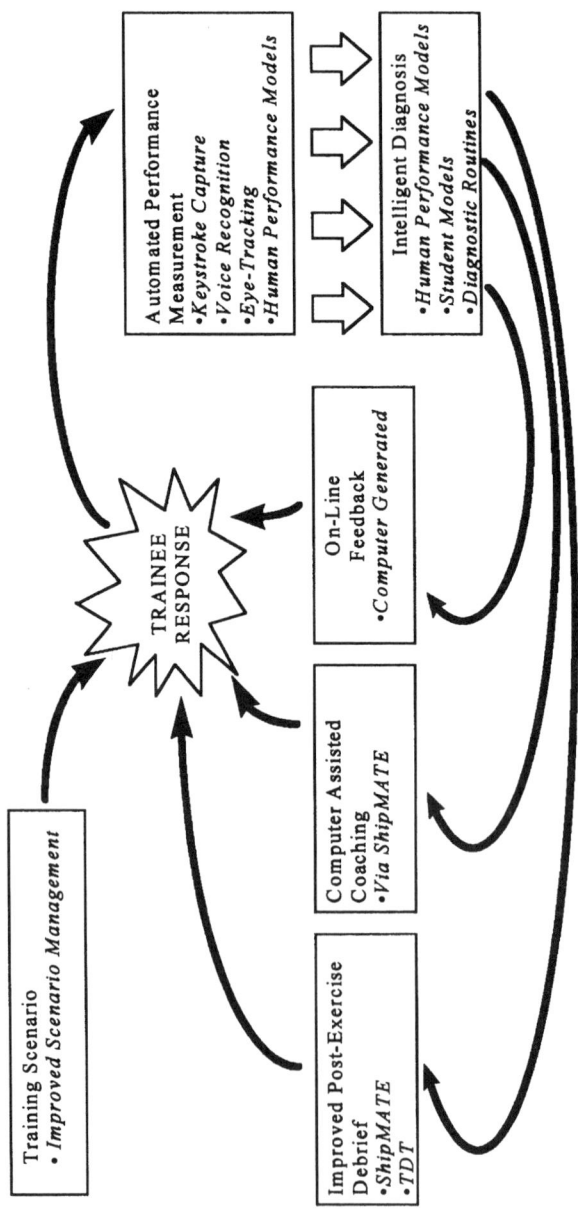

Figure 2. Advanced scenario-based training system.

Conclusion

As critical as automation is to improved training systems, perhaps the greatest return on this investment may be represented by a substantially enhanced ability to take advantage of naturally occurring learning opportunities. There is a growing consensus that dedicated training (i.e., the training event that is separate from the everyday operation) and on-the-job learning (i.e., learning that occurs naturally as a part of continuous operations) are points on the same continuum. As such, it behooves us whenever possible to implement advanced technologies with an eye toward both environments (Rostker, 1996).

Although we have described a vision for a scenario-based training system that supports dedicated training, we are cognizant of the need to support a more comprehensive approach to reaching and maintaining readiness. Our approach to taking advantage of this potential is to assess our methodologies and enabling technologies as they are being developed in order to determine if one or more of them might support training in everyday operations.

As we move into the 21st century, we will see the development and deployment of new ship classes with substantially reduced staffing requirements and substantially increased automation. Although it is certain that these new ships will represent capabilities well beyond anything we could have imagined just a few short years ago, careful consideration of the twin factors of reduced staffing and increased automation leads us to believe that we must provide for training that is as sophisticated as these new ships. The work we have described here, based on efforts begun under the auspices of the TADMUS program, lays a foundation for ensuring that training is ready to meet the need of the Navy's 21st century ships.

References

Cannon-Bowers, J. A., & Bell, H. R. (1997). Training decision makers for complex environments: Implications of the naturalistic decision making perspective. In C. Zsambok & G. Klein (Eds.), *Naturalistic decision making* (pp. 99–110). Hillsdale, NJ: Erlbaum.

Cannon-Bowers, J. A., & Salas, E. (1997). A framework for developing team performance measures in training. In M. T. Brannick, E. Salas, & C. Prince (Eds.), *Team performance assessment and measurement* (pp. 45–77). Hillsdale, NJ: Erlbaum.

Fowlkes, J., Dwyer, D. J., Oser, R. L., & Salas, E. (1998). Event-based approach to training (EBAT). *International Journal of Aviation Psychology, 8,* 209–222.

Oser, R. L., Cannon-Bowers, J. A., Dwyer, D. J., & Salas, E. (1997). Establishing a learning environment for JSIMS: Challenges and considerations [CD-ROM]. *Proceedings of the 19th Annual Interservice/Industry Training, Simulation and Education Conference* (pp. 144–153). Orlando, FL: National Training Systems Association.

Pruitt, J. S., Burns, J. J., Wetteland, C. R., & Dumestre, T. L. (1997). Shipboard mobile aid for training and evaluation. *Proceedings of the 41st Annual Meeting of the Human Factors and Ergonomic Society* (pp. 1113–1117). Santa Monica, CA: Human Factors and Ergonomic Society.

Rostker, B. (1996). *It's job performance, . . . ! A new paradigm for the training/job performance continuum.* Memorandum for the Chief of Naval Operations, Dept. of the Navy, Office of the Secretary.

Stretton, M. L., & Johnston, J. H. (1997). Scenario-based training: An architecture for intelligent event selection. *Proceedings of the 19th Annual Meeting of the Interservice/Industry Training Systems Conference* (pp. 108–117). Washington, DC: National Training Systems Association.

Zachary, W., Bilazarian, P., Burns, J., & Cannon-Bowers, J. A. (1997). Advanced embedded training concepts for shipboard systems [CD-ROM]. *Proceedings of the 19th Annual Interservice/Industry Training, Simulation and Education Conference* (pp. 670–679). Orlando, FL: National Training Systems Association.

16

Implications of Decision-Making Research for Decision Support and Displays

Jeffrey G. Morrison, Richard T. Kelly,
Ronald A. Moore, and Susan G. Hutchins

It is our premise that for human decision makers to be effective, they must foremost be able to access the data necessary to make a decision when it is needed, where it is needed, and in the form it is needed. The data must be integrated and organized so that they become useful as information to the user. Information must be meaningful, timely, and organized in a way that is consistent with how it is going to be used. The effects the operational environment may have on the human operators of a system complicate the understanding of how an operator uses information. The U.S. Navy has recognized the criticality of such factors in ensuring mission effectiveness and minimizing incidents of blue-on-blue engagements (mistaken attacks on friendly platforms) or blue-on-white engagements (mistaken attacks on neutral platforms).

As noted in chapter 1 of this book, the Tactical Decision Making Under Stress (TADMUS) program was initiated in response to one such incident: the accidental shooting down of an Iranian airbus by the USS *Vincennes* in 1988. The Congressional investigation of this incident resulted in the suggestion that emotional stress may have played a role in contributing to this incident, and the TADMUS program was established to assess how stress might affect decision making and what might be done to minimize those effects.

In any human–machine system, there are three possible approaches to addressing the human–machine system integration problem. One may select particular people on the basis of their skills and abilities. One may train people to enhance their skills and provide them with the knowledge needed to make optimal use of the system and their own capabilities and

We gratefully acknowledge his contributions as well as the substantial contributions of Jeffrey Grossman, Steve Francis, Brent Hardy, C. C. Johnson, Connie O'Leary, Mike Quinn, Will Rogers, Dan Westra, and our colleagues at the Naval Air Warfare Center Training System Division (NAWCTSD) during various phases of this project.

limitations. Finally, one may design the machine component of the system to accommodate and support the people using that system.

A fundamental premise of the TADMUS program was that the Navy had evolved a fundamentally effective system for selecting its commanders and tactical decision makers in the AEGIS combat system. It was plausible, however, that the Navy could make significant contributions toward improving the usefulness of these complex person–machine systems through the application of emerging theories of cognition and models of decision making to the areas of training and system design. Other chapters in this book address many of the training tools and techniques developed as part of the TADMUS program to facilitate effective decision making in real-time tactical systems.

This chapter addresses the ongoing work in developing improved human–system integration (HSI) for tactical decision makers. We begin by describing the design of integrated displays to support tactical decision making through recognition-primed and explanation-based reasoning processes. The methods and results of a study to evaluate these displays are then presented. The study placed expert Navy tactical decision-making teams in realistic simulations involving high-intensity, peacekeeping, littoral (coastal) situations. Teams used their current display system either alone or in conjunction with the enhanced, prototype displays. Finally, the lessons learned and design implications from this work are summarized, along with suggestions for using decision support displays in training.

Cognitive Analysis of Decision Tasks

Recent theories of decision making emphasize the importance of situation assessment for good decision making in naturalistic, event-driven situations. Moreover, they stress that decisions regarding actions to be taken are a by-product of developing the situation awareness that precedes action selection. Early TADMUS work focused on a family of cognitive theories that have come to be known as "naturalistic" decision making (Klein, 1989; Klein, Orasanu, Calderwood, & Zsambok, 1993). Naturalistic decision making differs from that found in the artificial intelligence–expert system literature in that these models are typically focused on emulating the outcomes of expert decision making by emulating the process a human decision maker might use in reaching the outcome. An example of these models is the analytic approach taken by many computers in playing chess. These programs typically consider all possible moves and countermoves in a computationally intensive manner and then make a move that represents the best solution from this exhaustive analysis. Human decision makers do not use this exhaustive, analytic approach, particularly not expert decision makers. The human expert typically looks at a situation and uses some general heuristic derived from his or her previous experience to choose a satisfactory action.

With regard to tactical decision making in a single-ship, peacekeeping mission, Klein found that the situation itself usually either determines or constrains the response options and that experienced decision makers make up to 90% of all decisions without considering alternatives. If the situation appears to be similar to one that the decision maker has previously experienced, the pattern is recognized and the course of action is usually immediately obvious. This has come to be known as *recognition-primed decision making* (RPD; Klein, 1989, 1993). On the other hand, if the situation does not seem familiar, a more complex form of decision making is involved, in which the decision maker considers general classes of explanations, selects from those that seem plausible to create a working hypothesis, and then rapidly adjusts this hypothesis after evaluating it. This less common form of decision making is referred to as *explanation-based reasoning* (EBR). In effect, this is decision making by telling a story to explain the discrepancies between expectations and what actually happens. As in RPD, the reasoning is not exhaustive but fairly short and concise, and the expert decision maker rapidly develops a reasonable hypothesis to explain the situation.

Additional support for these findings was found in an analysis of how experienced commanders make tactical decisions in realistic situations. Research was conducted to determine the decision requirements for command-level decision makers in the combat information center (CIC) of an AEGIS cruiser. Analysis of 14 actual incidents revealed 183 decisions; of these, 103 concerned situation assessments. Decision makers arrived at 87% of their situation assessments through feature matching, and the remaining 13% through story generation (Kaempf, Wolf, & Miller, 1993). Feature-matching strategies involve comparisons of observed data to sets of distinctive features or cues held in memory and based on the decision maker's training and experiences. When a match occurs, as a result of conscious or more automatic processes, the meaning of the observed situation becomes evident along with the appropriate responses. Story generation, on the other hand, involves more active EBR processes in which decision makers attempt to build a coherent story that accommodates the observed data, thereby providing a plausible interpretation of the situation. Story generation typically occurs when the patterns of observed data are unfamiliar or insufficient to allow a match to features of known situations held in memory.

The other 80 decisions that were identified from the analysis of real-world incidents involved course-of-action selection. Selecting courses of action involves determining what actions need to be undertaken to deal effectively with a particular situation. Kaempf et al. (1993) distinguished among strategies for course-of-action selection in which decision makers recognized the appropriate courses of action for the situation; selected a course of action from among multiple options; or generated a single, custom course of action that fit the details of the situation. These course-of-action decisions served a variety of functions, although relatively few were intended to end the incident. Twenty were intended as a final course-of-action decision, 14 were implemented to obtain more information, 22 to manage resources, and 24 to put the decision maker in a more favorable tactical position. A recognition-based strategy was used by decision makers to develop a final course of action most of the time. This strategy accounted for 95% of the actions taken in the 14 simulated incidents. The decision makers generated and compared multiple options in only 5% of the cases. In line with these findings, the TADMUS program adopted the position that decision support systems should assist in the decision-making *process* and focus on aiding the situation-assessment portion of the decision-making task.

Baseline tests in representative littoral scenarios further corroborated these analyses of tactical decision making among expert decision makers (Hutchins & Kowalski, 1993; Hutchins, Morrison, & Kelly, 1996). An analysis of communications patterns indicated a predominance of feature-matching strategies in assessing the situation, typically followed by the selection among preplanned response sets (tactics) that were considered to fit the situation. These tests also suggested that experienced decision makers were not particularly well served by current systems in demanding missions. Teams exhibited periodic losses of situation awareness, often linked with limitations in human memory and shared attention capacity. Environmental stressors such as time compression and presence of highly ambiguous information increased decision biases (e.g., confirmation bias, hypervigilance, and task fixation) and were correlated with tactical errors in executing the missions. Hutchins (Hutchins & Kowalski, 1993) cited several specific problems associated with short-term memory limitations, including the following:

1. Mixing up track (contact) numbers (e.g., track recalled as 7003 vs. 7033) and forgetting track numbers
2. Mixing up track kinematic data (e.g., track recalled as descending vs. ascending in altitude or as closing vs. opening in range) and forgetting track kinematic data
3. Associating past track-related events and actions with the wrong track and associating completed ownship actions with the wrong track.

Problems noted by Hutchins that related to decision biases included these:

1. Carrying initial threat assessment throughout the scenario regardless of new information that could be interpreted as indicating the initial assessment was inappropriate (e.g., framing error)
2. Assessing a track on the basis of information other than that associated with the track, such as old intelligence data, assessments of similar tracks, outcomes of unrelated events, and past experiences (e.g., confirmation bias)

Decision Support System Design

So that researchers could address the foregoing problems, the development of a prototype decision support system (DSS) became the focus of the TAD-MUS HSI development effort. The objective of this effort was to evaluate and demonstrate display concepts derived from current cognitive theory with expert decision makers in an appropriate test environment. The focus of the DSS was on enhancing the performance of tactical decision makers (viz., the commanding officer [CO] and tactical action officer [TAO] working as a team) for single-ship, air defense missions in high-density, ambiguous littoral warfare situations. The approach taken in designing the DSS was to analyze the cognitive tasks performed by the decision makers in a shipboard CIC, and then to develop a set of display modules to support these tasks on the basis of the underlying decision-making processes naturally used by the CO–TAO team.

Given that the CO and TAO decision makers were behaving in a manner consistent with that predicted by "naturalistic" decision-making theory (Klein, 1993), this theory became central to the design of a human–computer interface to improve tactical decision making. A prototype DSS was developed with the objectives of: (a) minimizing the mismatches between cognitive processes and the data available in the CIC to facilitate decision making, (b) mitigating the shortcomings of current CIC displays in imposing high information-processing demands and exceeding the limitations of human memory, and (c) transferring the data in the current CIC from numerical to graphical representations whenever appropriate. It was determined that the DSS should not filter or extensively process data; that is, it should support rather than aid (automate) decision making and leave as much decision making with the human decision makers as possible. The design goal of the DSS was to take the data that were already available in the system and present those data as meaningful information where, when, and in the form needed relative to the decision-making tasks being performed.

Decision Support System–1 Design

Version 1 of the DSS (DSS-1) was designed expressly for the evaluation of display elements to support feature matching and story generation (viz., RPD and EBR) with the goal of reducing errors, reducing workload, and improving adherence to rules of engagement. The design was significantly influenced by inputs from subject-matter experts to ensure its validity and usefulness for the operational community. It was implemented on a Macintosh computer to operate independent of, synchronized with, or linked to a scenario driver simulation. The design of DSS-1 was constrained by practical and research requirements. It was never intended to represent a display to be used aboard ship. The display was designed to complement

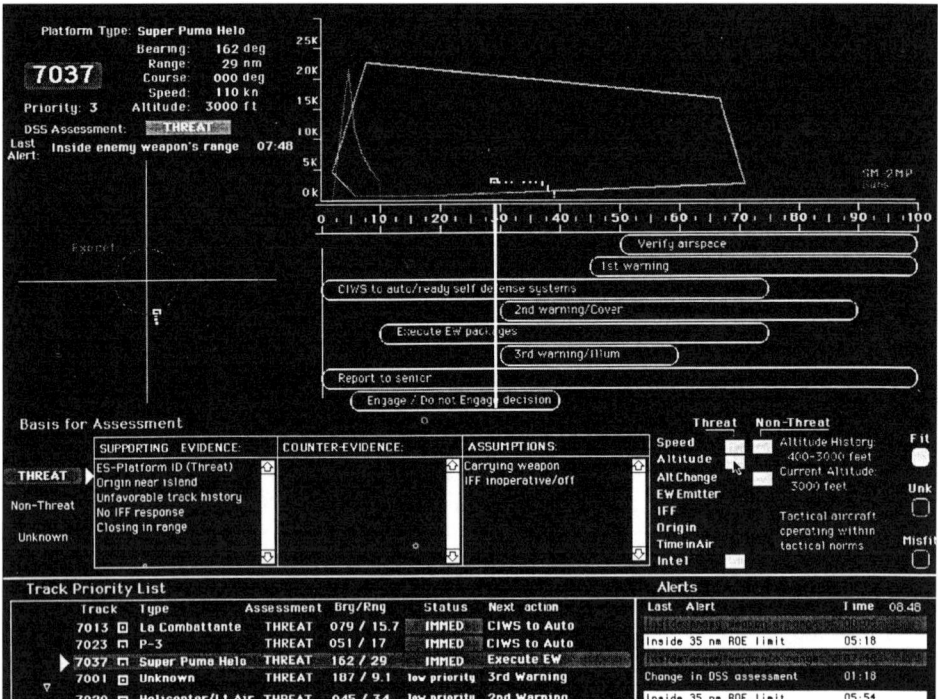

Figure 1.   Decision support system, Version 1.

an existing AEGIS geo-plot display so that performance could be evaluated with the AEGIS geo-plot as a baseline condition for comparison purposes.[1]

Figure 1 shows the DSS-1 prototype display. The DSS is a composite of a number of distinct display modules. Modules are arranged in a tiled format so that no significant data are obscured by overlapping windows. The DSS was conceived as a supplementary display to complement the existing geo-plot and text displays in current CICs. DSS modules have been discussed and demonstrated in detail elsewhere (cf. Moore, Quinn, & Morrison, 1996). Nevertheless, three of the modules are discussed in more detail as an illustration of how the DSS supports the information requirements of tactical decision-making tasks within the context of naturalistic decision-making theory.

Track profile. The track profile module consists of two graphical displays in the upper portion of the DSS that show the current position of a selected track in both horizontal and plan-form displays. Information requirements addressed by this module included the decision maker's need

---

[1]A geo-plot display presents a computer graphic representation of a geographic area, along with related information that is of concern to tactical decision makers. Geo-plot displays typically show land masses, political boundaries, and symbols that indicate the position of aircraft, ships, and submarines. Decision makers may choose to overlay other information on the geo-plot, such as commercial air lanes, navigation hazards, weapons envelopes, and exclusion zones.

to (a) see where the track is relative to ownship, (b) see what the track has been doing over time, (c) recognize whether the track can engage ownship, and (d) recognize whether ownship can engage the track. An important aspect of this display is that it shows a historical context of what the track has done in space and time (the history is redrawn each time the track is selected). This greatly reduces the short-term memory requirements on the CO and TAO in interpreting the significance of the selected track. This historical dimension of the display allows the decision maker to see what the track has done and primes his or her recognition of a likely mission for that track that could account for its actions. In addition, the profiles show ownship weapon and track threat envelopes displayed in terms of range and altitude so that the decision maker can visualize and compare mental models (templates) as he or she considers possible track intentions and own ship options.

Response manager. The response manager is located immediately below the track profile and is tied to it by a line indicating the track's current distance from own ship. It represents a Gantt chart type display showing a template of preplanned actions and the optimal windows in which to perform them. The display serves as a graphical embodiment of battle orders and doctrine, and it shows which actions have been taken with regard to the selected track. The display is intended to support RPD and serves the need to (a) recall the relevant tactics and strategies for the type of track being assessed, (b) recognize which actions need to be taken with the track and when they should be taken, and (c) remember which actions have been taken and have yet to be taken for the selected track.

Basis for assessment. This module is located in the lower left area of the DSS and is intended to support EBR (story generation). The basis-for-assessment module presents the underlying data used to generate the DSS's threat assessment for the displayed track. The display shows three categories of track assessment on which tactical decision makers focus: potential threat, nonthreat, and unknown. The decision maker selects the hypothesis he or she wishes to explore, and data are presented in a tabular format within three categories: supporting evidence, counterevidence, and assumptions. These categories were found to be at the core of all story generation in which commanders engage while deciding whether a track with the potential to be a threat is, in fact, a threat. This EBR related to threat assessment is also typically one of the decision-making tasks performed when one is deciding whether to fire on a track or not. The display was designed to present the data necessary for a commander to consider in evaluating all likely explanations for what a target may be and what it may be doing (i.e., "intent") through the generation of alternative stories to explain the available and missing data regarding the track in question. The display is also intended to highlight data discrepant with a given hypothesis to minimize confirmation and framing biases. Assumptions listed are the assumptions necessary to "buy into" the selected assessment. As a result, the basis-for-assessment module was expected to be particularly effective in helping sort out and avoid "blue-on-blue" (i.e., mistakenly shooting friendly contacts) and "blue-on-white" (i.e., mistakenly shooting neutral contacts) engagements.

Other Decision Support System–1 Modules

TRACK SUMMARY. The track summary module is located in the upper left of DSS-1. It summarizes current data related to a selected track. It is intended to support "quick look" RPD processes through data related to track kinematics and overall status (threat, neutral, or friendly) of the target. It was designed in response to the mental conversion and readability deficiencies of character readout (CRO) displays in the current CIC displays. This window also supports a drop-down track list for rapidly selecting individual tracks to be displayed, which are listed in order of priority.

COMPARISON TO NORMS. The comparison-to-norms module is located adjacent to the basis-for-assessment module. It determines how well the observed data for a track fit the established threat–nonthreat norms for that type of track. For instance, if the historical data seen for a track classified as hostile are consistent with established normal operating range (tactical norms) for a hostile track, there is a good fit of the data to the hypothesis with regard to the parameter. If the data are outside tactical norms for what the track would be expected to do given its classification but within its operational capabilities, it is a questionable fit. If the data are clearly inconsistent with that expected for that type of track with the current classification, there is a poor fit. The relative fit is shown by one of three colors in columns of color-coded "chips." Theoretically, the appearance of any misfit colors would signal the decision maker that there are ambiguous or conflicting data about the track, and cause him or her to consider alternative explanations or classifications for the track. In addition, this module supports EBR by allowing tactical decision makers to access detailed historical information summarizing the data categories for this track.

TRACK PRIORITY LIST AND ALERTS LIST. This module is located at the bottom of the DSS display and is intended to support the decision maker in the attempt to maintain general situation awareness (i.e., the "big picture"). It is intended to help the decision maker manage his or her own cognitive resources and attention. The track priority list presents the four highest priority tracks as well as a fifth track of special interest to the decision maker. The alerts list is linked to the track priority list. It normally displays the last alert issued for each displayed track. Clicking and holding the alert pulls up a list of all alerts in the system that have been issued for that track.

Decision Support System–2 Design

Figure 2 shows the current DSS-2 prototype display. The DSS-2 is a composite of several display modules that have been adapted from DSS-1 on the basis of user comments and performance tests. The DSS-2 is implemented with two 1024 × 768 CRT touch screen displays intended to simulate a single 2048 × 768 pixel display. The display features an integrated geo-plot and a variety of modules designed to solve specific decision-making problems encountered by the tactical decision maker. These modules are described in the following sections as an illustration of how the information requirements of tactical decision-making tasks were integrated with the cognitive processes described by naturalistic decision-making theory.

Geo-plot. The geo-plot occupies the left side of the DSS-2 display, shown in Figure 2. A close-up of the geo-plot is provided in Figure 3. The display uses variable-coded Navy Tactical Data System (NTDS) symbology (Nugent, 1996; Osga & Keating, 1994; Rausche, 1995) to represent the position of air, surface, and subsurface tracks over a geographic region. The module is intended to be the primary focus of decision makers and is designed for quick decision making associated with situation awareness and RPD processes. However, it also contains other geographic detail that could support EBR processes. Symbols are color and shape coded to indicate track identification and threat evaluation, and decision makers may toggle between NTDS symbols and track numbers. Track numbers are the "language" of the CIC and allow rapid location of various tracks. The map consists of several layers, which may be altered to suit particular mission requirements. Shown in Figure 3 is a desaturated map (Jacobsen, 1986; Van Orden & Benoit, 1994), which provides sufficient spatial reference for most tasks while minimizing screen clutter and excessive color for most decision-making problems. For strike or search-and-rescue missions, a topographical map may be overlaid to provide a more relevant context in which to frame tactical decisions. Likewise, overlays for infrastructure (e.g., highways, population densities, power grids) may be added when necessary to enhance the decision-making context. The use of a two-dimensional representation ensures that tracks can be located quickly with precision. An optional three-dimensional display format could be shown when appropriate for assessing general spatial relations. This three-dimensional format has not yet been fully implemented, and no data have been collected to validate its potential usefulness. Controls for altering the geo-plot are arranged along the left edge of the module. In addition to selecting map layers, scaling, and panning controls, the decision maker may supplement the display with velocity leaders to show the relative speed of all the tracks, with course histories that show a track's path over time relative to landmarks, with air corridors, and with other tracks. Weapon threat envelopes may be displayed for potential threat tracks, along with ownship, for rapid assessment of the criticality of a threat.

Whereas DSS-1 operated in conjunction with a geo-plot, that display

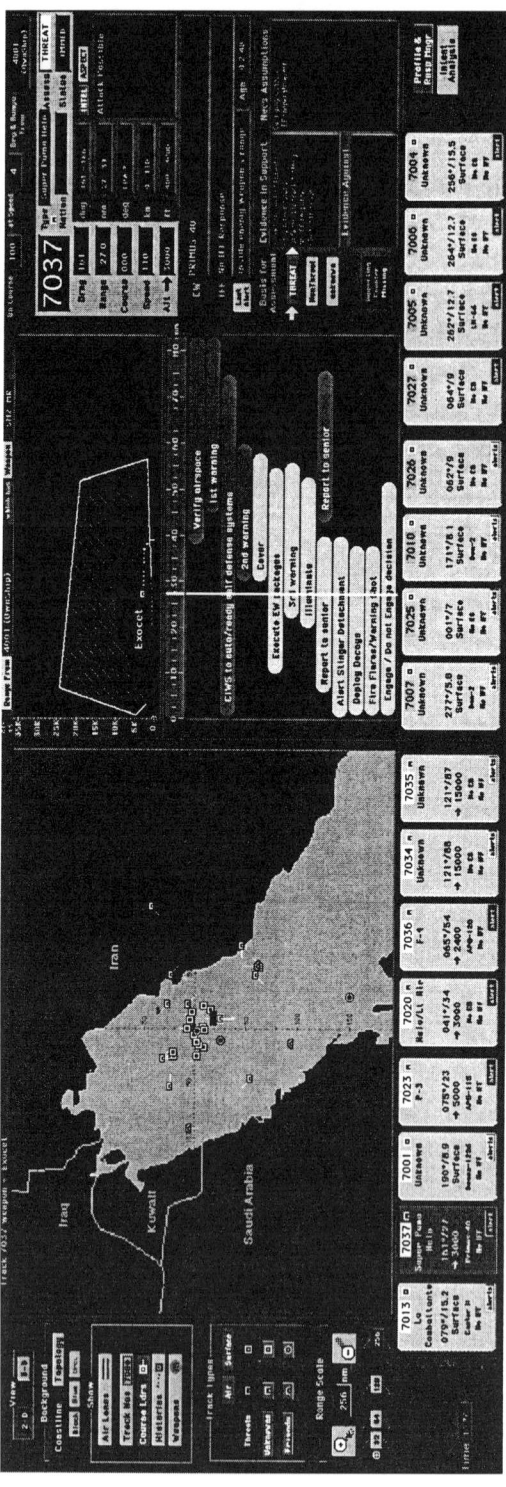

Figure 2. TADMUS decision support system integrated display. TADMUS = Tactical Decision Making Under Stress.

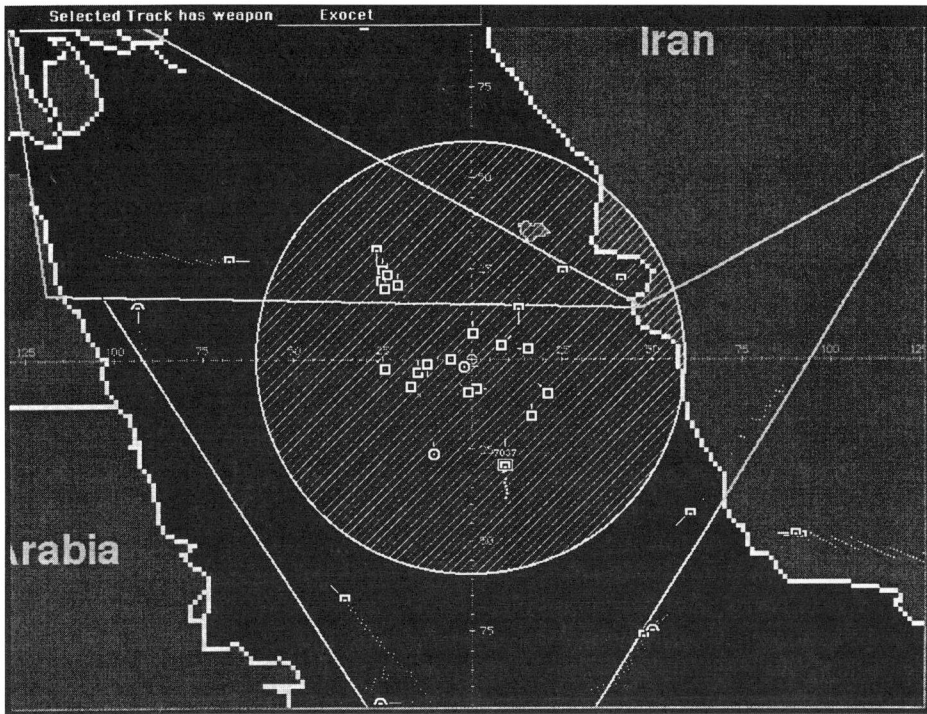

Figure 3. Geo-plot with desaturated map and variable coded symbology.

was operated independently of the DSS and had much less functionality, particularly with respect to overlays and alternate map types. On the basis of extensive comments and decision performance data from operational users, the DSS-2 was designed to incorporate the geo-plot as an integral part of the command display. This not only simplifies the human–computer interface dialogue by reducing the control actions required but also promotes better situation awareness by making it easier for decision makers to access and use all available tactical data displayed on the DSS.

Multi-character-readout access panel. Across the bottom of the DSS-2 display is a series of buttons for quick access to the highest priority tracks. A close-up view of one of these buttons is shown in Figure 4. The buttons serve as miniature character readouts (mini-CROs), displaying critical identification and kinematic information about the track and allowing the status of the most critical tracks to be monitored without additional interaction with the system. The buttons are arranged by a fairly simple algorithm in terms of their threat priority, with highest priority on the left and lower priority on the right. As situations evolve, the movement of the buttons relative to one another quickly draws attention to the changes, which helps preclude attention fixation to a single track or task when the decision maker is under stress. Thus, the collection of mini-CROs in this panel is intended to support RPD processes and to enhance situation awareness. The relative position of each mini-CRO, the color coding of

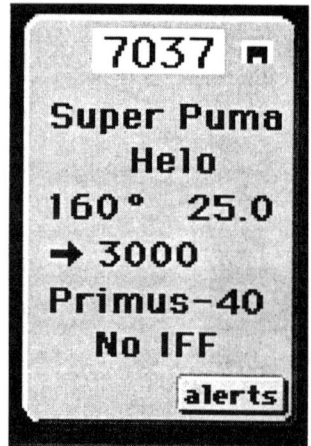

Figure 4. Sample mini-character readout. IFF = identification friend or foe.

threat level, and the summary of critical track information all support rapid feature matching by experienced tactical decision makers.

In the lower right portion of each track button is an alert button. This alert button is cyan colored when a new alert has occurred and gray when there are no new alerts. Pressing and holding the alert button generates a pop-up window to display a chronological list of alerts that have occurred for the selected track. Track age is shown rather than the time the alerts occurred (as in current systems) because decision makers are more interested in how old the alert is than when it occurred. This feature, which may be activated by users on demand, contributes to periodic need to engage in EBR processes. This implementation is an adaptation of the alerts list in DSS-1, which is based on feedback from operational users.

Track profile. The track profile, which is shown in Figure 5, is substantially the same as in DSS-1. The track profile module complements the geo-plot by showing a horizontal display of track altitude and range from own ship. Information requirements addressed by this module include the need for the decision maker to (a) see where the track is now, (b) see what the track has been doing over time, (c) recognize whether the track can engage ownship, and (d) recognize whether ownship can engage the track—all at a glance. The track profile also shows ownship weapon and track threat envelopes displayed in terms of range and altitude so that the decision maker can visualize and compare mental models (templates) as he or she considers possible track intentions and ownship options. Thus, the track profile supports RPD processes by experienced tactical decision makers.

To facilitate these evaluations further, the DSS-2 track profile incorporates two pull-down lists. The "perspective picker" allows the decision maker to change the viewpoint of the DSS to any other friendly (including air, surface, or land) unit in the area. This allows the decision maker to assess the possibility that other assets are the target of interest to a po-

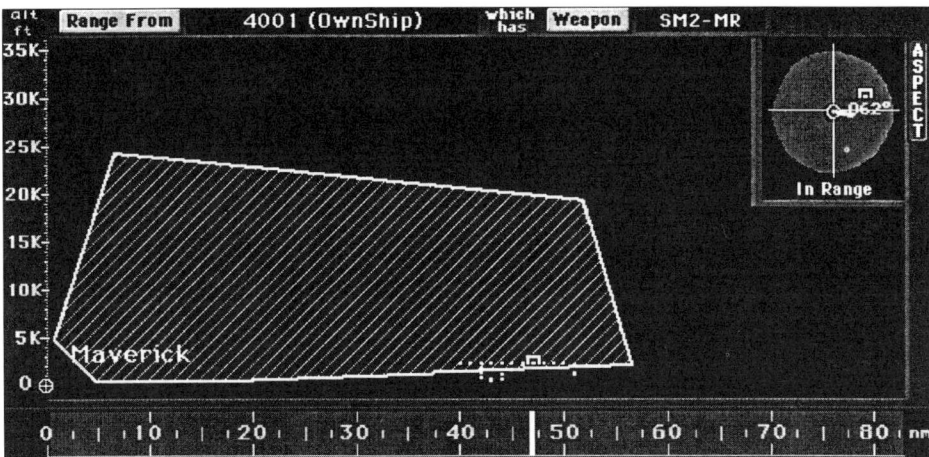

Figure 5.   Track profile with aspect inset.

tentially hostile track and to assess whether those friendly units could assist in engaging a prospective threat. When an alternative perspective is chosen, all modules in the display reflect the perspective and capabilities of the chosen asset. The second pull-down list is a "weapons picker" that further elaborates on capabilities and limitations of ownship and other friendly units showing how different weapons could be used against the selected track. In the event that a weapon system goes off-line, such information would automatically be reflected in the weapons displayed in the pull-down list. These capabilities are expected to play a significant role in joint (multiservice) and coalition (multinational) operations.

Embedded within the track profile is an inset window that shows ownship heading relative to the selected track. The display quickly shows radar cross section and weapons cutouts for assessing whether ownship should be maneuvered to optimize these parameters.

Response manager. The response manager is located immediately below the track profile and is linked to it by a line indicating the contact's current distance from ownship. As shown in Figure 6, it provides a Gantt chart type display, similar to that used in DSS-1, showing a set of preplanned actions and the optimal range windows in which to perform them. The display serves as a graphical embodiment of battle orders and doctrine, and shows which actions have been taken with regard to the selected track. The display is intended to support RPD and serves to (a) recall the relevant tactics and strategies for the type of target being assessed, (b) recognize which actions need to be taken with the target and when they should be taken, and (c) remember which actions have been taken and have yet to be taken for the selected target.

Track summary. The track summary module, shown in Figure 7, provides a more detailed summary of current and historical data for the currently selected track than was provided in DSS-1. Revisions were moti-

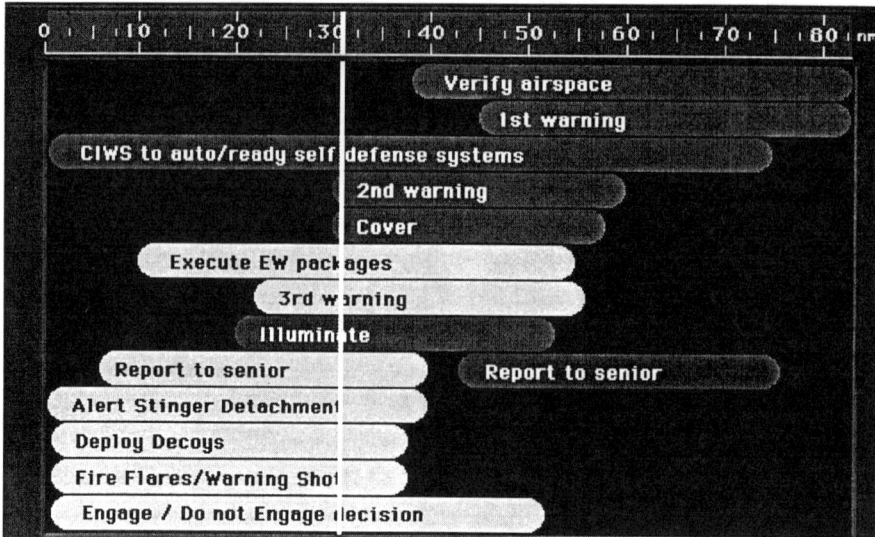

Figure 6. Response manager. EW = electronic warfare; CIWS = closed-in weapon systems.

vated by feedback from operational users concerning the DSS-1. The revised display provides a quick look at the track's kinematics as well as ancillary data, such as available intelligence, electronic warfare (EW), and identification friend or foe (IFF). Current information is shown in a cyan color, whereas historical or supplementary data are grayed-out. Kinematics (quantitative data) are read down from the track number, whereas track identifications (verbal data) are read across. Embedded within all

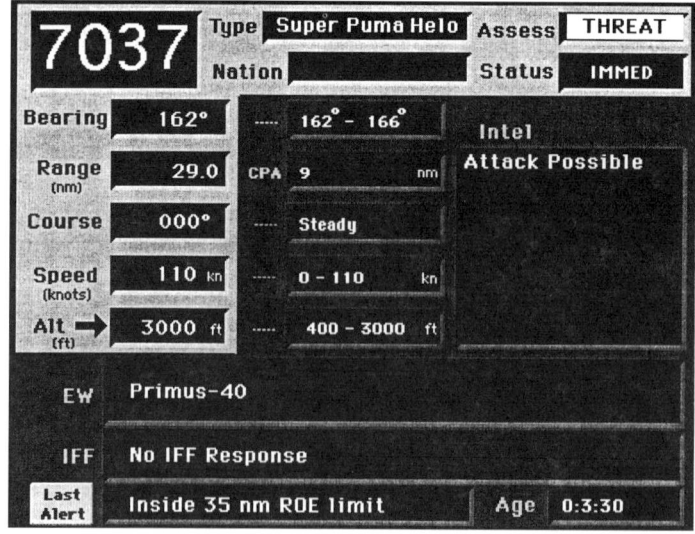

Figure 7. Track summary. EW = electronic warfare; IFF = identification friend or foe; ROE = rules of engagement.

CROs is a unique feature of the DSS: an altitude trend arrow that shows increasing, constant, or decreasing altitude. This is a critical feature of assessing threat intent that must be inferred by the tactical decision maker using conventional systems. There is a large pop-up window that may be accessed to provide a larger view of (possibly) more detailed alerts. This module is expected to be used when more detailed information is required for EBR.

Basis for assessment. The basis-for-assessment module is shown in Figure 8. It is similar in format to DSS-1, although substantial revisions are still under development. The basis-for-assessment module was explicitly designed to support EBR by providing a detailed listing of evidence for and against the current assessment of the selected track. It also presents important information that is unknown or unavailable and implicit assumptions made in accepting the assessment, which is described as potential threat, nonthreat, or unknown. This module supports EBR (story generation) by allowing the decision maker to explore alternative hypotheses and to see how the available data do or do not support them. The basis-for-assessment module presents the underlying data used to generate the DSS's threat assessment for the displayed track. The display was designed to present the relevant data necessary for a commander to consider and evaluate all likely explanations for what a target may be and what it may be doing (i.e., assess "intents") through the generation of alternative stories to explain the available and missing data regarding the target in question. The display is also intended to highlight data discrepant with a given hypothesis to minimize confirmation and framing biases. Assumptions listed are those necessary to "buy into" the selected assessment. Furthermore the assumptions are intended to prompt the decision

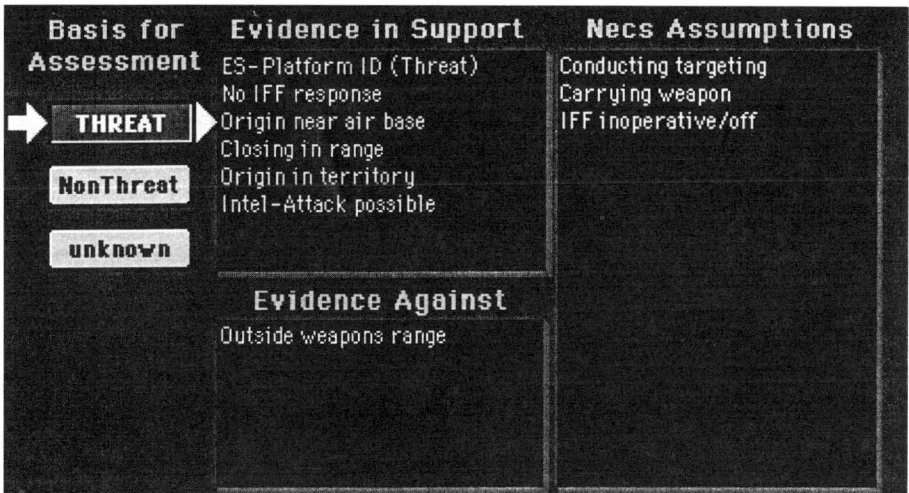

Figure 8. Basis for assessment. IFF = identification friend or foe; ES = electronic support.

maker to consider ways to resolve ambiguity. For instance, to assess a track as a threat, it may be necessary to assume that the track is carrying a weapon. The decision maker could then use organic assets such as friendly aircraft in the vicinity to fly out to the track to assess whether this is the case. As a result, the basis-for-assessment module is expected to be particularly effective in helping sort out and avoid blue-on-blue and blue-on-white engagements.

Decision Support System Evaluation Experiment

The ultimate goal of any display design is to influence positively the performance of the person—machine system of which it is a part. Therefore, a study was performed to examine how the DSS affected the decision making of COs and TAOs relative to performance in a traditional CIC in a medium-fidelity simulation. Although the contributions of individual display modules could not be assessed objectively owing to resource limitations, overall effects of the DSS on decision performance were examined in terms of a variety of performance criteria.

The results reported here address DSS-1 design issues. Other findings from this experiment have been reported elsewhere (Kelly, Hutchins, & Morrison, 1996; Morrison, Kelly, & Hutchins, 1996). Again, the DSS-1 did not feature an integrated geo-plot, but it relied on a geo-plot similar to that used in current tactical systems, provided as part of the Decision Making Evaluation Facility for Tactical Teams (DEFTT) simulator. The DSS-1 also had a comparison-to-norms module, which provided color-coded squares to show how well a set of critical parameters for the selected track fit a template for known threats and support pattern matching. This module was not well liked and was not used as had been intended. It was, therefore, dropped from the DSS-2. The DSS-2 is a refined version of the one tested, and it was refined on the basis of the results of this study.

As previously discussed, there is substantial evidence that experienced tactical decision makers employ feature matching and, to a lesser degree, story-generation strategies. Moreover, various errors observed during tactical scenarios and exercises have been linked with basic cognitive limitations (e.g., memory, attention). To build on the naturalistic decision strategies that experienced COs and TAOs use and to help overcome cognitive limitations, a series of decision support modules were developed. Because these decision support modules were developed with a user-centered design perspective, we expected that they would be effective in reducing decision maker errors. Also, we expected that COs and TAOs would consider these modules to be useful and easy to use, because they were consistent with the strategies these officers use in processing information and making decisions.

Method

Participants and support team. Sixteen active-duty U.S. Navy officers participated in this study as eight CO–TAO teams. These officers were highly experienced in air warfare tactical decision making, and several made up shipboard CO–TAO teams. The participants had completed the necessary training courses to be TAO qualified and had extensive shipboard experience standing watch as a TAO. Those participating as the CO were of appropriate ranks for this role (two captains and six commanders), had an average of 20.4 years in the Navy, and had been part of an average of 5.9 deployments. Those participating as the TAO also held appropriate ranks (five lieutenant commanders and three lieutenants), had served in the Navy for an average of 12.8 years, and had deployed an average of 4.0 times, mostly to the western Pacific and Persian Gulf.

A standard team of enlisted personnel served as the support staff for each CO–TAO team. These personnel served as electronic warfare supervisor, identification supervisor, tactical information coordinator, and anti-air warfare coordinator. They worked at consoles adjacent to the CO and TAO consoles within the DEFTT lab. The support staff were trained to use their displays and to provide information in a consistent manner across CO–TAO teams, which controlled a major source of extraneous variability.

Materials. This study was performed in the DEFTT laboratory at the Space and Naval Warfare Systems Center, San Diego, using dual-screen workstations for officers participating as the CO and the TAO. For each, one screen presented a standard geo-plot, and the other screen presented the DSS-1 display. Two large screen displays, comparable to those in many shipboard CICs, were also available.

Two training scenarios and four test scenarios were used. All scenarios were set in the Persian Gulf and involved a similar mix of fixed-wing aircraft, helicopters, and surface ships. The test scenarios, in particular, were constructed to simulate peacekeeping missions with a high number of targets to be dealt with in a short period of time (i.e., they were time compressed) and with a significant number of highly ambiguous tracks regarding assessment and intent. Analyses of the test scenarios indicated that although they differed in many details, they were roughly equivalent in decision workload overall. Each of the test scenarios had a duration of approximately 20 minutes.

Procedures. On arriving at the DEFTT laboratory, participants were given a prebriefing on the objectives and procedures for this study, oriented to the laboratory equipment and staff, and provided with necessary reference materials. Criterion-referenced training with the baseline DEFTT display system and with the DSS was then provided, and two practice scenarios were run before the test session was begun.

Participants were given appropriate geopolitical and intelligence briefings before each test run and were encouraged to take as much time as necessary to set up their displays the way that they wanted and to familiarize themselves with the tactical situation. This was done to reduce the artificial feeling of being dropped into the middle of a tactical situation that normally would have developed over several hours. During the scenario runs, the activity and communications from both the support team and various external sources were scripted. At various points in each scenario, CO–TAO teams were requested to report their primary tracks of interest. Subjective workload was assessed immediately following each test scenario using the National Aeronautics and Space Administration (NASA) Task Load Index (TLX; Hart & Staveland, 1988). Short breaks were provided between scenarios.

At the conclusion of the last test scenario, each participant was asked to complete a questionnaire that involved ratings of the usage, usefulness, and usability of each of the modules and of the DSS overall. Then, a brief structured interview was used to solicit comments about the strengths and weaknesses of the DSS, including suggestions for changing the displays and the information provided within them.

Experimental design and performance measures. A within-subject factorial design was employed across the four test scenarios such that each team performed two scenarios with the DSS and two scenarios without it. In all scenarios, the CO–TAO team had the use of geo-plot and CRO displays comparable to those in use in current ship CICs. The order of the scenarios and DSS conditions was counterbalanced using a Latin square procedure.

In addition to collecting objective data on tactical actions, display usage, control inputs, and voice communications, we solicited subjective assessments, through questionnaires and a structured interview, from each CO and TAO at the conclusion of the test session. The voice communications for each test scenario run were transcribed and analyzed to identify decision-making anomalies and communication patterns.

Results

Several classes of research questions were examined as part of this study. These concern usefulness and usability of the DSS, situation awareness, and team communications. Although the data discussed in the following sections were subjected to various statistical analyses, care should be taken in their interpretation. The small sample size and the simulated test environment used in this field research make it difficult to draw definitive conclusions about tactical decision-making performance. Nevertheless, these findings provide a strong indication of the potential usefulness of DSS displays for tactical decision making, particularly in littoral air warfare situations.

Usefulness. If COs and TAOs considered the DSS to be useful for tactical decision making, we would expect them to make use of it during the test scenarios when it was available. We also expected them to report that the information provided by the DSS was useful for their decision processes.

At 1-minute intervals throughout the test scenarios, subject-matter

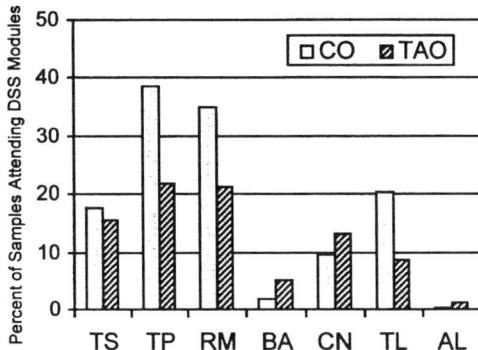

Figure 9. Mean percentage of time that the commanding officer (CO) and tactical action officer (TAO) attended to decision support system modules. TS = track summary; TP = track profile; RM = response manager; BA = basis for assessment; CN = comparison to norms; TL = track priority list; AL = alerts list.

experts recorded whether or not the COs and TAOs were attending to the DSS. On the average, participants were observed to be attending to the DSS for 66% of the time samples. For the remainder of the time samples, the COs and TAOs were observed attending to the geo-plot, which was on a separate display. In many cases, it was possible to determine to which DSS module or modules they were attending; these data are shown in Figure 9. It can be seen that the COs tended to use the DSS somewhat more than the TAOs, particularly for purposes of maintaining awareness of the behavior of individual tracks, the responses completed and pending, and the relative priority of active tracks. TAOs, on the other hand, tended to spend more of their time using the DSS to acquire quantitative track status and sensor data. Both COs and TAOs tended to use the upper half of the DSS (i.e., track summary, track profile, and response manager modules) more often. Self-reports of module use from the participants corroborated these findings.

At the end of the test session, participants were asked to complete a questionnaire that called for a variety of ratings of the DSS. Figure 10 shows the average ratings (on a 7-point scale) of the COs and TAOs concerning the usefulness of the information provided in the DSS modules. The ratings indicate that most modules were considered to be quite useful for tactical decision making, particularly those parts of the DSS designed to support quick decision making. COs and TAOs noted that these modules enabled them to extract key information rapidly and to visualize track behavior easily. Participants considered that the DSS overall was highly useful (average rating = 5.97 on a 7-point scale) for tactical decision making in littoral warfare situations.

Feedback from the CO–TAO teams who participated in this experiment indicated that the DSS provided them with an excellent summary of the overall tactical situation as well as of key data for individual tracks. In particular, COs and TAOs considered that both the track profile and the basis-for-assessment modules provided important information not

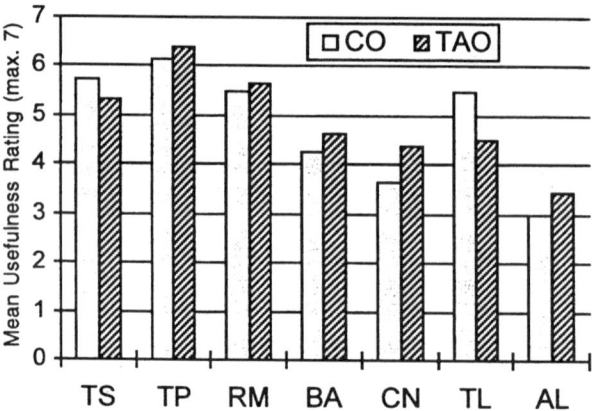

Figure 10.   Mean rating of the usefulness of Decision Support System (DSS) modules. TS = track summary; TP = track profile; RM = response manager; BA = basis for assessment; CN = comparison to norms; TL = track priority list; AL = alerts list; CO = commanding officer; TAO = tactical action officer.

readily available in present-day systems. Because the track profile module supported feature matching, which is the most commonly used decision strategy, its high rating was anticipated. However, when the track data are conflicting or ambiguous and when the decision maker has time available, the basis-for-assessment module was rated as helping substantially. By encouraging decision makers to consider the full range of available evidence along with various explanations for it, this module reduces the likelihood of mistakenly engaging friendly or neutral tracks and was rated highly with regard to avoiding blue-on-blue and blue-on-white engagements.

Situation awareness. Awareness of the tactical situation was examined through several performance measures. Specifically, it was predicted that if the CO–TAO team was more aware of the tactical situation in a peacekeeping mission, they would do the following:

- Identify the critical contacts earlier and more accurately.
- Take more of the defensive tactical actions required by the rules of engagement earlier, and take more of the provocative (offensive) actions later; that is, the window of time during which decision makers had to evaluate and handle a track would increase.
- Ask fewer questions to clarify previously reported track data and the relative locations of tracks.

CRITICAL CONTACTS.   During the scenario runs, the CO–TAO team was probed at prespecified times to identify the tracks that were considered to be of greatest tactical interest at that time. Their responses were contrasted with those of an independent group of five subject-matter experts. As shown in Figure 11, significantly more of the critical contacts were identified when the DSS was available. Significant differences (*p* <

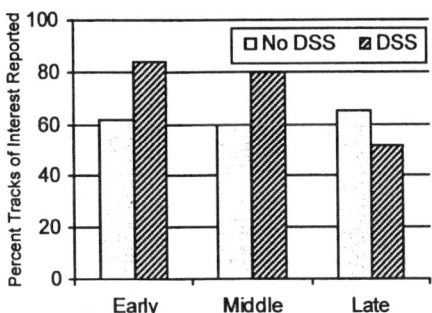

Figure 11. Percentage of critical contacts reported as tracks of interest. DSS = Decision Support System.

.05) were noted at both the early and midscenario probes; performance was comparable at the late probe, however. Late in the scenario, the critical tracks may become more obvious even without the DSS. Nevertheless, recognition of critical tracks earlier in the scenario affords decision makers a broader array of response options and permits more effective coordination of response actions.

TACTICAL ACTIONS. Using the rules of engagement as a benchmark for decision performance in the scenarios, a group of subject-matter experts assessed whether the CO–TAO teams warned or illuminated threat tracks at specified times and took appropriate defensive actions. A modified form of the Anti-Air Warfare Team Performance Index (ATPI) was used for scoring tactical performance (after Dwyer, 1992), and these data are summarized in Figure 12. In scenarios for which the DSS was available, CO–TAO teams were significantly more likely to take defensive actions in a timely manner against imminent threats ($p < .05$). This indicates that the DSS promoted an earlier recognition of the emerging risks of the tactical situation. By contrast, no difference was observed in the number of tracks that were warned or illuminated (i.e., provocative actions) when the DSS was available. However, several subject-matter experts contended

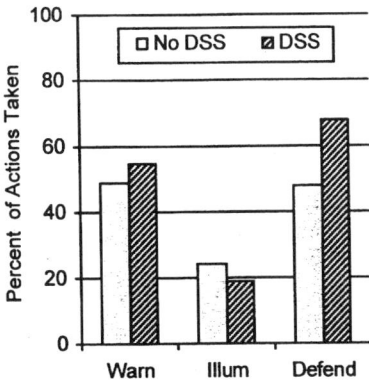

Figure 12. Team performance of tactical actions required by the rules of engagement. Illum = illuminate; DSS = Decision Support System.

that warnings and illuminations may not be diagnostic performance indices in these scenarios because they represent provocative tactical actions that commanders may consider to be inappropriate against certain tracks in a littoral situation. Not taking provocative actions would be appropriate and expected if commanders had assessed that the track was not an imminent threat and felt comfortable with deferring those actions because they had a good tactical picture—as would be expected if the DSS was effective in meeting its design objectives.

Team communications. Although the DSS was primarily designed to support an individual decision maker, the information that it provides could be expected to influence the team's collective decision process. One way that influence could be observed is through changes in the team's communications. Therefore, the communications rate, the pattern of communication, and the content of the communications were compared with and without the DSS.

COMMUNICATION RATE AND PATTERN.   It was hypothesized that when using the DSS, teams would have less need to exchange data verbally and would therefore communicate less often. To test this prediction, all voice communications that requested or provided information were tabulated for each of the 32 test runs. Because the lengths of the scripted test scenarios differed, the total number of voice communications observed was divided by the scenario duration to arrive at a communications rate.

Figure 13 shows the mean rate of communications originating with the CO, TAO, other members of their team, and others external to the ship's combat center (e.g., the battle group commander, the bridge). A general decrease in the rate of communications was observed with the DSS. This decrease remains fairly consistent regardless of who originated the communication. In fact, the pattern of communications was unaffected by

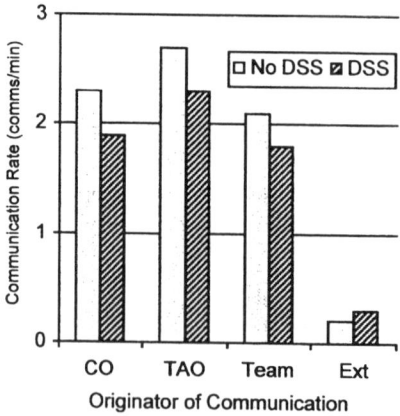

Figure 13. Voice communications rate by message originator and decision support system condition. CO = commanding officer; TAO = tactical action officer; Ext = external.

the presence of the DSS. About 40% of the communications occurred between the CO and TAO, and another 35% occurred between the TAO and the team. Each of the remaining links accounted for about 5% or less of the total communications. The decrease in communications across positions suggests that the DSS supported the entire team by providing basic data about tracks, thereby reducing the team's need to request or provide such data verbally.

COMMUNICATIONS CONTENT.   Whereas the pattern of communications was not found to be affected by the presence of the DSS, it was possible that the content of the teams' communications was altered by the DSS. That is, without the DSS, teams might need to spend more time exchanging basic track data, whereas teams using DSS might spend the bulk of their time assessing track intent or evaluating alternate courses of action. To explore the possibility of a qualitative trade-off in communications, voice communications were coded by their message content according to the following scheme:

- *Information*—exchange of sensor-based data
- *Status*—exchange of procedure-based data
- *Clarification*—redundant communication to elucidate, interpret, or correct other communications
- *Correlation*—association of two or more data
- *Assessment*—discussion of expected track behavior, likely intent, or future actions
- *Orders*—commands to perform an action

Figure 14 shows the overall average proportion of communications observed for each of these content categories. The largest proportion involved informational communications, in which sensor-based data were exchanged. This, of course, is not surprising, because these data effectively drive the decision processes. The *rate* of these communications, however, was found to be significantly lower when the DSS was available ($p < .05$).

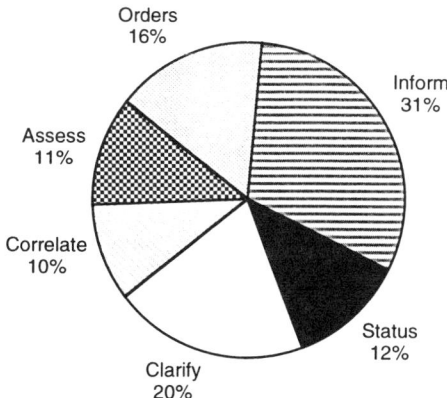

Figure 14.   Mean proportion of communications by content category.

Because the DSS provides much of these data, there was less need for verbal exchanges among the team. Similarly, fewer correlation communications were observed when the DSS was used. Although decision makers were less likely to ask about or report correlation data with the DSS (because much of it is displayed automatically), they were somewhat *more* likely to talk about correlations in the data that they observed on the DSS. This effect is consistent with the intended purpose of the DSS in supporting situation assessment and track evaluation. No differences between DSS and No DSS runs were observed in the other communications content categories.

CLARIFICATION COMMUNICATIONS.   Overall, about 20% of the communications were for clarification purposes, reflecting uncertainty about track location, kinematics, identification, status, or priority. Figure 15 shows the type of information that was discussed during clarification communications. The relative percentage of each type of clarifying communication is shown for when the DSS was available and when it was not available. It can be noted that clarification communications about track location (e.g., locating the symbol on the geographic display that corresponds to the track of interest) and track status (e.g., response to warnings) were equally likely whether or not the DSS was available. In contrast, clarification communications about track kinematics (e.g., speed or altitude), EW information (e.g., IFF and emitter signature), and tactical picture (e.g., track identity and relative position) were less likely when the DSS was available. Clarifications regarding ambiguous orders (e.g., incorrect track number) were somewhat more likely to occur when the DSS was available.

The pattern of these findings is quite revealing concerning the information in the DSS that was used by tactical decision makers. Because the multitrack geo-plot display and the DSS were not linked in this study, the DSS by itself provided little help in locating tracks in relation to each

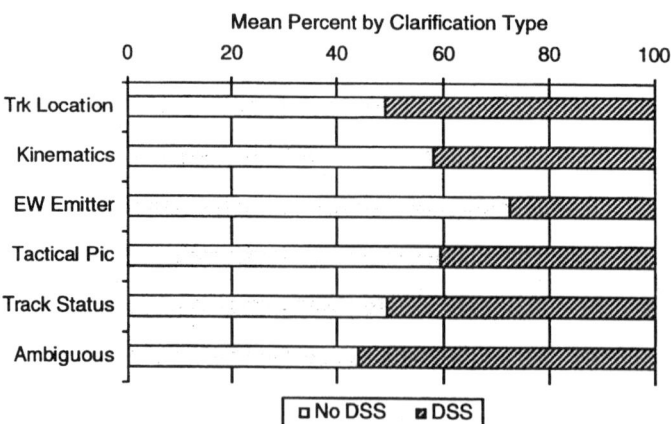

Figure 15.   Relative percentage of clarification communications with and without the decision support system. DSS = Decision Support System; Trk = track; Pic = picture; EW = electronic warfare.

other. The DSS also provided no information to decision makers about the status of actions taken or about tracks' responses to warnings. Therefore, no difference between DSS and No DSS conditions was expected for these types of clarifications, and the data confirmed this prediction.

In contrast, the DSS track summary module showed available track kinematic data; the supplementary data readout in the comparison-to-norms module summarized EW information; and the track profile and track priority list modules helped decision makers prioritize and maintain awareness of the overall tactical picture. Therefore, the DSS was expected to reduce the need to clarify these types of information. Again, these predictions were confirmed, especially for the EW information, which was consulted frequently in making threat assessment decisions.

Although it was not statistically significant, the tendency for there to be more clarification communications about ambiguous orders when using the DSS was interesting. In fact, we observed a tendency for decision makers to be more precise in referencing tracks when using the DSS. This increased precision thereby encouraged the team to ask for clarification about which track was being referenced when ambiguous orders occurred.

This detailed analysis of clarification communications confirmed that decision makers did indeed use the information that was displayed in the DSS, even when their experience with it was limited. This analysis also revealed ways in which the DSS might be enhanced to reduce further the burden of clarification communications. For example, linking the geo-plot display with the DSS such that selecting a track on either display would highlight it on the other would probably reduce the communications required to clarify track location. This enhancement was incorporated into the DSS-2 design, as previously described. Similarly, enhancements to the response manager module to show the status of actions and responses by tracks would be likely to reduce the need for decision makers to ask repeatedly about such track status information. Current work is exploring these and other enhancements to the response manager module.

COMMUNICATIONS ABOUT CRITICAL CONTACTS.   The tracks to which the teams' communications referred were also examined under the DSS and the No DSS conditions. The hypothesis was that the DSS would enable teams to focus on the critical contacts more quickly, resulting in a greater proportion of their communications focusing on those tracks. The average proportion of communications about the critical contacts was slightly greater (but not significantly greater) when the DSS was used. It is not particularly surprising that these teams concentrated the bulk of their communications on the critical contacts regardless of whether or not they were using the DSS. After all, these were highly experienced tactical decision makers who were accustomed to functioning effectively with their current (non-DSS) systems. Greater effects might be obtained with less experienced decision makers.

Usability. It is not sufficient that the DSS provide useful information, promote better situation awareness, and facilitate team communications;

it must also be usable. That is, decision makers should consider it easy to learn, easy to understand, and easy to use.

Participants were asked to complete a questionnaire following the test session in which they rated the usability of the DSS and its modules. Comments about the DSS interface were also solicited in the questionnaire and during a follow-up interview. The overall rating of DSS usability was high (average rating of 4.16 on a 5-point scale). Also, most modules were considered easy to use, as shown in Figure 16. The modules that promoted "quick-look" assessments of track status, location, and priority were rated as more usable. The modules that were predominantly text based, particularly the alerts list, were rated as less easy to use.

The COs and TAOs offered many valuable suggestions for improving the DSS to make it more useful and usable for tactical decision making, and these were incorporated into the DSS-2. Frequently heard suggestions included the need to (a) integrate the DSS display with the geo-plot to simplify track selection procedures and to display multiple tracks; (b) allow user-customizable display areas and content, particularly for control window size, range scale, and response manager actions; (c) allow command–override of track priorities and threat assessments; and (d) provide an expanded track priority list that shows more tracks in a graphic–spatial format.

Although several of the usability suggestions offered by experienced tactical decision makers concerned improvements to the "look and feel" of the DSS interface, others had substantial implications for the understanding of their underlying naturalistic decision processes. The call to integrate the DSS and geo-plot displays and to provide more tracks in the track priority list reminded us that feature-matching decision strategies involve evaluating tracks within the context of other related tracks and events. Decision makers' requests to permit them to customize their displays and to override system defaults have implications for their story generation and explanation-based reasoning strategies. Decision makers seem to use a "stepping-stone" approach whereby they use available data to reach an

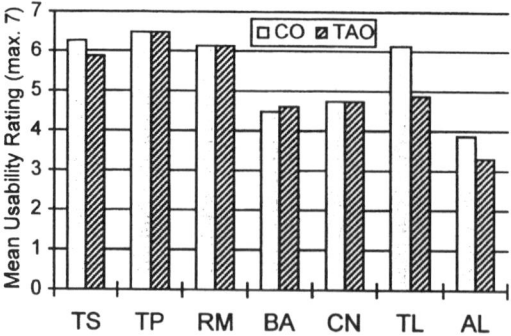

Figure 16. Mean rating of the usability of decision support system modules. TS = track summary; TP = track profile; RM = response manager; BA = basis for assessment; CN = comparison to norms; TL = track priority list; AL = alerts list; CO = commanding officer; TAO = tactical action officer.

intermediate conclusion about a track, such as its priority. On the basis of that intermediate conclusion, they continue to use other data to explore further implications for that track, given its priority. To support such a process, it is clear that decision makers would want to have the ability to override and customize their DSS displays.

Lessons Learned

Lesson 1

Experienced tactical decision makers' reactions and performance indicated that the DSS provided them with useful information in a readily understood form. An advantage of the DSS over current systems is that it supports "quick-look" decision processes (i.e., RPD), which are typically used in tactical situations. Naturalistic decision processes are further supported by the presentation of data in an operationally relevant context, which involves historical, geopolitical, and tactical doctrine components.

Implications of Lesson 1

A user-centered approach to decision support and display design is important for achieving user acceptance and ecologically relevant performance enhancements. A thorough understanding of the users' information requirements and decision processes provides a critical foundation for successful design efforts. This approach allows data to be viewed in relation to a specific decision situation (e.g., embedding a track's kinematic data in graphical threat templates).

Lesson 2

User reactions to the DSS indicated that tactical decision makers appreciated access to data that were not heavily filtered or preprocessed. Algorithms that were implemented (e.g., the track prioritization algorithm) were kept simple and understandable. The DSS makes use of tactical data already available in the CIC without fusion into more complex, abstract concepts, which gives the COs and TAOs greater confidence in their decisions, because they can maintain conceptual links to basic physical data (e.g., a contact's kinematics—range, speed, altitude). The DSS organizes these data around key tasks that tactical decision makers need to perform. This display organization makes it easier for experienced decision makers to get the information that they need quickly and to note inconsistencies or ambiguities in the data. Because the display is structured around specific operational tasks, decision makers were able to navigate the display efficiently and extract the required information. For less experienced decision makers, the DSS structures their information search so that they are better able to understand the relationships among the data and how the data can be used. As a result, less experienced decision makers function more like experts.

Implications of Lesson 2

Decision support and display design efforts should involve careful consideration of how automation is incorporated. Human decision makers require access to the underlying data to enable situation assessment on the basis of the patterns detected in these data. A promising design approach is to organize archival sensor data into modules that support critical decision-making tasks.

Lesson 3

DSS display modules that used graphics and that integrated information from several sources were used most often and were considered the easiest to learn and use.

Implications of Lesson 3

Time stress, multiple concurrent task requirements, and situation uncertainty characterize operational decision making. Displays that enable decision makers to recognize data patterns quickly provide substantial support under these conditions. Whenever possible, designers should make use of graphic displays and user-selectable overlays designed to enhance the decision-making context of tactical data, with color, shape, size, and position coding schemes.

Lesson 4

Early TADMUS research suggested that the user interface to existing tactical display systems required the decision makers to spend excessive time interacting with displays rather than making decisions. Field studies and analyses have revealed that with some current tactical systems, the user interface dialogue places an enormous burden on operators, requiring as much as a thousand control actions an hour (Osga, 1989). It was concluded that time spent interacting with the systems detracts from time available for making tactical decisions.

Implications of Lesson 4

Decision support systems should be designed to minimize the amount of interaction required of the decision maker to extract information. Interfaces that require the user to take overt actions to obtain information (e.g., selecting between overlapping windows, activating pop-up windows, using pull-down menus) interfere with the user's ability to process information and should be minimized. The use of tiled windows, with distinct modules organized around specific decision-making problems, is preferred when time is compressed, or the data being evaluated are complex and ambiguous.

Lesson 5

Development of the DSS did not follow a simple or straight path. The design was the product of extensive discussions among staff from many different disciplines and backgrounds. In addition, many forms of testing and evaluation of prototype displays were employed, from tabletop critiques to formal evaluation experiments.

Implications of Lesson 5

The design of decision support displays needs to draw on the unique skills of professionals in many complementary fields, particularly human factors engineers, subject-matter experts, and cognitive scientists, brought together as an integrated product team. Iterative testing with representative users needs to be conducted, in different forms, throughout development. In the early phases, testing of concepts can employ storyboards and static mock-ups. As the design matures, structured feedback from users can be obtained through use of scripted scenarios with a dynamic mock-up. Finally, more formal laboratory and field studies can and should be performed with prototype systems to evaluate its effectiveness.

Implications for Future Research Into Training Applications

In addition to its use as a real-time decision support tool, the DSS may have substantial value for training the complex cognitive skills that characterize expert tactical decision makers. Because the DSS organizes and presents tactical data in a form that is consistent with experts' usage patterns, it guides decision makers through the huge amount of tactical data available in the CIC. Essentially, the DSS is expected to help an intermediate decision maker quickly access relevant data, identify important patterns, and develop a higher level of expertise. In this regard, the DSS may be useful not only for initial training but also for refresher training.

In a training mode, we envision that the DSS could be central to teaching the application of tactical decision-making skills and developing expertise. Because the DSS has been designed to support expert decision makers, it is reasonable to expect that it would prove useful in shaping the decision-making strategies of novice decision makers and developing expert decision-making skills. There is the implication, however, that there would be a need to call up from a library particular scenarios that are appropriate for the tactical decision-making skills to be trained. There would need to be a comprehensive training strategy for developing expert decision-making skills as well as appropriate training objectives and performance criteria. The scenarios presumably would be sequenced to fit within an appropriate curriculum, leading trainees from more basic skills to more advanced skills under more demanding tactical situations (i.e., greater workload, more ambiguity). Nevertheless, the system could be flexible enough to enable detection of a trainee's level of expertise and particular training needs and adjustment of the curriculum accordingly. During training, the decision maker could be guided through the use of tactical data shown on the DSS as the scenario unfolded. Intelligent agents, wizards, balloon help, or other such HSI tools could "pop up" at appropriate points in the scenario to advise trainees on data relationships, requirements to shift attention, or specific applications of other key tactical skills. At the conclusion of each training scenario run, key indicators could be calculated to assess decision-making performance and provide rapid feedback to the trainee. The DSS would also need to include other standard features that support debriefing and evaluation following training exercises, such as the ability to replay selected parts of the scenario, to view selected points in time, and to annotate the scenario with comments.

To extend the DSS to support this type of tactical decision-making training, several research and development activities are required, which are being actively pursued as part of the ongoing TADMUS project. These include:

1. Identification of key tactical decision-making skills, knowledge, and abilities
2. Development of a training curriculum for producing (or sharpening) these skills
3. Development of a comprehensive library of scenarios that exercise these skills under various workload demands
4. Design and testing of intelligent agents, wizards, and other HSI tools that could be incorporated into the DSS as "over-the-shoulder" advisors or synthetic expert instructors
5. Investigation of diagnostic decision performance measures that could be compiled automatically to indicate whether the trainee mastered the skills

Conclusion

Operational decision making predominantly relies on feature-matching strategies. To a lesser extent, when faced with conflicting or ambiguous data, decision makers employ story generation or explanation-based reasoning strategies. Displays that are consistent with these naturalistic decision-making strategies provide the most useful support to commanders, facilitating the rapid development of an accurate assessment of the situation. Displays that support both feature-matching and explanation-based reasoning are recommended for complex decision-making tasks. Although the feature-matching displays will likely be used far more often, the explanation-based reasoning display is of substantial value under certain circumstances, particularly with less experienced decision makers.

The DSS was developed for application to Navy tactical decision making on a single ship in support of air warfare in dense, fast-paced littoral settings. With some adaptation, it could support other military decision situations, including concurrent decisions involving other warfare areas; higher level, supervisory decisions involving multiship battle groups; and even collaboration among tactical decision makers in joint service or multinational (coalition) operations. Several new research projects are underway to explore these applications. In addition to these direct applications to support military decision making, the decision support and display principles identified through this effort are relevant to other complex decision-making settings, such as nuclear power control, flight control, process control, and disaster relief planning. Furthermore, researchers are looking at developing derivative displays that reflect emerging theories of decision making, extension of the DSS concepts to other workstations within the CIC, and better integration of DSS modules with shipboard data-processing systems.

References

Dwyer, D. J. (1992). An index for measuring naval team performance. In *Proceedings of the Human Factors Society 36th annual meeting* (pp. 1356–1360). Santa Monica, CA: Human Factors Society.

Hart, S. G., & Staveland, L. E. (1988). Development of NASA-TLX (Task Load Index): Results of empirical and theoretical research. In P. A. Hancock & N. Meshkati (Eds.), *Human mental workload* (pp. 139–183). Amsterdam: North-Holland.

Hutchins, S. G., & Kowalski, J. T. (1993). Tactical decision making under stress: Preliminary results and lessons learned. In *Proceedings of the 10th Annual Conference on Command and Control Decision Aids*. Washington, DC: National Defense University.

Hutchins, S. G., Morrison, J. G., & Kelly, R. T. (1996). Principles for aiding complex military decision making. In *Proceedings of the Second International Command and Control Research and Technology Symposium* (pp. 186–203). Washington, DC: National Defense University.

Jacobsen, A. R. (1986). The effects of background luminance on color recognition and discrimination. *Society for Information Display Technical Digest*, 78–81.

Kaempf, G. L., Wolf, S., & Miller, T. E. (1993). Decision making in the AEGIS combat information center. In *Proceedings of the Human Factors and Ergonomics Society 37th annual meeting* (pp. 1107–1111). Santa Monica, CA: Human Factors and Ergonomics Society.

Kelly, R. T., Hutchins, S. G., & Morrison, J. G. (1996). Decision process and team communications with a decision support system. In *Proceedings of the Second International Symposium on Command and Control Research and Technology* (pp. 216–221). Washington, DC: National Defense University.

Klein, G. A. (1989). Recognition-primed decisions. In W. R. Rouse (Ed.), *Advances in man–machine systems research* (Vol. 5, pp. 47–92). Greenwich, CT: JAI Press.

Klein, G. A. (1993). A recognition-primed decision (RPD) model of rapid decision making. In G. A. Klein, J. Orasanu, R. Calderwood, & C. E. Zsambok (Eds.), *Decision making in action: Models and methods* (pp. 138–147). Norwood, NJ: Ablex.

Klein, G. A., Orasanu, J., Calderwood, R., & Zsambok, C. E. (Eds.). (1993). *Decision making in action: Models and methods.* Norwood, NJ: Ablex.

Moore, R. A., Quinn, M. L., & Morrison, J. G. (1996). A tactical decision support system based on naturalistic cognitive processes. In *Proceedings of the Human Factors and Ergonomics Society 40th annual meeting* (p. 868). Santa Monica, CA: Human Factors and Ergonomics Society.

Morrison, J. G., Kelly, R. T., & Hutchins, S. G. (1996). Impact of naturalistic decision support on tactical situation awareness. In *Proceedings of the Human Factors Society 40th annual meeting* (pp. 199–203). Santa Monica, CA: Human Factors and Ergonomics Society.

Nugent, W. A. (1996, September). Comparison of variable coded symbology to a conventional tactical situation display method. In *Proceedings of the 40th annual meeting of the Human Factors and Ergonomics Society* (pp. 1174–1178). Santa Monica, CA: Human Factors and Ergonomics Society.

Osga, G. A. (1989, March). *Measurement, modeling, and analysis of human performance with combat information center consoles.* (NOSC Technical Document 1465). San Diego, CA: Naval Ocean Systems Center.

Osga, G. A., & Keating, R. L. (1994, March). *Usability study of variable coding methods for tactical information display visual filtering* (NRaD Technical Document No. 2628). San Diego, CA: Naval Command, Control and Ocean Surveillance Center, Research Development Test and Evaluation Division.

Rausche, D. P. (1995, March). *Variable coded symbology and its effects on performance.* Unpublished master's thesis, Naval Post Graduate School, Monterey, CA.

Van Orden, K. F., & Benoit, S. (1994, March). *Color recommendations for prototype maritime symbolic displays* (NSMRL Tech. Rep. No. 1192). Groton, CT: Naval Submarine Medical Research Laboratory.

Part V

Reflections on the TADMUS Experience

17

Lessons Learned From Conducting the TADMUS Program: Balancing Science, Practice, and More

Eduardo Salas, Janis A. Cannon-Bowers, and Joan H. Johnston

Over the last 10 years, the military research and development (R&D) community has invested considerable resources in understanding the nature of team decision making in complex environments. Investment of the magnitude involved in this program brings high visibility that is both a blessing and a curse. On the one hand, doors are opened and people interested because they recognize that such an investment can yield important and groundbreaking contributions. On the other hand, the pressure is on program personnel to show the program's value even before any work has been done. For example, sponsors, operational personnel, and scientists who became aware of TADMUS consistently challenged us to answer the following questions: What would be the payoff? How would we know that we had succeeded? What kinds of products would be generated? How would transition or implementation of the products be accomplished? Is the field of decision making mature enough to enable completion of this program? Would the fleet understand the objectives and would they collaborate? Is there anything new that would advance the science and practice of decision making under stress? Of course, it was impossible to answer all of these questions completely; however, it can be seen from the chapters within this book that this program of research has made a significant contribution to the science of decision making under stress and to the operational Navy community.

Our message in this chapter is that getting here was neither easy nor straightforward; it was a challenge. Our purpose here is threefold: First, we discuss our observations and reflections about studying decision making in operational environments. A second class of observations is more general; these observations relate to the challenges associated with conducting a large-scale program of applied research. As we go along, we distill from these experiences a set of lessons learned that emerged from our experiences. Finally, we offer some concluding remarks.

Studying Team Decision Making "in the Wild"

As we reflect on the events that led us to the present, we must confess that the problem that we undertook to study—team decision making in command and control—was much more complex than we had originally understood. Our scientific training did not fully prepare us to face the challenges associated with designing or executing real-world research. Of course we understood experimental design (and quasiexperimental design); we could apply psychometrics; and we understood data analysis techniques; what we were not prepared for was the complexity we would confront in designing a viable, user-accepted test bed; deciphering the jargon and acronyms that characterized the environment; understanding the decision-making tasks themselves; and building measures that were valid, useful, and acceptable to our participants.

Lesson 1: Do not underestimate the complexities of conducting meaningful field research.

This statement implies, among other things, that researchers have to be fully committed to understanding the tasks and people they are studying, which requires a sizable investment. It also implies that at least part of the research team has to be made up of task experts (either through a "crash course" for the scientists or by finding task experts to join the research team). In fact, our strategy was to find "translators" in the fleet who could act as liaisons between the scientists and the job incumbents. We also had to spend a lot of time on ships and in training commands, which leads to our second lesson.

Lesson 2: Do not be afraid to get your "hands dirty"; it will pay off in the end.

A second benefit of getting one's "hands dirty" is that it has a tendency to buy credibility. One "talks the talk and walks the walk." In our case, this meant that because it was clear to the uniformed personnel that we were willing to make the investment in understanding the task, even if they disagreed with us they could not criticize our knowledge of the domain.

Lesson 3: Credibility enables the creation of partnerships between researchers and users.

The importance of partnerships cannot be underestimated. Partnerships allow users to buy into and create a sense of ownership. Nothing pleases (or gratifies) us more than to see our words in briefings given by the fleet without attribution. This means that our customers believe so completely in what we have said that they incorporate the thoughts into their own way of thinking. This partnership, however, does not come overnight. It requires careful managing of agendas, misunderstandings, egos, and personalities. Mainly it results from building trust and respect. If we can show that we can help improve our partners' work environment, they will help us. It is worth the effort.

Lesson 4: Partnerships create ownership of the problem and solution.

Ownership on the part of users is a double-edged sword. On the one hand, we need users to be supportive, as described in the preceding lesson. On the other hand, we have often been in the (some would argue enviable) position of having our products pulled away from us before they are fully ready for implementation. The problem is that if products undergo transition too soon, they face a higher chance of failure. In addition, R&D sponsors tend to think that anything that has already undergone transition should not be funded further. The danger is that an otherwise promising solution is never brought to fruition because it undergoes transition too soon.

Lesson 5: Do not be afraid of getting the users involved, but manage the process.

Overall, then, we conclude that studying decision making in the wild—or any complex phenomenon—requires a sizable, long-term investment on the part of the research team. It requires a motivation to understand and solve problems. It requires an ability and desire on the part of the scientists to handle ambiguous information and often contradictory messages from users and sponsors. A major (side) benefit of allis is that the task can be extremely gratifying. When the scientist–practitioner model works, it is an exciting and fulfilling experience.

Surviving a Large-Scale Program of Research

Besides the lessons we learned regarding the study of decision making in natural environments, we also learned a few that relate to the conduct of any large-scale R&D effort.

We begin by warning that within any highly supported, visible program, there are bound to be people who do not believe. These folks usually fall into one of four categories: (a) those who do not believe that the appointed researchers can do it, (b) those who believe that they can do it better, (c) those who believe that it has already been done, and (d) those who think it is impossible. Depending on where they sit, these people can be either mild irritants or dangerous adversaries.

Lesson 6: Do not listen to people who say it cannot be done.

This leads to a related notion: In a large program there are many constituencies—people who have a vested interest (positive or negative) in the results. The problem is that one program cannot possibly satisfy everyone's requirements. The project team needs to figure out what can reasonably be expected and then accept the consequences when one or more stakeholders are unhappy. One must think through carefully whose needs are most important and whose needs may be sacrificed to keep crucial stakeholders on board.

Lesson 7: No one can please everybody; one must pick stakeholders carefully.

Another reason to select stakeholders carefully is that the nature of behavioral research makes it difficult to represent the findings and products in a tangible, meaningful way. Even when one convinces supporters that one has something good to offer, it is not easy to couch one's progress in terms that they will appreciate. We struggled extensively with this issue and had to develop creative ways to represent our findings (see Salas, Cannon-Bowers, & Blickensderfer, 1997). We also invested in products that could be shown as "live" demonstrations. These products made our ideas, concepts, and theories come to life. Users and sponsors could see the possibilities, and they supported the ideas or concepts. This step required resources that we did not always wish to expend, but in the long run, it served us well.

Lesson 8: Do not underestimate the value of a good demonstration.

Perhaps the best thing that a demonstration provides is the time in which one can arm oneself with the most effective weapon—data. From a scientific standpoint, this gives the community something to criticize and debate, and it lends a measure of scientific credibility. From an operational standpoint, the quality movement (i.e., movement in industry toward continuous quality improvement and total quality management) has made salient the need for objective assessments of R&D, training interventions, and the like. We are almost at a point at which our customers are demanding that we prove that our research will have a measurable impact. Of course, this is not always easy, as we have explained previously. Data that illustrate the "value" of what has been done should be shown. We adopted a "show them, not tell them" perspective, which paid off.

Lesson 9: Do not underestimate the value of data.

Given the difficulties associated with complex research and of collecting meaningful data, we also conclude that being impatient is not useful. More often than not, progress is slow. Sometimes, one must celebrate small successes—"just noticeable differences" in progress. Otherwise, the process can quickly become discouraging.

Lesson 10: Be happy with small increments of progress; they eventually lead to the goal.

Overall, managing a large R&D effort is as much about the politics and marketing as it is about science. These are not always easy lessons to learn. Scientific training does not fully prepare students to understand or confront these challenges.

Concluding Remarks

As we look back over the project, a few additional thoughts are worth noting. First, the demands of a program as complex and difficult as TADMUS require an extraordinary degree of passion—for solving problems, for succeeding, for making things happen. We were fortunate to have working with us a group of people who believed with all their hearts that we were doing the right thing. These people—government employees, contractors, and operational personnel—never ceased to amaze us with their dedication, energy, and inspiration. Moreover, our success was achieved through teamwork. Again, we were fortunate that a common vision pervaded the research team, which allowed us to survive even when we disagreed (this happened more times than we care to remember).

Our final thought is that in addition to tangible products, there is a more subtle outcome of successful applied research, stated in the last lesson.

Lesson 11: One can change people's minds.

By this we mean that perhaps our strongest and most long-lasting legacy has been to make believers out of people who did not know or care about human performance issues in training or decision support. In the process of conducting the research, we have given our customers a new way of looking at the problem, a better way to handle decision making under stress, or a different way to frame the problem. We have given them (and others) new targets of opportunity.

What we have seen is a slow infiltration of ideas into the operational community. At least some of them, as a result of our efforts, have already changed—and will continue to change—fundamentally the way the Navy thinks about training decision makers. This is gratifying.

Reference

Salas, E., Cannon-Bowers, J. A., & Blickensderfer, E. L. (1997). Enhancing reciprocity between training theory and practice: Principles, guidelines, and specifications. In J. K. Ford, S. W. J. Kozlowski, K. Kraiger, E. Salas, & M. S. Teachout (Eds.), *Improving training effectiveness in work organizations* (pp. 291–322). Hillsdale, NJ: Erlbaum.

18

When Applied Research Works: Lessons From the TADMUS Project

William C. Howell

When I was asked in 1990 to help provide technical oversight for the Tactical Decision Making Under Stress (TADMUS) project I agreed, but not without serious misgivings. It was, I thought, an extremely high-risk venture—not only for the Navy, which was about to commit a substantial portion of its behavioral research budget for the next 5 years to this project, but also for the entire psychological science community.

For one thing, its origin in congressional reaction to the highly publicized USS *Vincennes* tragedy ensured that the project would receive an unprecedented level of attention. For another, its primary goal—providing a demonstrably effective, theory-based set of interventions to prevent future *Vincennes*-type accidents—seemed extremely ambitious in light of existing knowledge. A high-profile failure would have been heralded by critics as just another example of how useless social and behavioral science is and how the Defense Department squanders the taxpayers' money.

Former Senator William Proxmire is no longer around to dispense his "Golden Fleece Awards," but his spirit lives on in the Congress. Publicity-seeking lawmakers are overjoyed when they run across federally funded research that they can criticize as wasteful and seek to abolish. Their favorite targets are drawn from the social and behavioral sciences, largely because such research projects often have trivial-sounding titles that neither the lawmakers nor their constituents understand. In this climate, a proven failure of the magnitude of a TADMUS project would represent a political windfall. And the possibility of failure is a vital part of any meaningful research, because if the outcome were certain there would be little point in doing it!

Of course, it was also clear that were it to achieve even a modest level of success, the TADMUS project would yield important payoffs. High visibility is a two-edged sword. In the present case, demonstrating that training or aiding concepts derived from the basic cognitive literature can indeed be put to good use in something as important as the national security would represent a kind of validation opportunity rarely afforded behavioral science. Federally supported research has significantly advanced our understanding of how the human mind works in selecting, processing, interpreting, and acting on information. But the bulk of this knowledge has been gained through laboratory experiments designed to isolate particular—often rather simple—cognitive elements rather than the intact functions people actually use in coping with their complex and disorganized world. If it turned out that these laboratory-derived principles generalize to a context as rich and demanding as a shipboard air defense center, critics would have a difficult time arguing that investment in fundamental research is wasteful.

More important, any success achieved by the project would translate directly into improved performance of complex systems that we rely on to maintain our military preparedness and protect human lives. Although perhaps not as newsworthy as failure, success defined in these terms can hardly be challenged as trivial.

Fortunately, as the foregoing chapters so clearly illustrate, the naysayers were deprived the opportunity to criticize. By almost anyone's standard, TADMUS has turned out to be an unqualified success. Already, the Navy has incorporated many principles derived from the project into its training of combat information center (CIC) operators, and significant improvements in the information displays used by these teams to make critical decisions are in development. Eventually, we will see the evolution of a shipboard "embedded-training" capability for refining team skills and maintaining proficiency that will draw heavily on the knowledge provided by TADMUS research. Moreover, because science is a cumulative endeavor, the job of which is never finished, ideas spawned by this work will undoubtedly lead to further research, further refinements, further applications, and a deeper understanding of how teams of decision makers make crucial decisions under demanding circumstances and how best to equip them for it.

Lessons Learned—the Up Side

Why, then, did the project succeed when it might so easily have failed? Undoubtedly there were many contributing factors, including the skill and dedication of the research team as well as some fortuitous developments that would have been hard to predict and even harder to replicate. But I strongly suspect that at least part of the success is attributable to features of the project infrastructure that were designed explicitly to foster a good marriage between the competing requirements of good science and useful application. A review of these features may thus suggest some valuable lessons for the design and management of future applied research (e.g., 6.2) projects.

Advisory Panel Characteristics

The decision to appoint an independent, standing advisory body (the technical advisory board [TAB]) was an important initial step. Peer review, of course, is a time-honored mechanism for assuring quality in research, and advisory panels are commonplace in military research and development (R&D). But in both composition and function the TAB was a bit different. As for composition, it included representatives from both the relevant scientific and user (Navy operational and training) communities, each of which brought a unique—but salient—perspective to the table. Hence the inevitable tug between the demands of good science and context-specific application was played out continuously within the TAB, sometimes accompanied by vigorous debate. This prevented the project from drifting too far in either direction, an all too common tendency that can seriously erode the value of otherwise well-conceived research.

From a functional standpoint, the TAB maintained close contact with the project throughout its entire life span and provided feedback on a regular basis, concurrently, to both the project managers and the researchers. Meeting at 6-month intervals with the managers and lead scientists enabled the TAB to keep abreast of the project's progress as well as the inevitable problems that arise in any scientific endeavor of this magnitude. The frank discussion of these issues in open forum often produced on-the-spot solutions and course corrections. Two examples illustrate this point.

Early in the project, one of the investigators was criticized for failure to use more highly skilled personnel as participants in a particular study. When the investigator explained that she was having difficulty soliciting the cooperation of the responsible commanding officers, a Navy TAB member was able to intervene and help obtain the necessary support. Once established, this support base helped in soliciting the cooperation of other key fleet organizations. The net result? Increased external validity for all subsequent data collected.

The second example occurred when project managers enlisted contractor expertise in the design of experiments to evaluate specific interface features. The TAB considered the contractor's approach far too complicated to provide the desired information—particularly in view of the practical limitations on data collection—and as a consequence, the experimental design was greatly simplified and the studies yielded useful results.

In retrospect, one can identify a number of such actions that grew out of these regular, multifaceted discussions. It might be argued that the TAB contribution was not always positive, and the consequent actions not always well-advised. And, indeed, negative examples can undoubtedly be found. Viewed collectively, however, it is hard to escape the conclusion that the project benefitted significantly from the TAB process.

Written TAB Reports

In addition to the face-to-face discussions between the TAB and project personnel, documentation of the issues discussed and decisions reached at each TAB meeting was preserved in a brief report submitted by the TAB chair and reviewed by all concerned. Not only did this step enhance mutual understanding of the feedback, but also it served as a interim planning document for the project managers and a point of departure for the subsequent review. This helped maintain continuity and keep the reviews focused. It also reduced the likelihood of critical issues getting lost or reinterpreted between reviews. It enabled both the TAB and the managers to maintain contact with the most important developments throughout the project.

Customer/User Involvement

From the outset, the project management recognized the importance of gaining the support and encouragement of the Navy training and operational communities. It is inevitable that a project of this sort will encounter some initial resistance or suspicion from fleet personnel. For one thing, researchers often fail to understand the real-world systems they seek to improve well enough to justify confidence in whatever recommendations their work produces. For another, the prospect of change is always cause for apprehension among those most directly affected, and not just in the Navy.

Recognizing these realities, the project managers devoted extraordinary effort to developing relationships with the relevant Navy organizations, building a solid understanding of the relevant systems and system contexts, explaining the goals and potential value of the planned research to all those affected, and establishing their own credibility. Moreover, they insisted that TADMUS contractors adopt the same attitude. As a result, the initial resistance gradually melted away, cooperation grew, and by the end of the project, "customers" became the strongest advocates for the products and changes that evolved from the research. In some cases, the demand for these products actually preceded the empirical demonstration of effectiveness—users were convinced that the innovations would help them do their jobs better, even without the "hard evidence" the research would eventually provide.

Not only did customer involvement help overcome resistance to change, it also was essential for incorporating a high level of "external validity" into the research vehicles that were created and the research products that were generated. Navy content experts were relied on heavily in the development of the Decision Making Evaluation for Tactical Teams (DEFTT) simulator and the task scenarios, as well as the analysis and evaluation of individual and team performance. Trainers, trainees, system operators, and retired senior officers were consulted on various portions of the work as it evolved, often on a continuing basis. As a result, research findings were more likely to generalize to the operational setting, and training improvements were more likely to produce lasting transfer effects, than would otherwise have been the case. So when the empirical evidence supported these expectations, it came as no surprise.

Theoretical Grounding and Strategic Planning

The research literature abounds with fundamental knowledge that would appear relevant to the problem of decision making under stress. In the past several decades there has been a virtual explosion in research on human cognition and its role in the performance of various kinds of tasks (Medin, 1990; Pick, van den Broek, & Knill, 1992). Much of it has served to elucidate basic processes involved in concept formation (Medin, 1989), including mental models (Johnson-Laird, 1983) and situation awareness (Gilson, 1995); attention, including cognitive capacity and automaticity (Damos, 1991; Wickens, 1992); memory (Squire, Knowlton, & Musen, 1993; Tulving, 1985); judgment and decision making (Klein, Orasanu, & Calderwood, 1993; Yates, 1990), and problem solving (Sternberg, 1994).

Less well understood, but certainly far from ignored, are the processes responsible for stress effects (Gopher & Kimchi, 1989; Hancock & Warm, 1989) and multiperson (team or group) performance (Swezey & Salas, 1992). And last but by no means least is work on skill acquisition and maintenance (i.e., learning and training), which in recent years has also focused on underlying cognitive processes (Gagne & Driscoll, 1988; Goldstein, 1993; Howell & Cooke, 1989).

One might think that a knowledge base as rich as this would lead directly to practical solutions, but unfortunately such is rarely the case. There are two main reasons. First, as noted earlier, the knowledge is largely derived from laboratory research, and its generalizability to any specific real-world application thus involves a considerable leap of faith. Second, what it offers is by no means a coherent, complete picture of mental functioning. Rather, the "cognitive literature" consists of a host of relatively circumscribed theoretical perspectives, each supported by its own unique body of evidence. Therefore, when faced with a practical problem such as the deterioration of decision performance under stress, one encounters a whole array of theory-based options for addressing it. But there is very little, if any, compelling evidence to support the choice of one over the other on a conceptual basis, or to suggest how any of them should actually be implemented and, if implemented, whether they would produce any practical benefit.

Because of these limitations, the best existing knowledge often fails to make its way into practical applications. Instead, one of two things happens. Either the fundamental work is dismissed as irrelevant, and completely new (generally atheoretical) research is undertaken to answer the question at hand (i.e., the "quick-and-dirty" approach), or one theoretical perspective is plucked from the storehouse of basic knowledge and simply assumed, without further verification, to generalize to the applied context of interest.

Each of these alternatives presents severe drawbacks. The quick-and-dirty approach may provide an acceptable answer to the specific question posed, but it is unlikely to be the best answer, and when a similar question arises in another context, another complete research effort must be undertaken. The favored-theory approach, on the other hand, risks generating answers that don't fit the problem simply because the context contains features that unwittingly invalidate the assumptions of the theory.

How, then, did the TADMUS project cope with this age-old dilemma? Basically, in two ways: by considering a variety of theoretical perspectives in the selection of contractors and by conducting the research at several levels of abstraction concurrently. I next consider each in turn.

The typical approach for selecting research contractors is a standard acquisition process in which a fairly specific statement of the problem is written, bids are submitted, and choices are made on the basis of considerations such as the expectation that the proposed work will answer the question, the bidder's judged ability to carry out the work, and the cost. This process generally results in a research program built around the orientation of the successful bidder (i.e., the favored-theory approach cited earlier).

In the present case, the solicitation encouraged a variety of perspectives on both training and aiding approaches, and an unusually large number of contractors were selected to represent different theoretical orientations. Some, such as those representing the naturalistic decision making perspective, were chosen with the idea of testing the applicability of specific cognitive models to the CIC environment; others focused on more abstract components of the total problem, such as how to measure cognitive performance, what task elements are good candidates for automaticity training, and how best to manipulate stress.

Heterogeneity of expertise, however, is helpful only to the extent that it can be channeled in an appropriate direction, and that was accomplished by the overall research strategy adopted by the project managers. The plan was to develop a high-fidelity simulator as the core research vehicle and to link the simulation studies with field work on the "applied" side and laboratory experimentation on the "basic" side. This enabled the project managers to test conceptual issues under controlled conditions before adapting them to the more complex simulation environment, and to do some "reality testing" or shaping of the simulation work more or less "on the fly." The net result of this strategy was that the core simulation research remained anchored in both science and application throughout.

Resource Allocation

To this point, I have identified four principal characteristics of the way TADMUS was designed and managed that I feel contributed significantly to its success. None of the four, of course, is entirely unique. Many other behavioral research efforts have pursued one or another of them in varying ways with varying degrees of success. Rarely, however, has it been possible to adopt them all at once because of the cost involved—sustained investment at this level is rare indeed! So perhaps a superordinate lesson to be derived from all of this is that if a problem involving human performance is important enough to warrant serious attention, it is best addressed by a sustained, coordinated effort with adequate funding.

Bluntly put, political and budgetary constraints usually operate to prevent behavioral research from providing conclusive answers to important problems. Instead, investment tends to be piecemeal, sporadic, and insufficient given the difficulty of most human-performance questions. The implicit logic seems to run somewhat in the following vein: "Since we don't really trust behavioral science, but we need to do something about problem X, we'll give them a little money and hope for a miracle. In the likely event that the miracle doesn't materialize, we can at least say we tried."

This is not to suggest that all investment in mission-oriented research should be of the "big science" variety—indeed, without the considerable investment the Navy and other funding agencies have made over the years in basic cognitive research (i.e, small science), a TADMUS-like project would have been inconceivable. Obviously, we need both. What I'm suggesting is that when it comes to investing in "applied research," meaning research designed to provide the best available answer to a real-world problem, the tendency is to seriously underestimate the complexity of the human-performance issues involved and hence to seriously underfund whatever research is procured. Consequently, the answers are often incomplete or spurious, and everybody loses. Although I am sure other disciplines, such as the physical sciences, register similar complaints, and any failure can be blamed on insufficient funding, the contrast between the expectations applied to the behavioral versus the more traditional sciences seems inescapable. Somehow, the calibration of those who estimate the scope of behavioral problems and decide what resources are appropriate for addressing them needs to be brought into line with that applied to problems of a more physical nature—and into line with reality!

More Lessons Learned—The Down Side

No effort on the scale of a TADMUS project can be expected to come off without a hitch, and there are in fact some noteworthy hitches. And these, too, provide useful information for future reference.

Dual Project Management

In recognition of the fact that training and aiding represent two of the most plausible approaches to improving decision making under stress, and these approaches define the mission of two major Navy R&D organizations, NRaD and Naval Air Warfare Center Training Systems Division (NAWCTSD) assumed shared responsibilities for the respective components of the research.

It is difficult enough to coordinate closely the efforts of individual scientists and groups at the same site, but almost impossible when they are structurally, physically, and to some extent conceptually separated. Not surprisingly, therefore, the two organizations operated more independently than was desirable throughout the project. Consequently, opportunities to build on each other's progress were occasionally missed, and some duplication of effort occurred. For example, different approaches to cognitive task analysis, scenario generation, and performance measurement were developed, and different positions within the simulated CIC environment were adopted as the focal point of the respective research efforts. Training implications of aiding concepts incorporated into the display design were not considered by their developers until very late in the project. Likewise, the display requirements for different stages of individual and team training were not communicated adequately to the developers.

For all of their acknowledged contributions, neither the TAB process nor the overarching management afforded by a common funding source proved sufficient to overcome the barriers to integration that are inherent in distributed responsibility. This is not an indictment; it is merely a normal consequence of bifurcated management.

The lesson here seems to be that if a project requires participation of multiple management organizations, concerted effort should be devoted at the outset to devising an effective coordinating mechanism. For example, one unit could be accorded primary responsibility and the others supporting roles. It might even be worthwhile in some instances to have units from the supporting organizations collocated with the principal unit, at least during the development of the overall research strategy and at key points during its execution. In the present case, what coordination did occur happened only after the bulk of the work by each unit was already completed.

Contractor Diversity

Although having all of the theoretical advantages cited in the last section, there was a down side to the involvement of many perspectives and contractors in the work. Again, this reduces to a coordination and integration problem. It was apparent from the beginning that in some areas, such as development of task analysis tools or performance measures, there would be some redundancy, conflict, and seemingly wasted effort. And it was also predictable that some contractors would have a research agenda that was not entirely consistent with that of the project. Because it was not possible to predict which would prove most fruitful, it was necessary to explore many leads concurrently even at some cost in efficiency.

As the project evolved, however, the gravity of this problem declined dramatically—thanks, at least in part, to the sustained attention accorded it by the TAB and the project management. For the most part, the project came together admirably. Still, it required a great deal of effort, and no little time and expense, to winnow the initial diversity of research activities down to a reasonable number. It is possible that some of this effort could have been spared.

How, on the other hand, is not so easily specified. That is, there does seem to be a lesson here, but what it is isn't altogether clear. One possibility is that the solicitation might have been preceded by a much smaller, preliminary effort of some sort to narrow the scope of the main project and permit a bit clearer specification of promising theoretical and methodological parameters. The logic is much the same as that for investing in "pilot studies" prior to conducting a major survey or experiment. Whether such an approach would have proved more efficient than what transpired in this case—or indeed, whether any approach would have done so without sacrificing the valuable diversity in perspectives—is debatable. It would, however, seem worth exploring in future endeavors of a similar nature.

Consistency Within the TAB

Managers and contractors complained on occasion, with some justification, that the TAB feedback was not always consistent within or between review sessions. For one thing, there was considerable turnover in TAB membership, and for another, the ongoing tug between the various scientific and applied interests led to occasional perceived or actual inconsistencies in direction. For example, the recommendation to seek all possible advice from the ultimate customers was countered, in at least one instance, by a recommendation to avoid relying too heavily on objections raised by operational personnel! In another case, a directive to work toward a joint NRaD/NAWCTSD experiment was interpreted as more specific and time sensitive than the TAB had intended.

Obviously, the main lesson here is that insofar as possible, continuity within the advisory group should be maintained. If advisors turn over, as they inevitably do in the course of a long project, pains should be taken to sustain some consistency and institutional memory in the body. Written reports can help in this regard, but it is typically the case that once they are written, they are filed and forgotten.

It would perhaps be worthwhile, therefore, to have built into the process a requirement for systematic review of at least selected past reports by the advisory body. And to help promote mutual understanding, it might be well to include in the face-to-face debriefing following each review session a written list of explicitly stated major points and action items. As the current process was implemented, individuals took notes during the session, but these were not fully compared or reconciled until the written report was prepared and distributed. Thus subtle—and at times not so subtle—changes in interpretation were allowed to creep in, and misunderstandings to develop.

As I noted earlier, the TADMUS TAB process did, for the most part, seem to provide a good model for future technical oversight. Like any process, however, it can probably be improved on, and anything that can aid in capturing and communicating the advisors' collective judgment, the advisees' collective reactions, and any reconciliations that were reached, would have obvious usefulness.

Conclusion

The TADMUS project has served as a prime example of applied behavioral research that worked. It has illustrated that when properly planned, theoretically grounded, carefully managed, competently executed, and adequately funded, behavioral science is as capable of yielding solutions to significant real-world problems as are the physical or biological or any other sciences. This should go without saying, considering that science represents a generic approach to understanding and problem solving that is blind to disciplinary distinctions. But in reality, research and development policy makers consistently undervalue and underfund behavioral work. They expect little from it, hence invest little in it, thereby confirming their expectations. Demonstrating that significant payoffs can result from a significant investment in behavioral science thus has profound policy as well as practical implications.

To the extent that it serves to encourage future investment in projects of similar scope, TADMUS also provides a case study of infrastructure features that contribute to a successful outcome. As one of those formally charged with technical oversight for the project, I have had reason to ponder these features as the work unfolded. This chapter represents one deeply involved observer's attempt to draw useful conclusions from this case study and should be interpreted with that limitation in mind.

References

Damos, D. L. (Ed.). (1991). *Multiple-task performance.* London: Taylor & Francis.

Gagne, R. M., & Driscoll, M. P. (1988). *Essentials of learning for instruction* (2nd ed.). Englewood Cliffs, NJ: Prentice-Hall.

Gilson, R. D. (Ed.). (1995). Situation awareness [Special issue]. *Human Factors, 37,* 3–157.

Goldstein, I. L. (1993). *Training in organizations.* Belmont, CA: Brooks/Cole.

Gopher, D., & Kimchi, R. (1989). Engineering psychology. *Annual Review of Psychology, 40,* 431–455.

Hancock, P. A., & Warm, J. S. (1989). A dynamic model of stress and sustained attention. *Human Factors, 31,* 519–537.

Howell, W. C., & Cooke, N. J. (1989). Training the human information processor: A look at cognitive models. In I. Goldstein (Ed.), *Training and development in work organizations* (pp. 121–182). San Francisco, CA: Jossey-Bass.

Johnson-Laird, P. N. (1983). *Mental models.* Cambridge, MA: Harvard University Press.

Klein, G. A., Orasanu, J., & Calderwood, R. (Eds.). (1993). *Decision making in action.* Norwood, NJ: Ablex.

Medin, D. L. (1989). Concepts and conceptual structure. *American Psychologist, 44,* 1469–1481.

Medin, D. L. (1990). *Cognitive psychology.* Fort Worth, TX: Harcourt, Brace Jovanovich.

Pick, H. L., Jr., van den Broek, P., & Knill, D. C. (Eds.). (1992). *Cognition.* Washington, DC: American Psychological Association.

Squire, L. R., Knowlton, B., & Musen, G. (1993). The structure and function of memory. *Annual Review of Psychology, 44,* 453–495.

Sternberg, R. J. (Ed.). (1994). *Thinking and problem solving.* New York: Academic Press.

Swezey, R. W., & Salas, E. (Eds.). (1992). *Teams: Their training and performance*. Norwood, NJ: Ablex.

Tulving, E. (1985). How many memory systems are there? *American Psychologist, 40,* 385–398.

Wickens, C. D. (1992). *Engineering psychology and human performance* (2nd ed.). New York: Harper-Collins.

Yates, J. F. (1990). *Judgment and decision making*. Englewood Cliffs, NJ: Prentice-Hall.

Appendix A ───────────

Technical Advisory Board

Attendance List

Dr. Martin A. Tolcott, Chair
Washington, DC

Dr. Robert Angus
Defense and Civil Institute of
    Environmental Medicine
1133 Sheppard Avenue West
P.O. Box 2000
Downsview, Ontario, Canada
    M3M 3B9

CAPT Douglas M. Armstrong
Commanding Officer
Tactical Training Group Atlantic/Dam
    Neck
Virginia Beach, VA 23461-5596

CAPT Dallas Bethea
Commanding Officer
Tactical Training Group, Pacific
53620 Horizon Drive
San Diego, CA 92147-5087

CDR William Boulay
Johns Hopkins University
Applied Physics Laboratory
Laurel, MD 20723-6090

CAPT James L. Burke
Commanding Officer
Tactical Training Group, Pacific
200 Catalina Boulevard
San Diego, CA 92147-5080

CDR Charles T. Bush
AEGIS Program Office
Washington, DC

LTCOL James B. Bushman
Technical Training Research Division/
    Armstrong Laboratory
Air Force Lab/Human Resources and
    Training
Brooks AFB, TX 78235-5352

CAPT Martin L. Chamberlain
Deputy Director, Planning and
    Assessment
Office of Naval Research
Arlington, VA 22217

CDR Kevin Ebel
AEGIS Program Office PMS 400 B32C
2531 Jefferson Davis Highway
Arlington, VA 22242-5165

LTCOL Martin L. Fracker
Air Force Armstrong Laboratory
7909 Lindburg Drive
Brooks AFB, TX 78235-5352

CAPT William A. Gaines
Tactical Training Group Pacific
San Diego, CA

Dr. Ellen Hall
Cognition and Performance Division
Air Force Research Laboratory/HRCC
7909 Lindbergh Drive
Brooks AFB, TX 78235-5352

CAPT E. J. Halley, Jr.
Commander
Surface Warfare Development Group
2200 Amphibious Drive
Norfolk, VA 23521-3229

Scott Lutterloh
AEGIS Training and Readiness Center
5395 First Street
Dahlgren, VA 22448-5200

Gary Mahler
AEGIS Training Center
Dahlgren, VA 22448-5190

CAPT Edward C. McDonough
Surface Warfare Officers School
  Command
446 Cushing Road
Newport, RI 02841-1209

CAPT John McHenry
Commanding Officer
Tactical Training Group, Pacific
San Diego, CA

LCDR Vic Mercado
AEGIS Program Office
PMS 400 B30C
2531 Jefferson Davis Highway
Arlington, VA 22242-5165

CDR John R. Midgett
Director of Training
Combat Systems Training Group
  Atlantic/NAB, Little Creek
Norfolk, VA 23521

Luke H. Miller
Technical Director
AEGIS Training Center
CODE ANTD
Dahlgren, VA 22448-5190

CAPT Tom Mooney
Tactical Training Group, Pacific
53620 Horizon Drive
San Diego, CA 92147-5087

CDR Kathleen Paige
AEGIS Baseline Manager
AEGIS Program Office
Washington DC

CDR Lee Pitman
Fleet Training Unit
US Atlantic Fleet/NAB Little Creek
Norfolk, VA 23521

CAPT D. Salinas, II
Tactical Training Group, Pacific
53620 Horizon Drive
San Diego, CA 92147-5087

CDR Deborah R. Stiltner
PEO Surface Combatants
AEGIS Program Office
PMS 400 B3C
2531 Jefferson Davis Highway
Arlington, VA 22242-5165

Dr. Michael Strub
Chief, Army Research Institute Field
  Unit
P.O. Box 6057
Ft. Bliss, TX 79906-0057

Dr. Martin A. Tolcott
Apartment 1617
3001 Veazey Terrace, NW
Washington, DC 20008

Dr. Willard S. Vaughan
Director, Cognitive and Neural
  Sciences Division
Office of Naval Research
800 N. Quincy Street
Arlington, VA 22217-5000

Dr. Wayne Waag
Crew Training Division
Armstrong Laboratory/Air Force Lab/
  Human Resources and Training
Williams AFB, AZ 85240-6457

Dr. Dennis White
AEGIS Program Office
PMS 400
2531 Jefferson Davis Highway
Arlington, VA 22242-5165

CAPT Rick Williams
Commander Third Fleet Staff
SPOAP 96601-6001

LCDR Paul Willis
AEGIS Program Office
PMS 400 B30C
Washington, DC 20362-5102

Appendix B

Selected TADMUS Publications

Journal Articles

Cannon-Bowers, J. A., Salas, E., Blickensderfer, E. L., & Bowers, C. A. (1998). The impact of cross-training and workload on team functioning: A replication and extension of initial findings. *Human Factors, 40,* 92–101.

Cannon-Bowers, J. A., Salas, E., & Pruitt, J. S. (1996). Establishing the boundaries of a paradigm for decision-making research. *Human Factors, 38,* 193–205.

Cohen, M. S., Freeman, J. T., & Wolf, S. (1996). Meta-recognition in time stressed decision making: Recognizing, critiquing, and correcting. *Human Factors, 38,* 206–219.

Driskell, J. E., & Salas, E. (1991). Group decision making under stress. *Journal of Applied Psychology, 76,* 473–478.

Inzana, C. M., Driskell, J. E., Salas, E., & Johnston, J. H. (1996). The effects of preparatory information on enhancing performance under stress. *Journal of Applied Psychology, 81,* 429–435.

Johnston, J., Driskell, J. E., & Salas, E. (1997). Vigilant and hypervigilant decision making. *Journal of Applied Psychology, 82,* 614–622.

Kirlik, A., Walker, N., Fisk, A. D., & Nagel, K. (1996). Supporting perception in the service of dynamic decision making. *Human Factors, 38,* 288–299.

Kraiger, K., Salas, E., & Cannon-Bowers, J. A. (1995). Measuring knowledge organization as a method for assessing learning during training. *Human Factors, 37,* 804–816.

Rouse, W. B., Cannon-Bowers, J. A., & Salas, E. (1992). The role of mental models in team performance in complex systems. *IEEE Transactions on Systems, Man, and Cybernetics, 22,* 1296–1308.

Salas, E., Bowers, C. A., & Cannon-Bowers J. A. (1995) Military team research: Ten years of progress. *Military Psychology, 7,* 55–75.

Salas, E., Cannon-Bowers J. A., & Blickensderfer, E. L. (1993). Team performance and training research: Emerging principles. *Journal of the Washington Academy of Sciences, 83,* 81–106.

Saunders, T., Driskell, J. E., Johnston, J. H., & Salas, E. (1996). The effect of stress inoculation training on anxiety and performance. *Journal of Occupational Health Psychology, 1,* 170–186.

Volpe, C. E., Cannon-Bowers, J. A., Salas, E., & Spector, P. (1996). The impact of cross training on team functioning. *Human Factors, 38,* 87–100.

Weaver, J. L., Bowers, C. A., Salas, E. & Cannon-Bowers, J. A. (1995). Networked simulations: New paradigms for team performance research. *Behavior Research Methods, Instruments, & Computers, 27,* 12–24.

Other technical reports and unpublished manuscripts are available by contacting Janis A. Cannon-Bowers or Eduardo Salas, Code 4961, NAWCTSD, 12350 Research Parkway, Orlando, FL 32826.

Book

Driskell, J. E., & Salas, E. (Eds.) (1996). *Stress and human performance.* Hillsdale, NJ: Erlbaum.

Chapters

Blickensderfer, E., Cannon-Bowers, J. A., & Salas, E. (1997). Fostering shared mental models through team self-correction: Theoretical bases and propositions. In M. Beyerlein, D. Johnson, & S. Beyerlein (Eds.), *Advances in interdisciplinary studies in work teams series* (Vol. 4, pp. 249–279). Greenwich, CT: JAI Press.

Cannon-Bowers, J. A., & Bell, H. R. (1997). Training decision makers for complex environments: Implications of the naturalistic decision making perspective. In C. Zsambok & G. Klein (Eds.) *Naturalistic decision making—Where are we now?* (pp. 99–110). Hillsdale, NJ: Erlbaum.

Cannon-Bowers, J. A., & Salas, E. (1997). A framework for developing team performance measures in training. In M. T. Brannick, E. Salas, & C. Prince (Eds.), *Assessment and measurement of team performance: Theory, research, and applications* (pp. 45–62). Hillsdale, NJ: Erlbaum.

Cannon-Bowers, J. A., & Salas, E. (1997). Teamwork competencies: The intersection of team member knowledge, skills, and attitudes. In H. F. O'Neil (Ed.), *Workforce readiness: Competencies and assessment* (pp. 151–174). Hillsdale, NJ: Erlbaum.

Cannon-Bowers, J. A., Salas, E., & Converse, S. A. (1993). Shared mental models in expert team decision making. In N. J. Castellan, Jr. (Ed.), *Individual and group decision making: Current issues* (pp. 221–246). Hillsdale, NJ: Erlbaum.

Cannon-Bowers, J. A., Tannenbaum, S. I., Salas, E., & Volpe, C. E. (1995). Defining team competencies and establishing team training requirements. In R. Guzzo & E. Salas (Eds.), *Team effectiveness and decision making in organizations* (pp. 333–380). San Francisco, CA: Jossey-Bass.

Coovert, M. D., Craiger, J. P., & Cannon-Bowers, J. A. (1995). Innovations in modeling and simulating team performance: Implications for decision making. In R. Guzzo & E. Salas (Eds.), *Team effectiveness and decision making in organizations* (pp. 149–203). San Francisco, CA: Jossey-Bass.

Driskell, J. E., & Salas, E. (1991). Overcoming the effects of stress on military performance: Human factors, training, and selection strategies. In R. Gal & A. Mangelsdorff (Eds.), *Handbook of military psychology* (pp. 183–193). London: Wiley.

Duncan, P. C., Rouse, W. B., Johnston, J. H., Cannon-Bowers, J. A., Salas, E., & Burns, J. J. (1996). Training teams working in complex systems: A mental model-based approach. In W. B. Rouse (Ed.), *Human/technology interaction in complex systems* (Vol. 8, pp. 173–231). Greenwich, CT: JAI Press.

Johnston, J. H., & Cannon-Bowers, J. A. (1996). Training for stress exposure. In J. E. Driskell & E. Salas (Eds.), *Stress and human performance* (pp. 223–256). Hillsdale, NJ: Erlbaum.

Johnston, J. H., Smith-Jentsch, K. A., & Cannon-Bowers, J. A. (1997). Performance measurement tools for enhancing team decision making. In M. T. Brannick, E. Salas, & C. Prince (Eds.), *Team performance assessment and measurement: Theory, research, and applications* (pp. 311–330). Hillsdale, NJ: Erlbaum.

Kozlowski, S. W. J., & Salas, E. (1997). An organizational systems approach for the implementation and transfer of training. In J. K. Ford, S. W. J. Kozlowski, K. Kraiger, E. Salas, & M. Teachout (Eds.), *Improving training effectiveness in work organizations* (pp. 247–287). Mahwah, NJ: Erlbaum.

Kozlowski, S. W. J., Gully, S. M., McHugh, P. P., Salas, E., & Cannon-Bowers, J. A. (1996). A dynamic theory of leadership and team effectiveness: Developmental and task contingent leader roles. In G. R. Ferris (Ed.), *Research in personnel and human resource management* (Vol. 14, pp. 253–305). Greenwich, CT: JAI Press.

Kozlowski, S. W. J., Gully, S. M., Nason, E. R., & Smith, E. M. (in press). Team compilation: Development, performance, and effectiveness across levels and time. In D. R. Ilgen & E. D. Pulakos (Eds.), *The changing nature of work performance: Implications for staffing, personnel actions, and development* (SIOP Frontiers Series). San Francisco: Jossey-Bass.

Kozlowski, S. W. J., Gully, S. M., Salas, E., & Cannon-Bowers, J. A. (1996). Team leadership and development: Theory, principles, and guidelines for training leaders and teams. In M. Beyerlein, D. Johnson, & S. Beyerlein, (Eds.), *Advances in interdisciplinary studies of work teams: Team leadership* (Vol. 3, pp. 253–291). Greenwich, CT: JAI Press.

McIntyre, R. M., & Salas, E. (1995). Measuring and managing for team performance: Emerging principles from complex environments. In R. Guzzo & E. Salas (Eds.), *Team effectiveness and decision making in organizations* (pp. 149–203). San Francisco, CA: Jossey-Bass.

Means, B., Salas, E., Crandall, B., & Jacobs, T. O. (1993). Training decision makers for the real world. In G. Klein, J. Orasanu, R. Calderwood, & C. E. Zsambok (Eds.), *Decision making in action: Models and methods* (pp. 306–326). Norwood, NJ: Ablex.

Orasanu, J., & Salas, E. (1993). Team decision making in complex environments. In G. Klein, J. Orasanu, R. Calderwood, & C. E. Zsambok (Eds.), *Decision making in action: Models and methods* (pp. 327–345). Norwood, NJ: Ablex.

Rogers, W., Maurer, T., Salas, E., & Fisk, A. (1997). Training design, cognitive theory, and automaticity: Principles and a methodology (pp. 19–46). In J. K. Ford, S. W. J. Kozlowski, K. Kraiger, E. Salas, & M. Teachout (Eds.), *Improving training effectiveness in work organizations*. Hillsdale, NJ: Erlbaum.

Salas, E., & Cannon-Bowers, J. A. (1997). Methods, tools, and strategies for team training. In M. A. Quiñones & A. Ehrenstein (Eds.), *Training for a rapidly changing workplace: Applications of psychological research* (pp. 249–279). Washington, DC: American Psychological Association.

Salas, E., Cannon-Bowers, J. A., & Blickensderfer, E. L. (1997). Enhancing reciprocity between training theory and practice: Principles, guidelines, and specifications. In J. K. Ford, S. W. J. Kozlowski, K. Kraiger, E. Salas, & M. Teachout (Eds.), *Improving training effectiveness in work organizations* (pp. 291–322). Hillsdale, NJ: Erlbaum.

Salas, E., Cannon-Bowers, J. A., & Johnston, J. H. (1997). How can you turn a team of experts into an expert team?: Emerging training strategies. In C. Zsambok & G. Klein (Eds.), *Naturalistic decision making—Where are we now?* (pp. 359–370). Hillsdale, NJ: Erlbaum.

Salas, E., Cannon-Bowers, J. A., & Kozlowski, S. W. J. (1997). The science and practice of training: Current trends and emerging themes. In J. K. Ford, S. W. J. Kozlowski, K. Kraiger, E. Salas, & M. Teachout (Eds.), *Improving training effectiveness in work organizations* (pp. 357–368). Hillsdale, NJ: Erlbaum.

Salas, E., Dickinson, T. L., Converse, S. A., & Tannenbaum, S. I. (1992). Toward an understanding of team performance and training. In R. W. Swezey & E. Salas (Eds.), *Teams: Their training and performance* (pp. 3–29). Norwood, NJ: Ablex.

Salas, E., Driskell, J. E., & Hughes, S. (1996). Introduction: The study of stress and human performance. In J. E. Driskell & E. Salas (Eds.), *Stress and human performance* (pp. 1–46). Hillsdale, NJ: Erlbaum.

Smith, E. M., Ford, J. K., & Kozlowski, S. W. J. (1997). Building adaptive expertise: Implications for training design. In M. A. Quinones & A. Dudda (Eds.), *Training for a rapidly changing workplace: Applications of psychological research* (pp. 89–118). Washington, DC: American Psychological Association.

Swezey, R. W., & Salas, E. (1992). Guidelines for use in team-training development. In R. W. Swezey & E. Salas (Eds.), *Teams: Their training and performance* (pp. 219–245). Norwood, NJ: Ablex.

Tannenbaum, S. I., Salas, E., & Cannon-Bowers, J. A. (1996). Promoting team effectiveness. In M. A. West (Ed.), *Handbook of work group psychology* (pp. 503–529). Sussex, England: Wiley & Sons.

Proceedings, Professional Conferences, and Presentations

Bailey, S. S., Johnston, J. H., Smith-Jentsch, K. A., Gonos, G., & Cannon-Bowers, J. A. (1995). Guidelines for facilitating shipboard team training. *Proceedings of the 17th Annual Interservice/Industry Training Systems and Education Conference* (pp. 360–369). Washington, DC.

Baker, C. V., Salas, E., Cannon-Bowers, J. A., & Spector, P. (1992). *The effects of interpositional uncertainty and workload on teamwork and task performance.* Paper presented at the Annual Meeting of the Society for Industrial and Organizational Psychology, Montreal, Canada.

Bisantz, A. M., Gay, P., Phipps, D. A., Walker, N., Kirlik, A., & Fisk, A. D. (in press). Specifying training needs in dynamic judgment tasks using a lens model approach. *Proceedings of the IEEE Conference on Systems, Man, and Cybernetics.*

Blickensderfer, E. L., Cannon-Bowers, J. A., & Salas, E. (1994). Feedback and team training: Exploring the issues. *Proceedings of the Human Factors and Ergonomics Society 38th Annual Meeting* (pp. 1195–1199). Santa Monica, CA: Human Factors and Ergonomics Society.

Blickensderfer, E. L., Cannon-Bowers, J. A., & Salas, E. (1994). Feedback and team training: Team self-correction. *Proceedings of the Second Annual Mid-Atlantic Human Factors Conference* (pp. 81–85). Washington, DC.

Blickensderfer, E., Cannon-Bowers, J. A., & Salas, E. (1997). *Training teams to self-correct: An empirical investigation.* Paper presented at the 12th Annual Meeting of the Society for Industrial and Organizational Psychology, St. Louis, MO.

Brown, K. G., & Kozlowski, S. W. J. (1997, April). *Self-evaluation and training outcomes: Training strategy and goal orientation effects.* Paper presented at the 12th Annual Conference of the Society for Industrial and Organizational Psychology, St. Louis, MO.

Brown, K. G., Mullins, M. E., Weissbein, D. A., Toney R. J., & Kozlowski, S. W. J. (1997, April). Mastery goals and strategic reflection: Preliminary evidence for learning interference. In S. W. J. Kozlowski (Chair), *Metacognition in training: Lessons learned from stimulating cognitive reflection.* Symposium conducted at the 12th Annual Conference of the Society for Industrial and Organizational Psychology, St. Louis, MO.

Burgess, K. A., Burke, C. S., Salas, E., & Cannon-Bowers, J. A. (1993). *Team performance under the gun: An examination of team leader behaviors under stress.* Paper presented at the First Annual Mid-Atlantic Human Factors Conference, Norfolk, VA.

Burgess, K. A., Riddle, D. L., Hall, J. K., & Salas, E. (1992). *Principles of team leadership under stress.* Paper presented at the 38th Annual Meeting of the Southeastern Psychological Association, Knoxville, TN.

Burgess, K. A., Salas, E., & Cannon-Bowers, J. A. (1993). *Training team leaders: "More than meets the eye."* Paper presented at the 8th Annual Conference of the Society for Industrial and Organizational Psychology, San Francisco, CA.

Burgess, K. A., Salas, E., Cannon-Bowers, J. A., & Hall J. K. (1992). *Training guidelines for team leaders under stress.* Paper presented at the 36th Annual Meeting of the Human Factors Society, Atlanta, GA.

Burke, C. S., Volpe, C. E., Cannon-Bowers, J. A., & Salas, E. (1993). *So what is teamwork anyway? A synthesis of the team process literature.* Paper presented at the 39th Annual Meeting of the Southeastern Psychological Association, Atlanta, GA.

Cannon-Bowers, J. A., & Salas, E. (1990). *Cognitive psychology and team training: Shared mental models complex systems.* Symposium presented at the 5th Annual Conference of the Society for Industrial and Organizational Psychology, Miami, FL.

Cannon-Bowers, J. A., Salas, E., & Baker, C. V. (1991). Do you see what I see? Instructional strategies for tactical decision making teams. *Proceedings of the 13th Annual Interservice/Industry Training Systems Conference* (pp. 214–220). Washington, DC: National Defense Industrial Association.

Cannon-Bowers, J. A., Salas, E., & Converse, S. A. (1990). Cognitive psychology and team training: Training shared mental models and complex systems. *Human Factors Society Bulletin, 33,* 1–4.

Cannon-Bowers, J. A., Salas, E., Duncan, P., & Halley, E. J. (1994). Application of multi-media technology to training for knowledge-rich systems. *Proceeding of the 16th Annual Interservice/Industry Training Systems Conference* (pp. 6–11). Washington, DC: National Defense Industrial Association.

Cannon-Bowers, J. A., Salas, E., & Grossman, J. D. (1991). *Improving tactical decision making under stress: Research directions and applied implications.* Paper presented at the International Applied Military Psychology Symposium, Stockholm, Sweden.

Cohen, M. S., & Freeman, J. T. (1996). Thinking naturally about uncertainty. *Proceedings of the Human Factors and Ergonomics Society 40th Annual Meeting* (pp. 179–183). Santa Monica, CA: Human Factors and Ergonomics Society.

Converse, S. A., Cannon-Bowers, J. A., & Salas, E. (1991). Team member shared mental models: A theory and some methodological issues. *Proceedings of the Human Factors and Ergonomics Society 35th Annual Meeting* (pp. 1417–1421). Santa Monica, CA: Human Factors and Ergonomics Society.

Coovert, M. D., Campbell, G. E., Cannon-Bowers, J. A., & Salas, E. (1995). *A methodology for a team performance measurement system.* Paper presented at the 10th Annual Conference of the Society for Industrial and Organizational Psychology, Orlando, FL.

Coovert, M. D., Cannon-Bowers, J. A., & Salas, E. (1990). Applying mathematical modeling technology to the study of team training and performance. *Proceedings of the 12th Annual Interservice/Industry Training Systems Conference* (pp. 326–333). Washington, DC: National Defense Industrial Association.

Driskell, J. E., Salas, E., & Hall, J. K. (1994). *The effect of vigilant and hypervigilant decision training on performance.* Paper presented at the 9th Annual Conference of the Society for Industrial and Organizational Psychology, Nashville, TN.

Driskell, J. E., Salas, E., & Johnston, J. (1995). *Does stress lead to a loss of team perspective?* Paper presented at the 10th Annual Conference of the Society for Industrial and Organizational Psychology, Orlando, FL.

Driskell, J. E., Salas, E., & Johnston, J. (1995). *Is stress training generalizable to novel settings?* Paper presented at the 10th Annual Conference of the Society for Industrial and Organizational Psychology, Orlando, FL.

Duncan, P. C., Burns, J. J., Frey, P. R., Cannon-Bowers, J. A., & Johnston, J. H. (1995). Enhancing team performance in tactical environments: The team model trainer. *Proceedings of the 17th Annual Interservice/Industry Training Systems and Education Conference* (pp. 262–271). Washington, DC: National Defense Industrial Association.

Duncan, P. C., Cannon-Bowers, J. A., Johnston, J., & Salas, E. (1995, June). *Using a simulated team to model teamwork skills: The team model trainer.* Paper presented at the World Conference on Educational Multimedia and Hypermedia, Graz, Austria.

Dwyer, D. J. (1992). An index for measuring naval team performance. *Proceedings of the Human Factors Society 36th Annual Meeting* (pp. 1356–1360). Santa Monica, CA: Human Factors and Ergonomics Society.

Dwyer, D. J. (1995). Training for performance under stress: Performance degradation, recovery, and transfer (CD-ROM). *Proceedings of the 17th Annual Interservice/Industry Training Systems Conference* (pp. 154–162). Washington, DC: National Defense Industrial Association.

Dwyer, D. J., Hall, J. K., Volpe, C. E., Cannon-Bowers, J. A., & Salas, E. (1992). *A performance assessment task for examining tactical decision making under stress* (Special Report No. 92–002). Orlando, FL: Naval Training Systems Center.

Gay, P., Phipps, D. A., Bisantz, A. M., Walker, N., Kirlik, A., & Fisk, A. D. (1997). Operator specific modeling of identification judgments in a complex dynamic task. *Proceedings of the Human Factors and Ergonomics Society 41st Annual Meeting* (pp. 225–229). Santa Monica, CA: Human Factors and Ergonomics Society.

Gully, S. M., & Kozlowski, S. W. J. (1996, August). The influence of self-efficacy and team-efficacy on training outcomes in a team training context. In J. George-Flavey (Chair*), Defining, measuring, and influencing group level efficacy beliefs.* Symposium conducted at the Annual Convention of the Academy of Management Association, Cincinnati, OH.

Hall, J. K., Driskell, J. E., Salas, E., & Cannon-Bowers, J. A. (1992). Development of instructional design guidelines for stress exposure training. *Proceedings of the 14th Annual Interservice/Industry Training Systems Conference* (pp.357–363). Washington, DC: National Defense Industrial Association.

Hall, J. K., Dwyer, D. J., Cannon-Bowers, J. A., Salas, E., & Volpe, C. E. (1993). Toward assessing team tactical decision making under stress: The development of a methodology for structuring team training scenarios. *Proceedings of the 15th Annual Interservice/Industry Training Systems Conference* (pp. 87–98). Washington, DC.

Hall, J. K., Volpe, C. E., & Cannon-Bowers, J. A. (1992, August). *Mitigating the effects of stress: A look at potential team training strategies.* Paper presented at the Centennial Convention of the American Psychological Association, Washington, DC: National Defense Industrial Association.

Johnston, J. H., & Cannon-Bowers, J. A. (1996). A conceptual framework and recommendations for integrating training with a decision support system. *Proceedings of the Command and Control Research and Technology Symposium* (pp. 233–239). Washington, DC: National Defense Industrial Association.

Johnston, J. H., Cannon-Bowers, J. A., & Smith-Jentsch, K. A. (1995). Event-based performance measurement system for shipboard command teams. *Proceedings of the First International Symposium on Command and Control Research and Technology* (pp. 274–276). Washington, DC: National Defense Industrial Association.

Kirlik, A., Rothrock, L., Walker, N., & Fisk, A. D. (1996). Simple strategies or simple tasks? Dynamic decision making in "complex" worlds. *Proceedings of the Human Factors and Ergonomics Society 40th Annual Meeting* (pp. 184–188). Santa Monica, CA: Human Factors and Ergonomics Society.

Kozlowski, S. W. J. (1995, March). *Training for adaptive expertise.* Paper presented at the Personnel Human Resource Research Group, Tulane University, New Orleans, LA.

Kozlowski, S. W. J. (1996, March). *TEAMS/TANDEM: An experimental platform for examining skill acquisition, adaptability, and effectiveness at individual and team levels of analysis.* Paper presented at the Personnel Human Resource Research Group, University of Florida, Gainesville.

Kozlowski, S. W. J., Ford, J. K., & Salas, E. (1994, July). Team development: Levels, process, and learning outcomes. In D. R. Ilgen (Chair), *Work team performance: Some critical issues.* Symposium conducted at the 23rd International Congress of Applied Psychology, Madrid, Spain.

Kozlowski, S. W. J., & Gully, S. M. (1996, August). TEAMS/TANDEM: Examining skill acquisition, adaptability, and effectiveness. In J. Vancouver & A. Williams (Chairs), *Using computer simulations to study complex organizational behavior.* Symposium conducted at the Annual Convention of the Academy of Management Association, Cincinnati, OH.

Kozlowski, S. W. J., Gully, S. M., & McHugh, P. P. (1993, April). Leadership and team effectiveness: A developmental-task contingent model. In J. A. Cannon-Bowers (Chair), *Optimizing team performance through team leader behavior.* Symposium conducted at the 8th Annual Conference of the Society for Industrial and Organizational Psychology, San Francisco, CA.

Kozlowski, S. W. J., Gully, S. M., Nason, E. R., Ford, J. K., Smith, E. M., Smith, M. R., & Futch, C. J. (1994, April). A composition theory of team development: Levels, content, process, and learning outcomes. In J. E. Mathieu (Chair), *Developmental views of team processes and performance.* Symposium conducted at the 9th Annual Conference of the Society for Industrial and Organizational Psychology, Nashville, TN.

Kozlowski, S. W. J., Gully, S. M., Salas, E., & Cannon-Bowers, J. A. (1995, June). *Team leadership and development: Theory, principles, and guidelines for training leaders and teams.* Paper presented at the Third Annual University of North Texas Symposium on Work Teams, Dallas, TX.

Kozlowski, S. W. J., Gully, S. M., Smith, E. M., Brown, K. G., Mullins, M. E., & Williams, A. E. (1996, April). Sequenced mastery goals and advance organizers: Enhancing the effects of practice. In K. Smith-Jentsch (Chair), *When, how, and why does practice make perfect?* Symposium conducted at the 11th Annual Conference of the Society for Industrial and Organizational Psychology, San Diego, CA.

Kozlowski, S. W. J., Gully, S. M., Smith, E. A., Nason, E. R., & Brown, K. G. (1995, May). Sequenced mastery training and advance organizers: Effects on learning, self-efficacy, performance, and generalization. In R. J. Klimoski (Chair), *Thinking and feeling while doing: Understanding the learner in the learning process*. Symposium conducted at the 10th Annual Conference of the Society for Industrial and Organizational Psychology, Orlando, FL.

Kozlowski, S. W. J., & Salas, E. (1991, April). Application of a multilevel contextual model to training implementation and transfer. In J. K. Ford (Chair), *Training as an integrated activity: An organization system perspective*. Symposium conducted at the 6th Annual Conference of the Society for Industrial and Organizational Psychology, St. Louis, MO.

Nason, E. R., Gully, S. M., Brown, K. G., & Kozlowski, S. W. J. (1995, May). *Skill acquisition and declarative knowledge: Where structural knowledge fits*. Paper presented at the 10th Annual Conference of the Society for Industrial and Organizational Psychology, Orlando, FL.

Radtke, P. H., & Frey, P. (1996). PC-based part-task training for team decision making: Recommendations and guidelines. *Proceedings of the Command and Control Research and Technology Symposium* (pp. 241–259). Washington, DC.

Riddle, D. L., Hall, J. K., & Patten, A. L. (1992). *Effects of controllability on the noise–performance relationship: A review and implications for stress exposure training*. Paper presented at the 36th Annual Meeting of the Human Factors Society, Atlanta, GA.

Salas, E. (1993). Team training and performance. *Psychological Science Agenda, 6,* 9–11.

Serfaty, D., Entin, E. E., & Volpe, C. (1993). Adaptation to stress in team decision-making and coordination. *Proceedings of the Human Factors and Ergonomics Society 37th Annual Meeting* (pp. 1228–1232). Santa Monica, CA: Human Factors and Ergonomics Society.

Smith, K. A. (1994, October). Toward optimizing the impact of developmental feedback in team training simulations. *Proceedings of the Human Factors and Ergonomics Society 38th Annual Meeting* (pp. 1200–1203). Santa Monica, CA: Human Factors and Ergonomics Society.

Smith-Jentsch, K. A., Cannon-Bowers, J. A., & Salas, E. (1997). New wine in old bottles: A framework for theoretically-based OJT. In K. A. Smith-Jentsch (Chair), *Toward a continuous learning environment*. Paper presented at the 12th Annual Meeting of the Society for Industrial and Organizational Psychology, St. Louis, MO.

Smith-Jentsch, K. A., Payne, S. C., & Johnston, J. H. (1996). Guided team self-correction: A methodology for enhancing experiential team training. In K. A. Smith-Jentsch (Chair), *When, how, and why does practice make perfect?* Paper presented at the 11th Annual Conference of the Society for Industrial and Organizational Psychology, San Diego, CA.

Travillian, K. K., Baker, C. V., & Cannon-Bowers, J. A. (1992). *Correlates of self and collective efficacy with team functioning*. Paper presented at the 38th Annual Meeting of the Southeastern Psychological Association, Knoxville, TN.

Travillian, K. K., Volpe, C. E., Cannon-Bowers, J. A., & Salas, E. (1993, October). Cross-training highly interdependent teams: Effects on team process and team performance. *Proceedings of the 37th Annual Human Factors and Ergonomics Society Conference* (pp. 1243–1247). Santa Monica, CA: Human Factors and Ergonomics Society.

Volpe, C. E., Travillian, K. K., & Cannon-Bowers, J. A. (1992). *Walking in each other's shoes: A case for mental model training.* Paper presented at the 36th Annual Meeting of the Human Factors Society, Atlanta, GA.

Walker, N., Fisk, A. D., & Phipps, D. (1994). Perceptual-rule-based skills: When rule-training hinders learning. *Proceedings of the Human Factors and Ergonomics Society 38th Annual Meeting.* Santa Monica, CA: Human Factors and Ergonomics Society.

Index

Event-based scenarios, 44
EWS (electronic warfare supervisor), 228
Experiential learning, 249
Experimenter control station (ECS), 46
Expertise. *See also* Knowledge
    classic decision making and routine,
        117–118
    naturalistic decision making and adap-
        tive, 118–120
Explanation-based reasoning (EBR), 377,
    381–383, 389
Explicit coordination strategies, 225
Eye contact, 283–284

Facilitators, 135, 248–250
Feedback
    in learning process, 252–256
    in Team Dimensional Training, 286–287,
        295
Feedback augmentation, 103–110
    dimensions of, 103–105
    experiment in, 105–108
Fogarty, W. M., 3, 4, 5
Fractionation, task, 95

Geo-plots, 383, 385
Georgia Institute of Technology Aegis Sim-
    ulation Platform (GT-ASP), 45, 48, 56,
    98–103, 105–110
Goal orientation, 29
Goals-operators-methods-selection rules
    (GOMS), 324, 327
Grossman, J. D., 9
GT-ASP. *See* Georgia Institute of Technol-
    ogy AEGIS Simulation Platform
Guided team self-correction, ?72, 287–288,
    294

Hall, J. K., 45
HCI. *See* Human–computer interaction
Hemphill, J. K., 303, 306
Higher level goals, 168–169
High-performance skills, 91–92
Holt, V. E., 8–9
Hostile-intent stories, 164–170
HSI. *See* Human–system integration
Hughes, S., 19
Human–computer interaction (HCI), 315,
    320, 339
Human–system integration (HSI), 375–
    376, 404
Hutchins, S. G., 378
Hypervigilant decision making, 206

Identification friend or foe (IFF), 386, 388
Identification supervisor (IDS), 228, 230
Implicit coordination, 28, 225–227, 299,
    300

Information
    instrumental, 200
    procedural, 199–200
    sensory, 199
Instrumental information, 200
Interdependence, team, 302–305
Internal knowledge, 319–322
Interpositional knowledge (IPK), 300–303,
    306–308
Interpositional uncertainty, 300
Intuition, 157
IPK. *See* Interpositional knowledge

James, William, 95
JMCIS (Joint Maritime Command Informa-
    tion System), 7
Johnston, J. H., 30
Joint Maritime Command Information Sys-
    tem (JMCIS), 7
Joint Operational Tactical System (JOTS),
    7

Kaempf, G. L., 377–378
Keyboard skills, 97
Klein, G. A., 376–377
Knowledge
    declarative, 323
    of display symbols, 97–98
    interpositional, 300–303, 306–308
    perceptual, 325
    procedural, 324
Knowledge representation, 319
Kozlowski, S. W. J., 249

Lambertson, D. C., 306
LaPorte, T. R., 224
Lateral communication, 238
Leaders, team, 122
    behaviors of, during pre-briefs and post-
        action reviews, 253–259
    as facilitators of team performance, 248–
        250
    as team instructors/developers, 134–138
    and team learning cycle, 250–253
    training of, 76, 259–264
Lecturing, 256
Levels issues, in team training, 121–122,
    130–133
Lewin, Kurt, 17
Localization, 169

Mastery orientation, 29
Measurement, performance. *See* Perfor-
    mance measurement
Measures of effectiveness (MOEs), 62
Measures of performance (MOPs), 62
Mental models, 26–28, 272–274, 289–291.
    See also Shared mental models

About the Editors

Janis A. Cannon-Bowers, PhD, is a senior research psychologist in the Science and Technology Division of the Naval Air Warfare Center Training Systems Division (NAWCTSD), Orlando, Florida. She received her master's and doctoral degrees in industrial and organizational psychology from the University of South Florida, Tampa. As the team leader for advanced surface training research at NAWCTSD, Dr. Cannon-Bowers has been involved in several research projects directed toward improving training for complex environments. These have included investigation of training needs and design for multioperator training systems, training effectiveness and transfer of training issues, tactical decision making under stress, the impact of multimedia training formats on learning and performance, and training for knowledge-rich environments.

Dr. Cannon-Bowers is an active researcher who has published more than 75 articles, book, and technical reports and has given numerous professional presentations. She is on the editorial board of *Human Performance*. Dr. Cannon-Bowers's research interests include team training and performance, training effectiveness, training complex decision-making skills, and the application of advanced technology to the design of embedded training systems.

Eduardo Salas, PhD, is a senior research psychologist and head of the Training Technology Development branch of the Naval Air Warfare Center Training Systems Division. He received his doctoral degree in industrial and organizational psychology from Old Dominion University in 1984. Dr. Salas has coauthored more than 100 journal articles and book chapters and has coedited 5 books. He is on the editorial boards of *Human Factors, Personnel Psychology, Military Psychology*, the *Interamerican Journal of Psychology*, and *Training Research Journal*.

Dr. Salas's research interests include team training and performance, training effectiveness, tactical decision making under stress, team decision making, performance measurement, and learning strategies for teams. He is a fellow of the American Psychological Association, a recipient of the Meritorious Civil Service Award from the Department of the Navy, and has courtesy appointments at the University of South Florida and the University of Central Florida.